# Maintenance Engineering and Management

*Precepts and Practices*

*Second Edition*

# Maintenance Engineering and Management

*Precepts and Practices*

*Second Edition*

D. R. Kiran

BSP BS Publications

CRC Press
Taylor & Francis Group
Boca Raton London New York

CRC Press is an imprint of the
Taylor & Francis Group, an **informa** business

Maintenance Engineering and Management - *Precepts and Practices, Second Edition*
*by D.R. Kiran*



*Published by*

BS Publications
A unit of BSP Books Pvt. Ltd.
4-4-309/316, Giriraj Lane, Sultan Bazar,
Hyderabad - 500 095, India.

*For*

CRC Press
Taylor & Francis Group, an informa business
6000 Broken Sound Parkway, NW
Suite 300, Boca Raton, FL 33487
711 Third Avenue
New York, NY 10017
2 Park Square, Milton Park
Abigdon, Oxon OX14 4RN, UK
www.taylorandfrancisgroup.com

*For distribution in rest of the world other than the India, Pakistan, Nepal, Myanmar (Burma), Bhutan, Bangladesh and Sri Lanka.*

ISBN: 978-1-138-04849-2



*British Library Cataloguing in Publication Data*
A Catalogue record for this book is available from the British Library

*This Book is Dedicated to*

All the persons who were associated with me during my 45 years long industrial and academic career and who helped me in bringing this book to this shape.

# Contents

## CHAPTER 6

## Condition Monitoring



## CHAPTER 7

## Logical Fault Location

## CHAPTER 8

## Wear Debris Analysis



## CHAPTER 9

## Reliability Based Maintenance

## CHAPTER 10

## Lubrication

## CHAPTER 11

### Cutting Fluids



## CHAPTER 12

### Maintenance Costs

## CHAPTER 13

### Failure Modes and Effects Analysis



## CHAPTER 14

### Lean Maintenance

## CHAPTER 15

## Scientific Inventory Management

## Chapter 16

# Spare Part Inventory Control

# Preface to Second Edition

At the outset the author wishes to thank the publishers and the readers for necessitating the release of a second edition of this book in conformance to the syllabi from around 20 Universities and Institutions of India as well as those of Boston University and University of Dar Es salaam. Several eminent academicians and industry leaders have given positive review about this book, and the author wishes to adapt the suggestions made by some of them. Topics like wear debris analysis, non-destructive testing, environmental impact, instruments for condition monitoring, logical fault location, tramp oils etc. which are specified by Universities but hardly covered in many other maintenance books, found special appreciation from the reviewers.

In view of the technical advancement in the recent times, the maintenance practices have widened their tentacles into the computerized systems, mobile applications and web-based applications needing specific mention. This books provides the preliminary understanding of this aspect.

While the basic text and treatment of the theme is maintained as in the first edition, some suggestions from the reviewers have been adapted and incorporated in the new edition. New and improved figures and diagrams together with case studies have been added. The chapter on Scientific Inventory Control has preceded that on the Spare Part Inventory Control to conform to the logical precepts of the inventory principles.

Consequently this book continues to be of immensive help to the busy maintenance engineer who needs instant answers to a wide variety of on-the-job problems at the same time providing a comprehensive knowledge to the engineering/management student who wants to build up a career in industry.

Several case studies are given from the practical experience of the author, both in India and abroad, presenting an authoritative treatment of maintenance engineering and management that can be found only in international reference books.

As stated in the first edition preface, though a separate book on Industrial Safety by this author is expected to follow soon, in deference to syllabi of some universities that specify basics of industrial safety in the maintenance management syllabus, two chapters on industrial safety that illustrate the basic principles are included.

**D. R. Kiran**

# Preface to First Edition

The economic liberalization of late twentieth century saw enormous industrial growth in India. As a result today's maintenance engineer faces new challenges in minimizing downtime for trouble free operation of the machinery and other equipment that are under his charge. He finds that he has to manage more than he can engineer the maintenance function. In other word she is as much responsible for efficiently managing his day-to-day performance of his functions, as for the sound physical knowledge and skill in maintenance of the production facilities including the expanding scope of modern maintenance that includes dimensions beyond the factory walls. As the modern management techniques change, the maintenance management techniques too change, and needless to say that for the best possible functioning, the maintenance engineer must be up-to-date in the latest management precepts and practices.

Under this background this book deals with the management principles and practices that govern the maintenance function apart from the engineering techniques for the effective and optimal systems for the maintenance of equipment. It gives the maintenance engineer, the latest developments in maintenance engineering techniques like wear debris analysis, preventive maintenance and condition monitoring as well as management concepts like reliability based maintenance, logical fault location and lean maintenance. Consequently this book is expected to be of immensive help to the busy maintenance engineer who needs instant answers to a wide variety of on-the-job problems at the same time providing a comprehensive knowledge to the engineering/management student who wants to build up a career in industry.

Syllabi from around 20 Universities and Institutions of India as well as those of Boston University and University of Dar Es salaam are collected to draw the outline for this book. Topics like wear debris analysis, non-destructive testing, environmental impact, instruments for condition monitoring, logical fault location, tramp oils etc. which are specified by these Universities but hardly covered in any other Indian book, are covered in detail in this book. Several case studies are given from the practical experience of the author, both in India and abroad, presenting an authoritative treatment of maintenance engineering and management that can be found only in international reference books.

The title of the book is hence aptly chosen as *Maintenance Engineering and Management - Precepts and Practices.*

Though a separate book on Industrial Safety by this author is expected to follow soon, in deference to syllabi of some universities that specify basics of industrial safety in the maintenance management syllabus, two chapters on industrial safety that illustrate the basic principles are added.

**D. R. Kiran**

# About the Author

Prof. D. R. Kiran, B.Sc., B.E., M.Sc. (Eng.), (Ph.D.), FIE(I), FIIIE, FIIProdE, FIIPlantE, FITTE, FISNT, has a rich practical experience of forty years both in industry and academy. Starting his career in 1968 with Larsen & Toubro, he held top positions like Planning Manager of Rallifan (CF division), World Bank Adviser/Instructor for Transport Managers in Tanzania and the Principal of PMR Institute of Technology, Chennai.

At industries in India and abroad, he did projects at L&T, National Bicycle Company etc., besides several Tanzanian industries, both as an internal and external consultant that included logical fault location, condition monitoring systems, WDA etc., besides ABC analysis, codification and other industrial engineering applications. He started his academic career in 1979, teaching Maintenance Engineering at B.E. level at the University of Dar Es Salaam. This rich experience in maintenance and planning helped him in introducing topics of special interest and specific use to the maintenance engineer.

In recognition of his services in the field of engineering education, he was presented with the coveted Bharat Excellence Award and Gold Medal for Excellence in Education in New Delhi in 2006. He is listed as an International Expert in Industrial Engineering and Management in the International Directory of Experts and Expertise. He is nominated for the post of Honorary Deputy Director General in India for International Biographical Center.

Earlier during the eighties, he was introduced to Dr. Julius Nyerere, the then President of Tanzania as a Pioneer of Work Study in that country. He was one among few non-political foreigners to be interviewed by the government newspaper of Tanzania.

He has authored three text books on Professional Ethics and Human Values (2nd edition), Maintenance Engineering and Management (2nd Edition) and Total Quality Management. He also published 23 papers in professional journals and seminar proceedings and was the Chief Guest in several Technical Meets. He is the Vice Chairman of Indian Institute of Plant Engineers, Tamilnadu Chapter.

He is widely travelled having visited over 30 countries and is a philanthropist.

# Abbreviations

| | |
|---|---|
| ABC control | A, B and C category control |
| ABCD PEN F2 PVC LOSS | Acronym for the 16 characteristics of lubricants |
| ADP | Automated Data Processing |
| AIIE | American Institute of Industrial Engineers |
| AKS | Asset Knowledge Science |
| ALARP | As Low as Reasonably Practicable |
| AMMS | Asset Maintenance Management System |
| AN | Alkylated naphthalene |
| ARR | Average Rate of Returns |
| ASTM | American Society of Tool and Manufacturing Engineers, now called Society of Manufacturing Engineers |
| BLS | Bureau of Labor Statistics |
| BSI | British Standards Institute's |
| BIS | Bureau of Indian Standards |
| CAM | Challenger's Adverse minimum |
| CASPA | Computer Aided systematic Particle Analysis |
| CAVE | Computer Aided Visual Engineering |
| CBM | Condition based maintenance |
| CCI | Construction Cost Index |
| CFC | Chlorofluorocarbon |
| CFMEA | Concept FMEA |
| CMMS | Computerized Maintenance Management System |
| CPFR | Collaborative Planning, Forecasting and Replenishment |
| CT | Computer Tomography |
| CVTWDA | Computer vision techniques for wear debris analysis |
| DAM | Defender's Adverse minimum, |
| DERA | Defense Evaluation and Research Agency |
| DFMEA | Design FMEA |
| DOD/UAL | US Department of Defense and United Airlines |
| DOE | Design for Environment |
| DOM | Design out maintenance |

| | |
|---|---|
| DRBFM | Design Review Based on Failure Mode |
| EAM | Enterprise asset management |
| EAMS | Enterprise Asset Management Systems |
| EBME | Electro Biomechanical Engineering |
| ECM | Environmentally conscious manufacture |
| EDM′ | Electric discharge machine |
| EFDC | Early Fatigue Detection Centre |
| EIA | Environment Impact Assessment |
| EMS | Environmental Management Systems |
| EOQ | Economic order quantity |
| EOT Cranes | Electric Overhead Traveling Cranes |
| ERP | Enterprise resource planning |
| ERR | Economic rate of return |
| ERV | Estimated Replacement Value |
| ET | Eddy current test |
| FAR | Fixed Asset Register |
| FMEA | Failure mode and effects analysis |
| FMECA | Failure mode, effects and criticality analysis |
| FSN | Fast moving, Slow moving or Non-moving |
| FTI | Fixed time inspection |
| FTM | Fixed time maintenance |
| GDP | Gross Domestice Produce |
| GHGs | Green house gases |
| GOK | God only knows |
| GOLF | Government, Local or Foreign |
| HML | Hyundai Motors Limited |
| HML | High cost, medium cost and low cost |
| HPS | High Pressure stabilizers |
| HRMS | Human resource management systems |
| HVAC | Heating, Ventilation and Air Conditioning |
| ILO | International Labor Organization |
| Imax system | Fixed Maximum Stock Ordering system |
| Imin system | Fixed Safety Stock Ordering system |
| IMMS | Integrated Maintenance Management System |
| IRR | Internal Rate of Returns |

## Chapter 17

### Total Productive Maintenance



## Chapter 18

### System Approach to Management Theory

## CHAPTER 21

### Work Order Systems



## CHAPTER 22

### Computerised Maintenance Management System



## CHAPTER 23

### Replacement of Machinery and Equipment

## CHAPTER 24

## Environmental Issues

## CHAPTER 25

## Industrial Safety

## CHAPTER 26

### Job Hazard Analysis

| | |
|---|---|
| ISNT | Indian Society of Non-destructive Testing |
| JHA | Job hazard analysis |
| JIPE' | Japan Institute of Plant Engineers |
| JIT | Just-in-time |
| JSA | Job safety analysis |
| LORA | Level of Repair Analysis |
| LT | Lead Time |
| MAPI | Machinery and Allied Products Institute |
| MDTR | Machine Down Time Reduction |
| MET | Machine Education and Training |
| MFM | Material Flow Management |
| MIMOSA | Machinery Information Management Open Systems Alliance |
| MIS | Machine Inspection Standards |
| MMSI | Maintenance Management System Improvement |
| MND | Moving, Non-moving or Dead items |
| MNG | Moving, Non-moving or Ghost items |
| MPP | maintenance prevention program |
| MRP | Material Resources Planning |
| MSA | Machine Safety Awareness |
| MT | Magnetic particle test |
| MTA | Machine Time Availability |
| MTBF | Mean Time Between Failure |
| MTTR | Mean Time To Repair |
| MWT | Mean Waiting Time |
| NCCM | National Conference on Condition Monitoring |
| NDE | non-destructive evaluation |
| NDT | Non-destructive Testing |
| NEWS FOR CD POP TV | Acronym for the 14 lubricant additives |
| NPC | National Productivity Council (India) |
| NPV | Net Present Value |
| NPW | Net Present Worth |
| ODP | Ozone Depletion Potential |
| OECD | Organisation for Economic Co-operation and Development |
| OM | Opportunity maintenance |

| | |
|---|---|
| OSHA | Occupational Safety and Hazards Administration |
| OTF | Operate to failure |
| P system | Fixed Period Ordering system |
| PAG | Polyalkylene glycols |
| PAO | Polyalpha-olefin |
| PCB | Pinted Circuit Board |
| PDF | Probability density function |
| PdM | Predictive maintenance |
| PEEP | A place for everything and everything in its place |
| Pf | Probability of failure |
| PFMEA | Process FMEA |
| PPE | personal protective equipment |
| PRM | Proactive Reliability Maintenance |
| PT | Dye penetrant test |
| Q system | Fixed Order Quantity system |
| QCFI | Quality Circle Forum of India |
| Qmax | Maximum Stock level |
| Qmin | Minimum stock level |
| RBM | Reliability based maintenance |
| RCFA | Root cause failure analysis |
| RCM | Reliability centered maintenance |
| RFT | Response time to failure |
| RIF | Reliability Improvement Factor |
| ROA | Return on Assets |
| ROL | Re-order Level |
| ROQ | Re-order Quantity |
| RPN | Risk Priority number |
| RT | Radiographic test |
| SAE | Society of Automobile Engineers |
| SAN | Strong Acid Number |
| SDE | Scarce, Difficult and Easily Available |
| SFC | Soft foot condition |
| SfFMEA | Software FMEA |
| SHC | Synthesized hydrocarbons |
| SQC | Statistical Quality Control |

| | |
|---|---|
| SREDDIM, | 7 step methodology fro hazard analysis |
| SrFMEA | Service FMEA |
| STOU | Super Tractor Oil Universal |
| SyFMEA | System FMEA |
| TAN | Total acid content |
| TBN | Total base content |
| TBO | Two button operations |
| TIPER | Tanzania Italy Oil Refineries |
| TM | Thermographic inspection |
| TPM | Total productive maintenance |
| TPS | Toyota Production System |
| TQM | Total Quality Management |
| UNCED | United Nations Conference on Environment Development |
| UNHCHS | United Nations Habitat Conference on Human Settlement |
| UT | Ultrasonic test |
| UTTO | Universal Tractor Transmission Oil |
| VED | Vital, Essential or Desirable |
| WDA | Wear Debris Analysis |
| ZDP | Zinc dithiophosphate |

# Chapter 1

# Introduction

*Chapter synopsis:* *Maintenance is an integral part of production for optimal achievement of company objectives. This is crux of today's Total Productive Maintenance. After giving the common definitions of the maintenance function, this chapter introduces the purpose and scope of maintenance and outlines the importance, scope and objectives of maintenance. The three dimensions are highlighted.*

*Keywords:* *Five M's of input, three dimensions of maintenance, level of technology, mechanization. Primary and secondary functions, objectives and responsibilities, economic and technical dependency, and social change, tero-technology Integrated Maintenance Management System (IMMS), expert system, intelligent system, Maintenance Management System Improvement.*

## 1.1 Dictionary Definitions

The dictionaries define maintenance in several forms as:

- The correction or prevention of faults in hardware by a programme of inspection and the replacement of parts
- The act of removal of existing faults and the modification of software in response to changes in specification or environment
- The act of maintaining or the state of being maintained
- A means of support; livelihood
- *A modifier* of or relating to the maintaining of buildings, machinery, etc., like the maintenance team
- A provision ordered to be made by way of periodical payments or a lump sum, as after a divorce for a spouse

In the context of maintenance applied to industry, we shall limit our definitions and explanation to the first three meanings.

## 1.2 Importance of Maintenance Function

Why does a factory need a maintenance department? The answer to this question is the basis for developing the general concepts, the philosophy, the policies, the techniques and the organization of the maintenance engineering facility. Basically the purpose of having a maintenance facility in an organization is to attend to the day to day problems in keeping and maintaining the machinery and other plant equipments and services in good working condition. Machine failures have far reaching effects and may result in loss of thousands of man hours.

Broadly speaking the term maintenance embraces all the activities involved in keeping the entire production systems or specific equipment in working order and to enhance its reliability. The reliability can be defined as the probability that the production system will function properly or a reasonable length of time after being put into use. The policy objectives from the management point of view are to minimize the total cost of maintaining the system at an optimal level of equipment reliability. In view of its significance, the term reliability and its related terms are discussed in detail in chapters 4 and 9.

The trend towards mechanization and automation has already increased the investment in capital goods. To obtain a fair return of the investment, high level of utilization is required, and to achieve this, the equipment must be kept in perpetual operating condition. The extent to which this objective is achieved depends largely upon the maintenance policy.

Manufacturing activities at any level of technology involve the five M's of input, viz, men, materials, money, management and more significantly the machines and tools as illustrated in Fig. 1.1. Obviously every machine cannot work continuously throughout their life time. The mechanization and automating of the manufacturing processes do increase the productivity, but at the same time increase the dependency on proper running of the equipment.

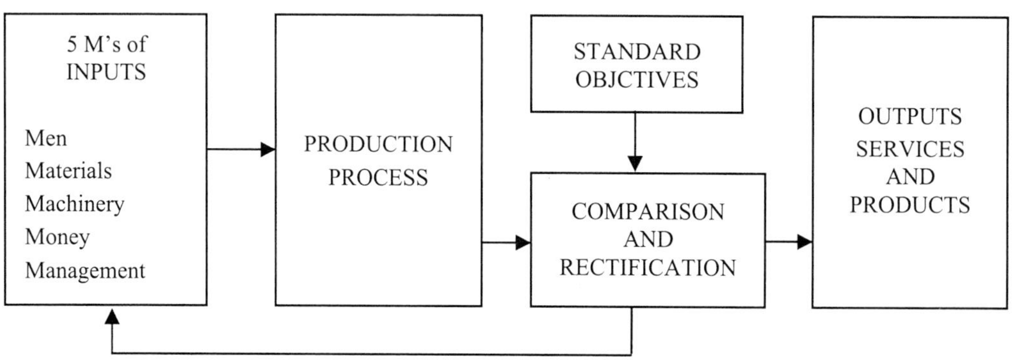


**Figure 1.1** The 5 M's of input.

The justification for maintenance engineering facility lies in its efforts to ensure the availability of the machines, buildings and services needed by the other wings of the organization for the performance of their functions with optimum return on investment, whether the investment is in terms of machinery, materials or men. The maintenance function should be considered an integral and important part of the organization.

Dependence of the operating personnel on the maintenance engineering facility is ever increasing with the complexity of machinery used in modern industry. The cost of maintenance has become a greater part of the total cost of manufacture and the maintenance unit of the industry. Regardless of the tremendous growth of the importance, cost and complexity of the maintenance function, it should be remembered that the function exists because it is an essential part of the whole plant operation, and is not a self sufficient unit. It is one part of the team that can be successful only when function in tandem.

## 1.3 The Three Dimensions of Maintenance

We thus see that preventive maintenance, which involves the act of estimating the probability of the possible breakdowns in the future and planning the course of maintenance action before they occur and this is an integral part of the production function and involves three major dimensions.

(a) *Level of technology:* Manufacturing is a technical process of physical and chemical transformation of materials. The function and organization of maintenance is obviously determined by the kind of technology (chemical, engineering etc.), and the level of technology (small, medium, large etc.). This relationship is illustrated in Fig. 1.2.

- *At cottage level:* All members of the family are involved in the manufacturing, using indigenous skills and very simple tools, which are made and maintained by the respective user.
- *At small industry level:* Technical division of the labor and mechanization is in small scale. Machine operators are relatively experienced and perform a lot of repair work themselves. Yet outside craftsmen with specialized skills in tool making and repair are often called in.
- *At large scale level:* Highly developed professional specialization exists in designing, tool making, machine operation, maintenance etc.

(b) *Organization:* Any function of manufacturing industry, including maintenance is based on the technical division of labor. Hence the three C's viz,

- Communication
- coordination and
- cooperation

are essential for the successful preventive maintenance program. The organization patterns that are most suitable for maintenance function are discussed further in chapter 19.

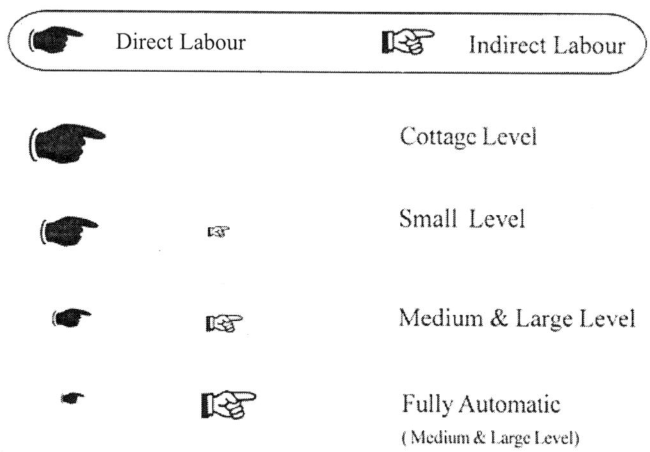


**Figure 1.2** Maintenance labor vs. production labor.

(c) *Economy:* Since the profitability of a maintenance program is essential to an organization, the following economics must be given consideration,

- Economy of time
- Economy of material resources,
- Economy of human recourses.

In other words, we find that, as mechanization increases,

- Machine operators tend to be less skilled
- Machine maintenance staff repairing in the respective facility need higher skills with formal specialized training, and
- In a highly mechanized situation, the cost of breakdown is so high that it is always essential to foresee when a breakdown would occur and set the weakening characters right before the breakdown occur. This is dealt more in detail in chapters 5 and 6.

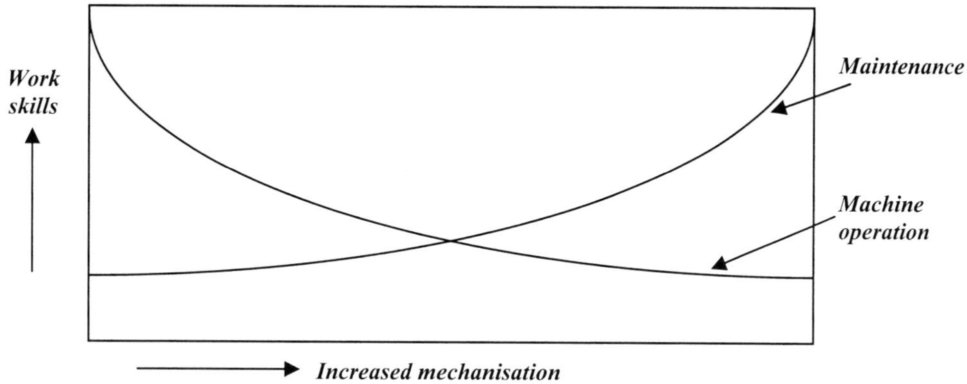


**Figure 1.3** Maintenance skills increase with increased mechanization.

## 1.4 Scope of Maintenance Function

Although in practice, the scope of the activities of a maintenance engineering department is different in different plants and is generally influenced by the plant size, type, company policy, industry category etc. It comprises of plant maintenance, construction, utilities, generation and miscellaneous services for the plant operation. While the basic function of the maintenance department is to provide the engineering and support facilities for the uninterrupted, efficient and safe operation of the plant, we can group these activities into two general classifications as *primary functions* that are included in the justification of the department and *secondary functions* most of which are included because of the experience, knowhow, precedence or because there is no logical division of the plant to which these responsibilities can be assigned or delegated to.

*Primary functions:*

1. Engineering, scheduling, execution and control of the planned maintenance and repair of the equipment
2. Condition monitoring of the plant and equipment
3. Maintenance of the building and other facilities of the plant
4. Equipment inspection and lubrication
5. Maintaining and controlling of the workshops
6. Generation and distribution of utilities like power, steam and compressed air
7. New installations and/or alterations to the buildings and equipment
8. Engineering and supervision of the construction and installation projects within the scope of the organization

9. Technical consultation on mechanical problems with the production department
10. Performing all the above functions in a safe and efficient manner.

**Secondary functions:**

1. Storekeeping, generally of maintenance items
2. Plant protection including fire protection
3. Industrial safety
4. Waste disposal and salvage
5. Insurance administration
6. Janitorial services
7. Property accounting
8. Pollution and noise control
9. Scrap yard control and maintenance
10. Vehicle maintenance and estate function
11. Any other function delegated by the management.

## 1.5 Objectives of Maintenance Function

1. To provide freedom from breakdowns during manufacturing operations
2. To maintain the equipment in a satisfactory condition for safe operation
3. To maintain the equipment at its maximum operating efficiency
4. To reduce to a minimum, the downtime due to breakdowns
5. Consistent with the above points to reduce to a minimum the cost of maintenance
6. To maintain a high level of engineering practices in the performance of the work handled by the department
7. To store and maintain sufficient and optimum quantity of spares consistent with the current demand and conditions
8. To investigate continuously the causes and remedies of emergency breakdowns
9. To keep abreast of the technological developments, new methods, equipment and materials
10. To have close cooperation with production personnel.

## 1.6 Responsibilities of a Maintenance Department

1. *Engineering and execution of planned maintenance*: The responsibility for the continued economical use of the equipment lies with the operating supervisor. It is however the responsibility of the maintenance staff, in cooperation with the operating supervisors to organize preventive maintenance schedules, to improve the equipment efficiency necessary to meet the production schedule requirements.
2. *Generation and distribution of power and other utilities:* The power engineer is responsible for the generation and distribution of the utilities like steam and compressed air in suitable quantities as also the purchase and distribution of electrical power, water etc. at minimum cost.
3. *Stand-by power generation:* Apart from purchase of the electric power from EB. If stand-by diesel generator is available in the factory, it is the maintenance staff that would be responsible for the operation of this generation and supply sufficient power during power cut periods.
4. *Administrative and staffing:* In order to handle mechanical work involved in carrying out the functions of maintenance department, an adequate skilled labor force provided with suitable equipment and properly supervision is necessary. The type and size of this group is determined on the basis of a balance between the costs of establishing and maintaining the group versus the cost of contracting for the maintenance work by outside firms, considering the factor if emergency and round the clock service. It is the responsibility of the head of the maintenance department to plan and coordinate work distribution among the crafts and to organize and carry out the programmers for the training of the supervisors and craftsmen.
5. *Engineering and supervision of construction projects within the scope of the group*: Maintenance engineering is a definite part of the preliminary and final design of the construction projects. It is the responsibility of the maintenance group to be familiar with all the projects under consideration or execution that will eventually become a part of the plant.
6. *Technical consultation:* The maintenance engineer would act as a consultant to the production department in solving the mechanical problem of the equipment.
7. *Plant safety activities and equipment*: Provision and control of all safety equipment, their maintenance and monitoring for the general observance of plant safety practices by all workforces in the plant, would be the responsibility of the maintenance department.

8. *Plant insurance:* Planning for the plant insurance and maintenance of all cornered records and documents as well as taking the necessary action in the event of accidents.
9. *Service facilities:* Engineering, administration and maintenance of various service facilities delegated to the departments like yards, roadways, washing rooms, sewers and waste disposal sections and employee welfare services within the budgets provided by the management.
10. Maintaining the property and all relevant records of all plant and equipment required for the annual accounts and audit.
11. Keeping the breakdown record for each and every machinery and equipment meticulously as required for future maintenance planning.
12. Last but not the least is to perform all the above functions in a safe and efficient manner.

## 1.7 Responsibilities of Maintenance Engineer

*Reliabilityweb.com* suggests the following functions for the maintenance engineer.

1. Ensures that the equipment is properly designed, selected and installed based on the life cycle philosophy.
2. Ensures that equipment is performing effectively and efficiently.
3. Establishes and monitors programs for engine and compressor analysis and vibration and other condition monitoring techniques.
4. Reviews deficiencies noted during corrective maintenance.
5. Provides technical guidance for Computerized Maintenance Management System (CMMS).
6. Maintains and advises on use and disposition of stock, surplus and rental equipments.
7. Promotes equipment standardization, recommends spare part levels and coordinated sharing of spare parts with other asset teams.
8. Becomes available for consultation with maintenance technicians.
9. Monitors new technology and keeps management and staff apprised of the new developments.
10. Champion's quality assurance services including shop qualifications for outside services.
11. Develops standards and procedures for major maintenance jobs.

12. Periodically makes cost/benefit review of maintenance management programs for areas of responsibility and exchanges information across asset reams.
13. Provides technical guidance for preventive and predictive maintenance programs.
14. Monitors competitors' activities in the fields of maintenance management.
15. Becomes the focal point for monitoring performance indicators for maintenance management programs.
16. Optimizes maintenance strategies.
17. Becomes focal point for analyzing equipment operating data, root cause analysis, ultrasonic and vibration analysis.
18. Imparts maintenance training.

## 1.8 Tero-Technology

In 1970, UK introduced the new concept of Tero-technology which is a combination of management, financial, engineering and other aspects of management related to physical assets in achieving economical life cycle costs. This includes the specification and design for reliability and maintainability of machinery, equipment, buildings and structures with their installation, commissioning, maintenance, modification and replacement.

## 1.9 Integrated Maintenance Management System (IMMS)

Integrated Maintenance Management System is the modern maintenance practice similar to Tero-technology but concentrates on the maintenance activities. All the maintenance tasks are planned and executed in a holistic approach. Of more significance is the expert system, where all data on the maintenance history of each and every asset is made available in the form of computerized reports. The name expert system or intelligent system, is derived from the fact that with the availability of full data as stated above, the maintenance engineer would be in a better position to understand, plan and execute the optimal maintenance tasks. The relevant software for this is Maintenance Management System Improvement (MMSI or MMIS) which is described more in detail in chapter 22.

## 1.10 Conclusion

We should look at maintenance as an integral part of production and should apply the respective principles of production planning, work study, stock control; etc., for the efficient fulfillment of this function.

Peter Drucker said, 'in the third world societies, the modern technology demand is a radical break with the social and cultural traditions and produces a fundamental crisis in the society'. Specific to Indian industry, the factors leading to such a crisis are:

- *Economic dependency*: In developing countries the mechanization is highly linked with the foreign aid given by the donor countries, as also the equipment supplied by the collaborators. Due to this and the limited foreign exchange situation, the Indian economy depends on the proper maintenance of the existing machinery and equipment.
- *Technical dependency:* This too follows the same argument as above. The more sophisticated the machinery is, the more we have to depend on the supplying country for maintenance of the plant and equipment.
- *Social change:* The imported mechanized technology is basically of the industrial culture and philosophy of the western countries. The fact that India has a rural oriented society necessitates a different approach of education and training.

The following chapters hence are devoted to the explanation of these basic principles of maintenance practices as can be applied to Indian industry.

# Chapter 2

# Principles of Maintenance Planning

***Chapter synopsis:*** *Since it is essential to have well-established maintenance systems for maximizing production and organizational efficiency, the significance of well-planned maintenance activities needs no second opinion. This chapter explains the basic principles behind sound planned maintenance, the what, how, and when the maintenance is to be planned and executed. It illustrates the procedure and benefits of planned maintenance. The working pattern and working cycle of maintenance activities are discussed in detail.*

***Keywords:*** *Maintenance planning, five to one thumb rule, developmental medium and long range plans, critical shutdowns, what, how, and when to maintain, working pattern, planning, analysis, execution and records control, analysis and assessment, reliability, maintenance staff and cost effectiveness.*

## 2.1 Planned Maintenance

Planned maintenance is a maintenance concept developed over a span of time. When machines were no more than simple aids to human muscles like wheelbarrows, the problems posed by these were few and the maintenance could normally be provided by the operator himself like an occasional drop of oil where the quantity or the frequency have any major significance. But modern machines and equipments which have grown in complexity call for more precise planning and attention. Planned maintenance is a program designed to improve the effectiveness of maintenance through the use of systematic method and plans. The primary objective of maintenance is to keep the equipment functioning in a safe and efficient manner to ensure the production targets are kept with minimum stoppage of machines and low operating costs. As Rushton says, *'you cannot maximize production or reduce costs without the use of an effective planned maintenance system'.*

## 2.2 Why Maintenance Planning?

The fundamental mission of any company is to generate profits to the stake holders and planned maintenance activities tend to increase profits in two ways:

- Increased production by reducing the wasteful down time
- Reduced maintenance costs (direct as well as indirect)

It becomes the responsibility of the maintenance department to continuously achieve higher production time by reducing the down-time of each and every equipment. A machine that is not running due to mechanical problem is producing nothing. Thus it is an unwritten fact that in most cases it is the maintenance department and not the production department that has the greatest opportunity to increase the overall production time.

## 2.3 Advantages of Well Planned Maintenance

1. Releases front line supervisors from major planning duties and gives them more effective concentration on crew supervision and executional planning.
2. Charts out the definitive procedures for planning, execution, monitoring and controlling of maintenance resources.
3. Reduces delays and waiting times for men, materials and tools once the maintenance job starts in progress.
4. Enables systematic collection of materials before the commencement of a maintenance job.
5. Provides procedures for implementing and continuing preventative maintenance program.
6. Provides a communication link between the maintenance, production and other departments.
7. Provides definite guidance for the hiring of daily wage labor and/or contractors.
8. Helps the field staff to coordinate with the production and project personnel.
9. Helps senior management to monitor the progress by the performance reporting.
10. Reduces time required for critical shutdown and overhauls.
11. Reduces overall maintenance costs.
12. Provides a tool for assigning priorities.
13. Reduces emergency breakdown requiring crisis management.

## 2.4 The Five to One Thumb Rule

This thumb rule states that for every one rupee you spend for proper maintenance planning, you save 5 rupees. Several case studies can be cited to show how organizations saved lakhs of rupees by spending marginally on improving the maintenance practices.

## 2.5 Types of Maintenance Plans

The plans for maintenance activities can be classified into developmental plans, medium range plans and short range plans.

- *The developmental plans* or long range plans are made for the purpose of surveying an appraising future maintenance requirements and needs for resources, especially during the development of new products or processes conceived by the organization. They can be meaningful if they include objectives that can be predicted within reasonable amount of certainty. They normally cover a period of 3 or more years and include, among others.
- *The medium range plans* cover those plans made for a period of one year, include annual preventive maintenance schedules, rehabilitation of machinery and equipment, needing repair and/or replacement. Maintenance of other facilities like boilers, air conditioning equipment or even building modifications. A typical medium range plan situation can be represented by the Fig. 2.1.

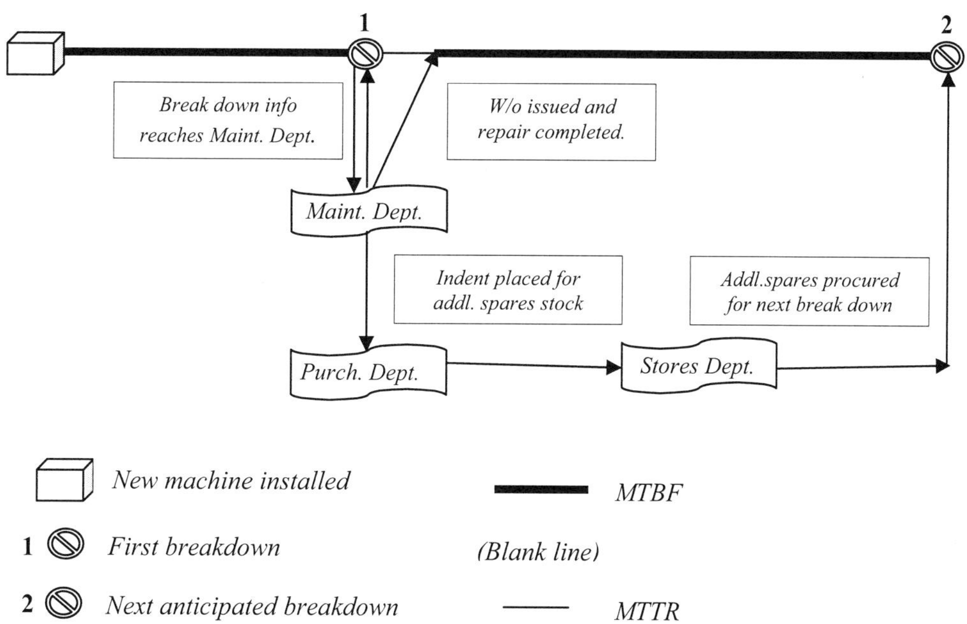


**Figure 2.1** A typical medium range plant situation.

- *The short range plans* or the near term plans generally cover one month and include the breakdown maintenance, routine maintenance like lubrication, condition monitoring, facility inspection and other day-to-day activities. For closer control, these plans are divided into weekly and even daily plans.

As time passes the long range plans become part of medium range plans and ultimately the near term plans.

## 2.6 The Essentials of Maintenance Planning

Maintenance planning is the application of maintenance procedures planned in a scientific manner. It is the comprehensive planning of the maintenance function. Each piece of the plant, each item of the machinery or equipment calls for different treatment. The most effective method of maintenance will vary according to the item concerned and its importance of the production process.

In essence, to establish planned maintenance, it is necessary to answer for each item of the plant the following questions:

- WHAT is to be maintained?
- HOW is it to be maintained?
- WHEN is it to be maintained?
- IS the maintenance cost effective?

## 2.7 What is to be Maintained?

As essential prerequisite to the organization of maintenance work is a comprehensive list of all equipment requiring maintenance. While the exact nature of the physical assets depends on the size and type of the industry concerned. There are certain essential items of information that is to be collected like:

- Make, model, age etc.,
- Location,
- Significance to production,
- Details of major operating components etc.

## 2.8 How is it to be Maintained?

For each item of the plant or major operating component, it is necessary to prepare a maintenance schedule. This is a comprehensive list of all the necessary maintenance tasks, and the intervals at which they should be carried out, as indicated earlier; there is no best way of carrying out planned maintenance that is universally applicable, but to adapt one or more of the four maintenance systems discussed more in detail in chapter 3. They are:

1. Run till the item breaks down and then replace or repair (Operate to failure or also called OTF)
2. Run till the specified, preplanned period and then replace (Preventive maintenance or PM),
3. Inspect for condition at specified intervals, adjust, repair or replace as required (Condition based maintenance or CBM), or,
4. Analyse the cause for the previous failure and take action to prevent similar failure in future. (Design out maintenance or DOM)

Factors that influence the decision about the method to be adopted and periods between examination and/or repairs are:

1. The manufacturer's recommendations,
2. Operating experience which indicates the actual deterioration rate and reliability of each item,
3. Results of planned inspections and test that provide useful guide to the deterioration rate,
4. The signification of the item or the effect of failure on production safety etc.

## 2.9 When is it to be Maintained?

Based on the physical asset register, the schedule of work to be performed on each item and the frequency of each task, a maintenance program can be prepared. This allocates each maintenance task to a specific time or period. There are four objectives to be achieved in formulating the program. These are:

1. The most efficient use of resources - men and machinery. Maintenance work calls for a wide variety of skills like fitters, and electricians, sometimes all being needed on a certain task. If all these are not available in the plant, outside contractors may have to be employed. Hence, careful planning is necessary to ensure that the right skill is available when required, without delay.
2. Adequate and economical provision of spares: The maintenance program and the individual maintenance schedules provide a sound basis for determining spares requirements. Proper records enable buffer stock levels to be kept low and so minimize cash tied up in stocks of unwanted items.
3. The maintenance program covers both the running and shut down maintenance. The latter can vary from a complete overhaul to a few minutes' job when an inspection is made and adjustments carried out. Close liaison with the user, particularly in the production areas is essential, without which, the

maintenance staff will not be able to gain the necessary access to the item with a complaint that the production will be interrupted.

4. There should be effective coordination with the production or other user departments.

## 2.10 Priority Allocation of Maintenance Activity

- *Emergency:* Safety of life or property threatened. Immediate serious impact on mission.
- *Urgent:* Continuous facility operation threatened. Impending serious impact on mission.
- *Priority:* Degrades quality of mission support. Significant and adverse effect on project.
- *Routine:* Redundancy available. Impact on services insignificant.

## 2.11 Objectives of Planned Maintenance

Planned maintenance is expected:

1. To involve and educate the operators in the functional concept, operation, and maintenance of equipment by delegating the responsibility,
2. To secure health of operators and others by reducing the chances of accidents,
3. To achieve good quality of goods and avoid scrap by keeping machines clean and maintaining their accuracy,
4. To assure full capacity utilization of equipment, and to avoid down time and loss of output by proper scheduling,
5. To prevent major repair work by forethought, watching wear out of the machines and performing minor repair work in due time,
6. To optimize volume of maintenance by proper planning and control of services and repair work in order to minimize maintenance costs,
7. To extend physical life-span of the equipment as long as it is economical to do so, in order to get an optimum return on the investment,
8. To implement reconditioning and replacement program by recording the history and costs regarding each equipment,
9. To improve future production and maintenance planning by evaluating experience and feedback for loading, budgeting etc.,
10. To reduce wasteful or unnecessary downtime to increase production and thereby the profits,
11. To achieve higher production and reduction of maintenance costs.

## 2.12 Benefits of Planned Maintenance

| *Objective achieved* | *Figs.* | *By means of* |
|---|---|---|
| 1. Production |  | (a) Ensured availability of the equipment at the needed time<br>(b) High utilization of plant and machinery<br>(c) Reduced down time<br>(d) Improved reliability<br>(e) Good housekeeping and safe methods and working conditions |
| 2. Cost Control |  | (a) Efficient utilisation of services like steam and compressed air<br>(b) Reduction of labour cast and overheads |
| 3. Working Capital | $ | (a) Reduced inventories<br>(b) Material submission<br>(c) Standardization |
| 4. Sales turnover | 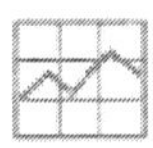 | (a) Improved quality and quantity of product<br>(b) Less selling expenses<br>(c) Company's prestige |
| 5. Human Relations |  | (a) Safer working conditions<br>(b) Skill development and job satisfaction<br>(c) Increased wage earnings<br>(d) Employee morale |
| 6. Research &Development |  | (a) Feedback information of all assets<br>(b) Design out maintenance to improve maintainability<br>(c) Problem analysis and modification |
| 7. Corporate planning |  | (a) Communication and information feedback systems |

**Figure 2.2** Benefits of planned maintenance.

## 2.13 Basic Steps of Planning of Maintenance Activities

1. Determine the objectives.
2. Decide the path to be taken that is the type and quantum of the resources, methods to be followed etc.
3. Schedule the activity - as and when these resources are to be used and in what manner.
4. Issue orders so that the resources are utilized according to the plans, and
5. Take corrective actions if any.

| ACTIVITY | WEEKS | REMARKS |
|---|---|---|
| A new machine breaks down in 12 weeks | 0 | 1. Machine wrongly selected<br>2. Machine manual not followed properly<br>3. Operator not instructed properly<br>4. Set practices not followed<br>5. Lack of lubrication |
| Maintenance engineer does fault analysis and locates the fault | 1 to 2 | 1. Loss of production time<br>2. Cost of labor and materials<br>3. High overheads |
| List of spares needed and indents made to the purchase department | 6 to 8 | 1. Spares not available in stock<br>2. No stock planning done<br>3. Recording of spares received with machinery not available<br>4. Lack of foreign currency<br>5. High cost of procurement due to air freighting |
| Either the production is stopped or transferred to another line | 8 to 12 | 1. Loss of production<br>2. Second line of production can give only 50%<br>3. Needs rerouting and extra handling<br>4. Additional cost of production<br>5. Disruption of regular production into related flow |
| Machine parts arrive but repair cannot start | 13 | 1. No feedback between the maintenance department, purchase department and the supplier<br>2. Maintenance department not prepared for the extra emergency workload<br>3. No proper manpower planning |
| Repair takes 2 to 3 weeks | 15 to 16 | 1. No maintenance planning<br>2. No formal training. Often an expert from the machine supplier is required. |

**Figure 2.3** An illustration of bad maintenance planning.

## 2.14 Working Pattern of Maintenance Activities

To attain the maintenance objective, it is important not only to carry out the right activities but to execute them optimally to achieve the best possible use of the resources. For this coordination and rationalization is important. Fig. 2.4 illustrates this process and its components.

***Planning***: By proper planning of the maintenance activities, many costly repairs could be avoided. In this way, the maintenance plan may be adapted to the market conditions and other external factors. Not only the preventive maintenance schedules etc. should be planned, but all activities connected with the maintenance activity like employment of personnel, stock control of spares etc. A significant symptom of bad companies is the poor and ineffective maintenance planning. The maintenance activities are carried out only in case of need, mostly on crisis management basis. These emergency jobs involve higher costs than the planned and preventive maintenance work, not only in direct maintenance costs but in indirect costs because such emergency jobs are difficult to plan and involve in loss of production time. Hence the concentration on systematic planning must be the first step for improving the maintenance activities in a company. Besides, all the related activities like selection of proper machines, corrective maintenance, employment and training of personnel, stock keeping of spares and consumption of maintenance materials shall be monitored.

***Execution***: Planning leads to execution activities. Even an emergency repair, where detailed preplanning is not possible, can be carried out better, if planning is done just before or during executing, with regards to tools, instruction etc. In this way emergency repair work can be done at lower costs.

***Recording and control***: A system for recording the plant operation and maintenance data should be set and well maintained. The maintenance system should include a recording section, from which we can easily obtain information like *what, when* and *why* the maintenance activities are carried out. In addition, observations in relation to the maintenance work, time consumed, spare part materials, the loss of production, etc. yield valuable guidance to the future planning. This information if used constructively would form the basis for expert systems in maintenance. The statistics collected and summarized during this assessment would enable optimal decisions.

***Analysis and assessment***: The recorded data has to be analyzed and assessed. The results of this analysis and assessment influence further planning considerably. Even at the project stage, that is while installing new plant or even equipment, recorded data should be used to ensure that the new plant is reliable. When the working cycle outlined above is completed, new information will enter system that is the maintenance plan gets modified. This is a continuous process in order to achieve optimal performance.

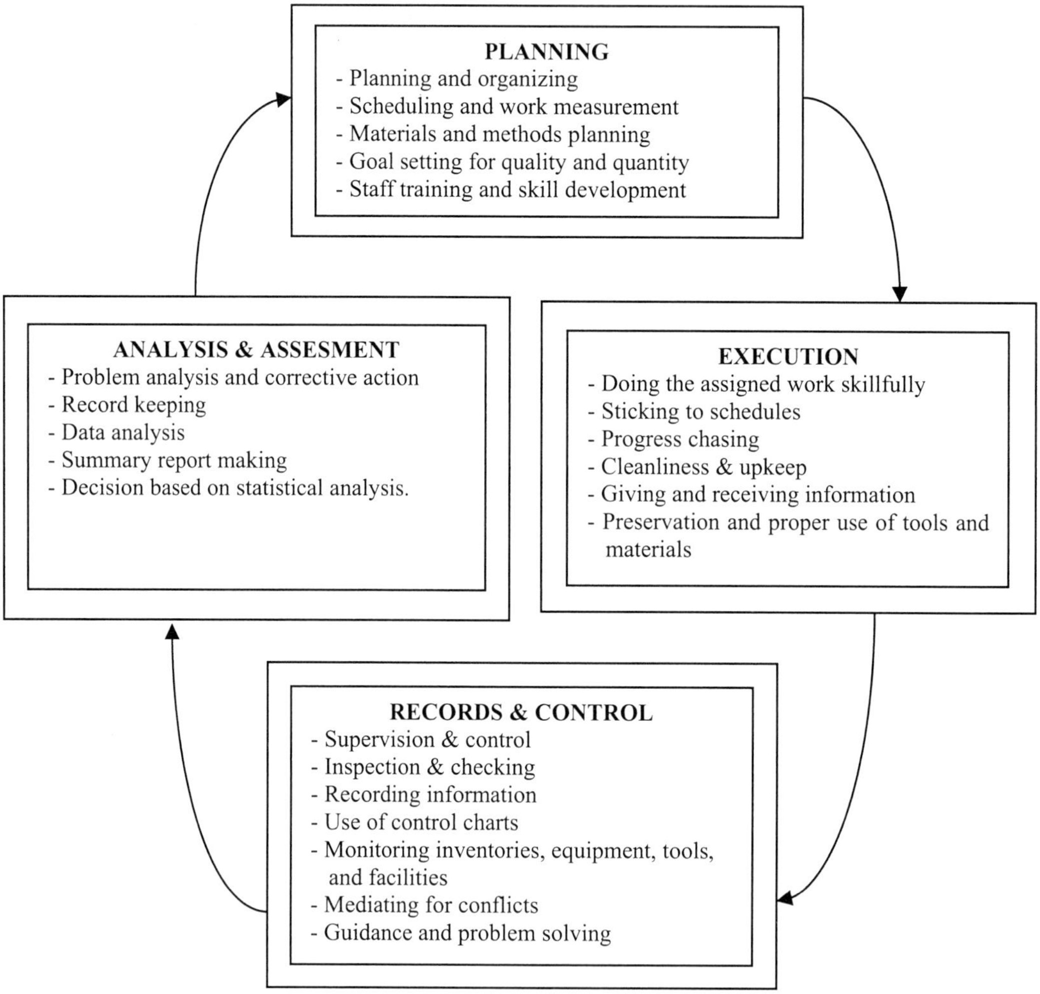


**Figure 2.4** Working cycle for maintenance activities.

## 2.15 Principles of Planned Maintenance Activity

Maintenance activities form major part of the production activity of the company, and hence contributes to the common goal of the company, viz maximizing profit.

1. *Production:* Maintenance activities should be carried out and adapted to enable the stipulated availability of the plant.
2. *Economy:* Maintenance activities should be carried out so that the total costs for maintenance resources will be kept at minimum for a specified attainment of availability. Costs for maintenance should not exceed the given objectives.

3. *Plant economy:* Maintenance activities should, where possible, be assessed and estimated with regard to the lowest cost of retaining the life cycle.
4. *Reliability:* The goal for all the maintenance activities should be to raise the reliability. This should be possible by first raising the functional reliability, secondly by raising the maintainability and thirdly by raising maintenance efficiency. These terms - reliability, maintainability and maintenance efficiency are elaborated further in chapter 4.
5. *Level of self supporting:* The Company's own personnel resources within the maintenance activity should cover atleast 85% of the maintenance volume of the plant. For the remaining part of the activities, especially if they involve specialized activities like boiler maintenance, external specialist's services may be sought for.
6. *Maintenance vis-à-vis production personnel*: Maintenance forms a natural part of a production system, which can function well if the production personnel are well acquainted with the maintenance requirements and methods. It is also necessary for the maintenance personnel to be acquainted with the production functions and requirements. Fig. 2.5 illustrates the relationship between the production and maintenance activities.

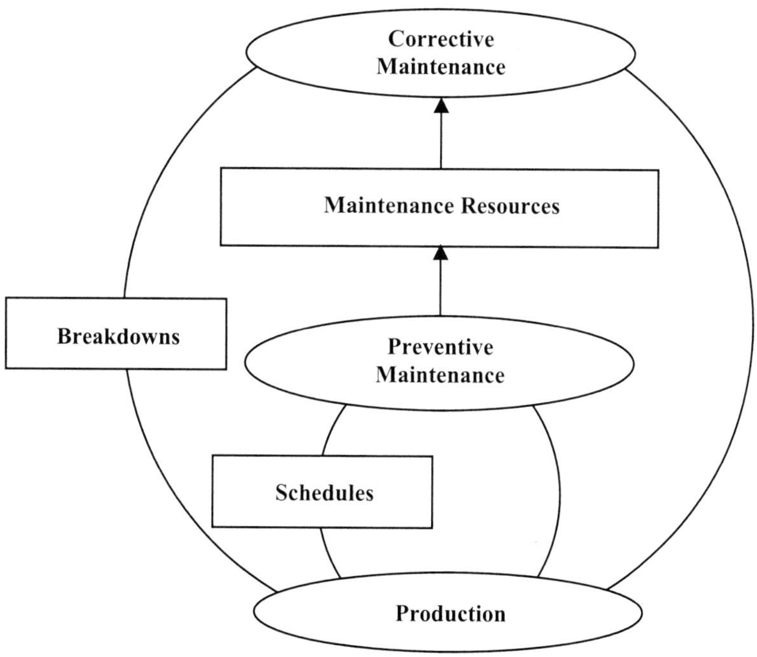


**Figure 2.5** Relationship between production and maintenance activities.

7. *Which type of maintenance to adapt*: Maintenance activities should be directed to foresee and counteract plant troubles (preventive maintenance) as well as to correct them (corrective maintenance). Whether the former or the latter to be adapted depends upon the situation, as well as the profitability to the company. Maintenance activities with intervals of two years or more and of major types like overhauls should be carried out when long term difficulties to attain a certain availability are observed or may be expected.

8. *Production of spares:* Production of spares, operational equipment etc. or subcontracting job to external companies should be carried out when the following are justified.
    - Clear cut profitability,
    - Objective of equalization of burden of work,
    - Delivery situation from the supplier.

9. *Investments:* Jobs of minor proportion in connection with investments may be carried out within the maintenance department, subject to the following conditions.
    - Specialized know how, experience and resources are available within the department,
    - Profitability would be influenced positively,
    - Up keeping of the maintenance condition does not decline.

10. *Carrying out maintenance jobs:* Maintenance jobs are carried out by the following units.
    - Local maintenance resources within the department,
    - Other resources within the plant like requesting the help of tool room,
    - Maintenance bodies in the area or district like boiler inspection agencies,
    - External maintenance bodies,
    - External suppliers.

    When placing maintenance job orders on external agencies, the following should be observed.
    - The best possible fulfillment of the objectives,
    - Clear-cut distribution of responsibility and organisation,
    - Making efficient use of resources,
    - Consideration for maximum overall profitability,
    - Long term development of maintenance activities.

11. *Personnel:* Maintenance activities need considerable contribution from the personnel. Cost for paying wages and salaries amount to normally 60% of the

direct maintenance costs. Hence for efficient utilization of manpower, the following points must be taken care of.

- Planning of available capacity,
- Adequate direct and indirect management,
- Well balanced guidance towards the achievement of the objectives,
- Continuous training of personnel and development of manpower resources,
- Making a budget and following up on the personnel costs.

12. *Responsibilities*: The responsibility for the maintenance resources like staff, materials, technology, administration and space lie on the head of the maintenance section. He will also be responsible for the rationalization and efficient planning and control of the maintenance activities.

13. *Supplies:* Maintenance supplies can be subgrouped into:

    - Consumption goods, like oil, cotton waste etc.,
    - Consumption spares, like bearings, electric bulbs etc.,
    - Specific spare component, identified by the drawing, and
    - Emergency spare parts.

    Consumption goods and consumption spares are regularly consumed items, and the supply is guided by statistical or the normal methods for purchase and stocking as adopted for the production items. Specific spare components and emergency spare items are taken into stock as a precaution against potential need as and when the maintenance demands, either in preventive or in corrective maintenance. These should be checked and assessed regularly.

14. *Technology:* Preventive maintenance, central lubrication, selection of materials etc., are examples of different technologies used in order to reduce the total need of maintenance resources. Guidance, checking, and assessment should be carried out in order to develop constantly and establish the essential new technologies.

15. *Administrative facilities:* Administrative facilities within the company should be made use of. For the maintenance activity, there is additional need for facilities. For example, with the maintenance instructions, machine history cards etc., These facilities should well cover the plant in order to reduce stoppage time and to reduce the need for direct supervision.

16. *Result of the maintenance activity:* The results of the maintenance activity is assessed by evaluating the availability and other indices, given in chapter 4, and comparing them with the cost incurred and/or productivity improved.

17. *Profitability:* For control purposes, all maintenance costs should be distributed on the different production equipment, so that productivity improvements for each equipment can be compared with the cost incurred.

Fig. 2.6 illustrates the impact of the principles on the cost effectiveness of a maintenance policy.

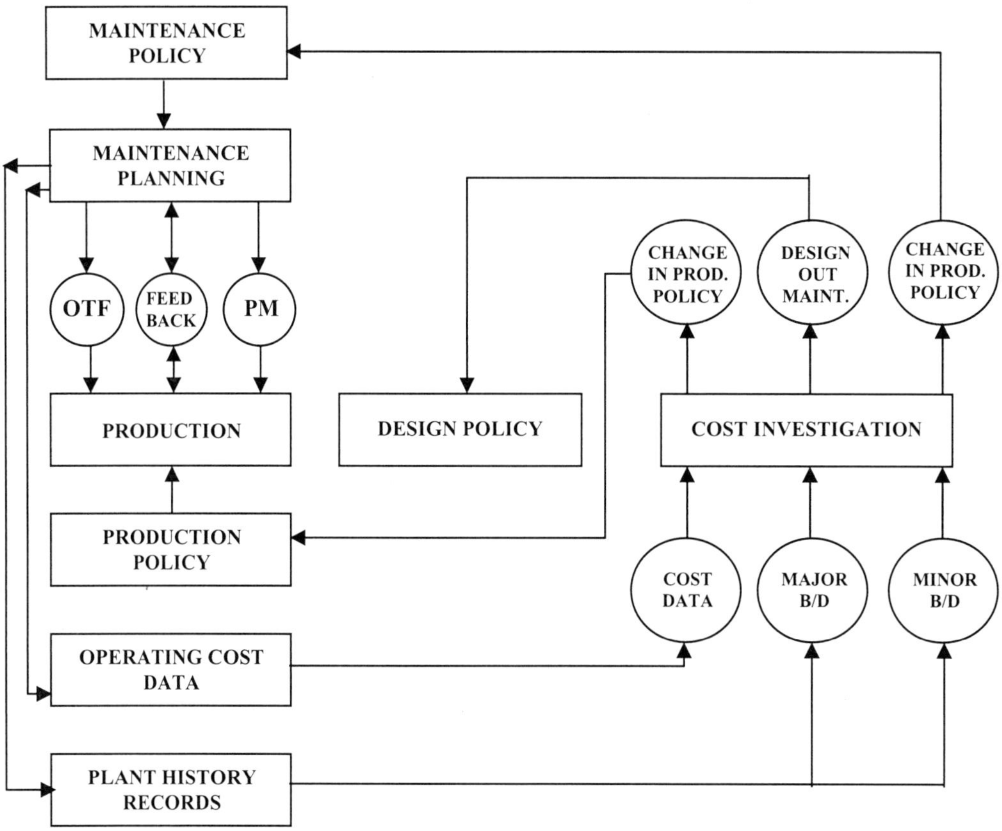


**Figure 2.6** Cost effectiveness of a maintenance policy.

## 2.16 Conclusion

It has become a foregone conclusion that effective maintenance planning is the backbone of manufacturing. If the fundamental principles explained in this chapter are understood and followed, the overall production efficiency could always improve irrespective of the type of industry.

# Chapter 3

# Categories of Maintenance

*Chapter synopsis:* *While citing the different categories of maintenance approaches, each of these approaches is defined and explained in detail with illustrative diagrams. These categories are discussed further in terms of their equivalent policies. The factors that determine the policy to be adapted in different situations are discussed. The scope and procedure for critical and complex replaceable items are highlighted.*

*Keywords:* *Planned and unplanned maintenance, basic categories and approaches, directly preventive maintenance, indirectly preventive maintenance, predictive maintenance, remedial maintenance, running maintenance, opportunity maintenance, design out maintenance, proactive maintenance, proactive reliability maintenance, operate to failure, sound maintenance management, maintenance policy, complex replaceable items, permanent items.*

## 3.1 Different Approaches to Maintenance

The different approaches for the maintenance practices can be better understood by the Fig. 3.1, outlining the basic categories and illustrating their interrelationships.

## 3.2 Definitions of the Categories

*Maintenance*: This covers all actions necessary to keep an item in or restore it to a specified condition. This basically falls into planned and unplanned categories. Planned maintenance covers all maintenance organized and carried out with forethought, control and use of records, to a predetermined plan. This category comprises of the preventive, predictive and to a large extent the corrective maintenance which refers to the work that is carried out after the failure, but for which advance planning and provision is usually made in terms of spares, materials, labor and equipment. Unplanned maintenance activities are necessitated due to unexpected breakdowns and/or plant shut downs.

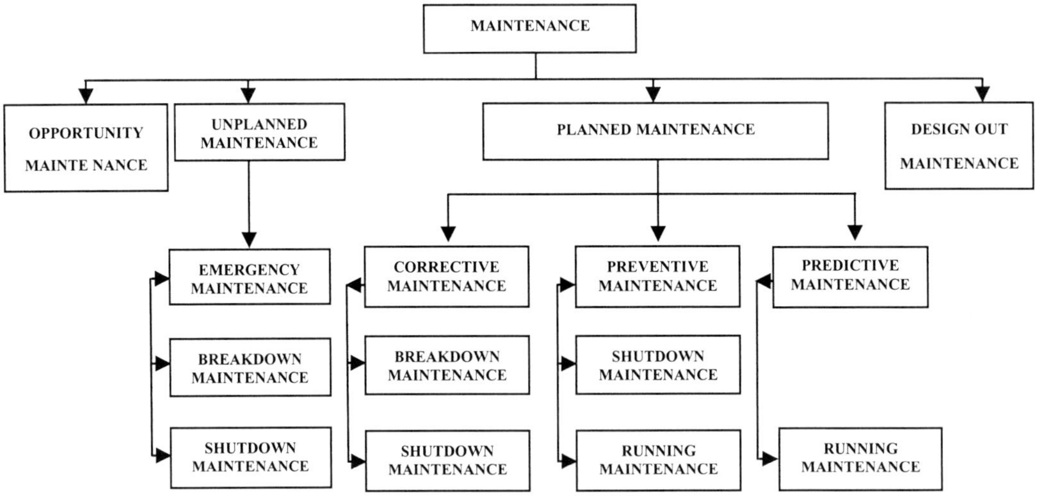


**Figure 3.1** Basic categories of maintenance.

***Preventive maintenance***: This covers all programmed activities carried out in order to prevent failures or to reveal the causes of failures before damage or disturbance occurs. Logical fault finding methods are adapted to identify the trouble spots. Preventive maintenance can either be direct or indirect.

*Directly preventive maintenance:* This covers those tasks or activities that can be classified under planned overhauls, either minor or major, that may be needed to bring the equipment back to a specified standard of to the level of required effectiveness. These tasks are always preplanned.

*Indirectly preventive maintenance:* This covers the routine or established attention tasks designed to preserve the equipment at a specified standard of effectiveness. These tasks include routine inspections, lubrication, and regular adjustment and where justified, running replacement of normal component parts like drive belts, tool post bolts and indicator lamps. In short, the routine preventive maintenance jobs assigned to the semiskilled staff come under this category. In these days of Total Production Maintenance, these jobs are being assigned to the machine operator itself.

***Predictive maintenance***: This is a planned maintenance, where the condition of all the vital components are continuously checked and assessed, either when the machine is running or when it is stopped for a short time. This is also known as condition based maintenance and is more popularly known as condition monitoring. Even though preventive maintenance and condition monitoring are two traditionally distinct types, in practice they supplement each other to such an extent that today the term preventive maintenance applies to all its techniques including condition monitoring. This aspect is illustrated in chapters 6 and 12.

*Corrective or remedial maintenance*: Maintenance undertaken to restore an item to a specified condition, after the failure has occurred, is called corrective maintenance. Many components of the plant and equipment system operate as individual items or separate from the actual manufacturing process components like the indicator bulbs or gear box covers. In some cases the companies have stand-by equipment so that breakdowns donot stop production and equipment can be repaired at scheduled times.

In these cases, their failure would not immediately or greatly affect the overall production process or constitute safety hazards. The cost of preventing their failures would be more than the loss due to the breakdown. In this particular situation it would be more optimal to continue to run the equipment till it fails before carrying out the major breakdown till maintenance action. Hence the components are allowed to work, they fail at random and then only they are replaced. This category comes under planned maintenance unlike the breakdown maintenance where the components that effect the production directly fail at random.

Automobile service provides a typical illustration of this situation wherein apart from minor preventive maintenance actions like oil-checking or oil changing or battery testing, detailed preventive maintenance actions are not undertaken for several running components, unless the driver starts hearing any abnormal sounds while the car is running. Another illustration is the replacement of fused out indicator bulbs or electronic components either on machines or in cars. These failures are minimal and occur at random without any pre-indication of the conduction. It would be cheaper to replace these after failure than during preventive maintenance. This type of running an equipment till it fails and then repairing it is called operate to failure.

Utmost care has to be taken to decide when corrective maintenance is to be adapted and has to be fully justified by the maintenance supervisor why he did not adapt preventive maintenance schedules in that particular situation. This is because, though corrective maintenance appears to be easy for the personnel, even a small provision for this in the policy would result in discretionary relaxation of the schedules of preventive maintenance. The maintenance personnel are too busily occupied with the preventive maintenance schedules planning and compliance. The more the corrective maintenance is accepted, the more are the chances that even unexpected breakdowns too would be considered as those planned for corrective maintenance. In practice it is surprising that even in many large organizations this system of letting the machine operate till failure in the name of corrective maintenance and then take fire fighting measures.

*Running maintenance*: This is the work that can be carried out when the facility is working and the facility need not be stopped atleast without affecting production.

Routing lubrication, testing and tensioning of the V-belts, V-belt cover tightening, tool post screw replacement, other minor adjustments during routine patrol maintenance etc. come in under this category.

***Opportunity maintenance***: This refers to the situation wherein, whenever a machine or a production line is stopped either due to failure of a certain item or for preventive maintenance or for condition based repair, opportunity is taken to direct the maintenance actions to items other than those which are the primary cause of failure or the stoppage, these actions being uneconomical otherwise. A simple example is, when the gearbox of an automobile is opened to repair a broken gear, opportunity is taken to check or repair parts in the clutch or other systems normally hidden behind the gearbox. Another common illustration is that when the plant is shut down for the repair of a primary machine, the secondary machines could be subjected to unscheduled preventive maintenance, since anyway these machines cannot work when the primary machine is stopped.

***Design out maintenance***: This aims to eliminate the cause of maintenance rather than the effect. Past experience is made use for the designing new machinery with lower probability of failures. Structural modifications of machines are sometimes possible, with a view to eliminate certain failure causing snags. Thus, in general, analysis of the design of the machine components with maintenance in view, is termed as design out maintenance. This design aspect ensures easy accessibility of items that need to be adjusted more frequently. An example of this is the location in the motor cycles of the spark plug which is the component that is to be accessed by the rider more often than any other component. On the other hand in some of the scooters, the spark plug is located deep inside, requiring the dismantling of the side panel and it becomes very difficult to clean the spark plug.

*Another illustration is the design of components for the mobile phones. Apart from the keyboard, the most common components handled by the user are the charger receptacle unit, the Micro SD card slot and the sim card receptacle unit (called tray) where the mechanical wear and tear caused by the frequent movement of the plugs or cards inside cause them to fail much earlier in comparison to the motherboard which is rarely touched. But since the these receptacles are integral part of the motherboard, the service personnel of a certain mobile manufacturing company say that these items, though minor, cannot be replaced without replacing the whole motherboard which costs almost as much as the whole handset. This is a bad example of product design without forethought and without applying the concept of design-out maintenance. This author wishes that all the mobile manufacturers would keep this in mind and come out with motherboards where these three components can be detached, replaced and reassembled by the service personnel without touching the motherboard.*

***Breakdown maintenance***: This falls under the unplanned category and is carried out after failure, but for which advance provision is usually not made. This advance provision can be considered as the chief criterion that distinguishes the corrective maintenance with the breakdown or emergency maintenance.

***Shutdown maintenance*:** This is the work which can be carried out only when the facility is out of service or is taken out of service.

As a summary we can say that reactive maintenance takes action when a pump breaks down, while in a *preventive* maintenance setting, the pump is cleaned, re-lubricated, etc., at regular intervals (x operating hours). In a *predictive* maintenance setting, the pump may be monitored on-line. When vibration levels increase, maintenance actions are initiated. Applying *design-out* maintenance means the pump is "redesigned" with a different type of lubricant or bearing, providing an easier accessibility to the internal components that fail more frequently.

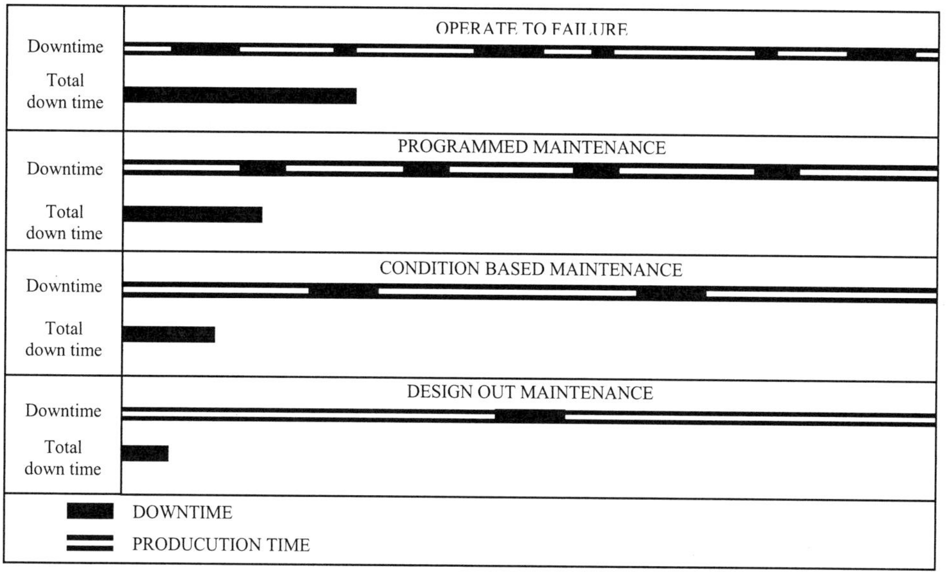


**Figure 3.2** Production downtime is influenced by the maintenance policy.

Fig. 3.2 illustrates how the production down times can be influenced by these different policies.

***Proactive Maintenance*:** The website www.skf.com/aptitudexchange/glossary.html indicates two more types of maintenance which are basically a combination of two or more categories. Proactive maintenance is the combination of operator-performed maintenance, preventive maintenance, and predictive maintenance activities whereby maintenance is conducted to prevent, eliminate, delay or reduce maintenance before failure.

**Proactive Reliability Maintenance (PRM),** based on four key steps – predictive maintenance, diagnostics and root cause analysis, key performance indicators, and operational review is a strategy that addresses failures and implements the processes necessary to prevent recurrence. PRM enables an organization to take complete control over what is happening on the plant floor. In industry, PRM is partly compared to design-out maintenance.

## 3.3 Maintenance Policies with Specific Reference to Maintenance Spares

From these definitions it can be inferred that there are basically 5 types of maintenance policies that can be adapted either individually or in combination, depending upon the situation.

1. *Operate to failure*: This is a maintenance policy wherein no predetermined action is taken to prevent a failure and is applicable for those items of the plant and equipment that operate as individual units or separate from the direct processing process. This category of maintenance is hence corrective maintenance, and is suitable where the cost of preventing a failure would be more than the loss due to the breakdown itself. Replacing of indicator bulbs, switches, relays etc. fall under this category.
2. *Fixed time replacement or preventive maintenance:* This is effective where the failure for a component is clearly time dependent, the item being expected to wear out within the life of the unit, and where the total costs, both direct and indirect, of such replacements are substantially less than those of failure replacement. The preventive maintenance can generally have two components, *fixed time maintenance* where adjusting and repairing or replacing is done at fixed preplanned periods, or *fixed time inspection*, where the inspection is done at equal or variable inspection periods, but adjustments repair or replacement is done as warranted by the condition. Basically those actions that are carried out before a failure are called preventive, while those that are carried out after failure are called corrective.
3. *Condition based maintenance:* Maintenance carried out in response to a significant deterioration of a unit as indicated by a change in the monitored parameter of the unit's condition or performance is called condition monitoring as defined above. This defers from the earlier two policies, in the manner in which the condition of any parameter of the unit is being monitored regularly, and suitable action is taken when a change in the condition so monitored indicates a necessity. This means that preventive maintenance added to the programmed replacements, overhauls, lubricating, cleaning etc., also includes systematic condition

monitoring. Preventive maintenance used in conjunction with condition monitoring, leads to maintenance works that may easily be planned and this in turn results in lower maintenance costs with higher production.

4. *Opportunity maintenance:* In general, since several machines work in continuous production line, stopping one will affect the working of all other machines. In this case whenever the main machine of the line is stopped for repair/adjustment, then this opportunity is taken to inspect/repair/replace the components of the less critical machines of the line or whenever there is a power shutdown.

5. *Design out maintenance:* As a clear contrast to the preceding policies, which aim to minimize the effect of the failure, this policy aims to eliminate the cause of failure itself. Clearly this requires engineering action than the maintenance action. Nevertheless it is often the responsibility of the maintenance department, especially, wherever there is a possibility to modify the existing structural or functional design of a unit to result in lower failure probability. For this, the past data on the breakdown history of the unit has to be analysed. Fig. 3.3 illustrates these alternative maintenance procedures.

| No. | What | When | What action to be taken |
|---|---|---|---|
| 1 | OTF | Operate to failure | Replace or repair after failure |
| 2 | FTM | Fixed time maintenance | Adjust/repair/replace at preplanned intervals. |
| 3 | FTI | Fixed time inspection | Inspect at equal or variable inspection periods and adjust/repair/replace as needed |
| 4 | CBM | Condition based maintenance | Inspect on continuous basis and adjust/repair/replace as needed |
| 5 | OM | Opportunity maintenance | Inspect an item during some other item's maintenance/inspection period and if needed adjust/repair/replace |
| 6 | DOM | Design out maintenance | Incorporate maintenance accessibility, ease of maintenance etc. during the design stage. |

**Figure 3.3** Alternative maintenance procedures.

## 3.4 Nomenclature used in SAE Standard JA1011

It may be brought to the notice of the reader that SAE released a standard No. SAE JA 1011, on Reliability based Maintenance, in which these policies are further explained, but with different nomenclature for the maintenance categories in its section of 5.9. This nomenclature used in SAE JA 1011 as against that used in this chapter is compared and illustrated in Table 9.2 in chapter no. 9.

## 3.5 The Importance of Sound Maintenance Management

1. Preventive maintenance efforts can interfere with production just as breakdowns can. Maintenance efforts must be managed effectively to keep either type of interference to an economical minimum.
2. Operating a maintenance department costs money, so it should be planned and controlled just as any other department. Maintenance activities affect a company's return on investments since they reflect in additional expenses.
3. Effective maintenance helps to enhance the productivity of a company by influencing the percentage of time that the equipment can operate.
4. It keeps the equipment in fit condition improving the quality of the system's output.
5. It effects the economic life time and the salvage value of the equipment.
6. The increase in downtime reflects in delayed or missed sales.

## 3.6 The Determination of a Maintenance Policy

The maintenance policy for a plant should be built up by selecting for each unit, the best combination of the policies outlined in paragraph 3.3, and then by coordinating these policies in order, to make the best use of the resources and time.

Ideally the preventive and corrective actions for each unit of the plant should be specified in some detail by the manufacturers. This is usually so in the case of simple replacement items, where maintenance is inexpensive and deterministic. This is rarely so for complex replaceable items where maintenance is costly and probabilistic. An explanation of these types of items is given in paragraph 3.6.

Many factors effect the selection of a policy appropriate for each item. This, together with the large number of items usually involved, gives rise to the need for some systematic procedure for determining the policy, which again would be valid for a particular length of time. These are generally reviewed annually. This procedure is outlined below.

- *Classification and identification of equipment:* The simplest of various systems of classification in some form of a number or mnemonic code, identifying the location, function, part and replaceability etc. of the machine and its part. A system of identification is discussed in detail in chapter 20.
- *Collection of Information:* Collection of all information that might be relevant to the maintenance policy planning is certainly essential to every unit of the plant.

Since maintenance is inseparable from production, the information of greatest relevance is the production pattern, whether continuous, fluctuating or intermittent. Such an information having been obtained it would then be possible to construct a schedule, for each unit and each decision period, of the expected time available for maintenance that will not involve production loss.

- *Other information*, much of which is normally provided by the manufacturer is:
  (a) Manufacturer's maintenance recommendations, like trouble shooting, required actions and frequencies,
  (b) Equipment factors, like failure characteristics, minimum life and reliability,
  (c) Economic factors, like cost of failure, cost of replacement prior to failure,
  (d) Safety factors, like internal, environmental, safety regulations etc.

## 3.7 Maintenance Procedure for Complex Replaceable Items

The best procedure for each item is determined by first identifying the procedures which are effective and then deciding which is the most desirable (Fig. 3.4).

The choice will depend upon many factors and the decision criteria are normally of minimum cost, safety (statutory or otherwise) etc. It is necessary to apply this selection process separately to the different categories of each item, which can be sub classified, with reference to the ease of maintenance and replaceability.

| ITEM CATEGORY | SIMPLE REPLACEABLE ITEMS | COMPLEX FREPLACEABLE ITEMS | PERMANENT ITEMS | |
|---|---|---|---|---|
| | | | MAIN | SECONDARY |
| MAINTENANCE PROCEDURE | Adjust or replace by FTM/CBM | Repair or replace by CBM/FTM/OTF/DOM | Repair by CBM | No action |
| PREVENTIVE PROGRAM | Inspection, routine lubrication & routine periodic service | Condition based replacement/overhaul | Routine inspection | None |
| CORRECTIVE POLICY | | Refer decision guidelines for corrective action | | |
| INVENTORY POLICY | Stock mostly 'A' items, where demand for some components is predictable | Stock one or two of most needed items (Use past experience since demand is difficult to predict) | No stocking except for insurance spares | |

**Figure 3.4** Maintenance procedures for complex replaceable items.

Fig. 3.4 illustrates the guidelines for the selection of maintenance procedures for complex replaceable items which is more elaborated in paragraph 16.3 of chapter 16.

## 3.8 Summary of Factors for Selection of the Policy

1. The fixed time replacement or preventive maintenance is usually most suitable for low cost simple replaceable items.
2. A condition based monitoring systems is generally most suited for high cost complex replaceable items.
3. Design out maintenance systems is suitable, where necessary and possible, for all high cost maintenance items, replaceable or permanent.
4. Where the preventive maintenance replacement action is not suitable or economically justified, the policy of operate to failure may be followed. Fig. 3.5 illustrates the alternative maintenance procedures.

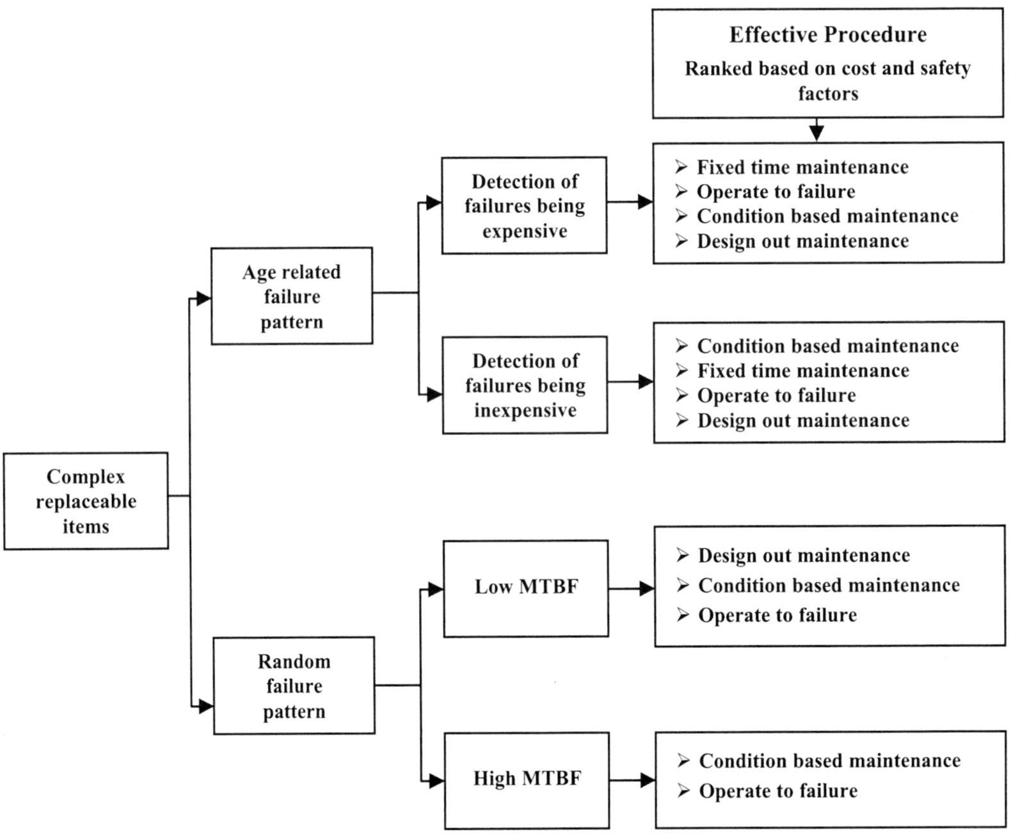


**Figure 3.5** Maintenance procedures for critical units.

## 3.9 Conclusion

This chapter clearly explains the different maintenance policies and how a specific policy should be followed to suit specific maintenance issue. It also emphasizes how two or more policies can be integrated like preventive maintenance together with condition monitoring as illustrated in Figure 12.2 of chapter 12.

# Chapter 4

# Reliability and Machine Availability

*Chapter synopsis: This chapter discusses the physical aspects of machine availability and reliability and the computational relationship between them so as to provide an initial understanding of reliability, which is dealt more in detail in a later chapter on reliability based maintenance. The several factors, on which the machine availability depends, are discussed. Numerical examples are worked out to provide illustrations on these factors and to make the concept clearer.*

***Keywords:*** *Reliability, availability, breakdown time distribution, bath tub characteristics, failure rate, performance curves, $T_{up}$, $T_{down}$, system effectiveness, machine capacity, functional reliability (MTBF), maintainability (MTTR), maintenance efficiency (MWT), probability of failures, reliability improvement factor, indices of maintenance control, Pareto principle, 80–20 rule, law of the vital few, principle of factor sparsity.*

## 4.1 Availability of a Machine

Maintenance can be considered as a combination of actions carried out in order to replace, repair, service or restore the components or an identifiable group of components of an equipment or machine or plant, so that it will continue to operate to a specified availability for a specified time. In short; the function of maintenance is to control and increase the plant availability.

Then what is this availability?

Availability is defined as the situation in which a component or a product or a system is performing its required function at a given point of time, when used under stated operating conditions. It is expressed as a percentage of time a component or product or a system is operating over a specified time interval.

In general

$$\text{availability} = \frac{\text{working time}}{\text{working time + down time}}$$

To understand the above equation better we shall first distinguish between working time and down time. For this, it is necessary for us to develop the following concepts.

## 4.2 Breakdown Time Distribution

The down time refers to the time a component or a system is down, which is some form of break down. The useful period is free from breakdown.

As we said above, the useful life period is not free from breakdowns, they occur at random intervals as illustrated in Figure 4.1, that shows atypical portion of the service life of an equipment, during which failures occur at random. Here only the breakdown maintenance system, via repairing the machine only after the breakdown occurs, is considered. A comparison of a similar case with other systems is given in chapter 2. The lower part of the figure is a hypothetical representation of the total time lost due to the breakdowns during the period considered.

In fact the breakdown time treated as in Fig. 4.1 has 2 components, the waiting time before the repair can start and the internal repair time from the time the repair, team commences the identification of the fault, till it is rectified.

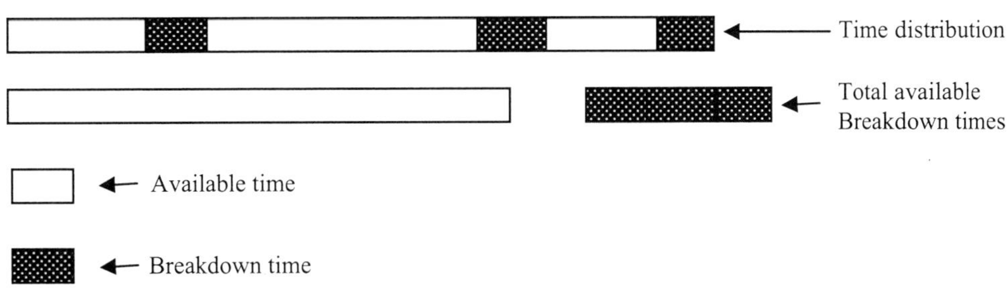


**Figure 4.1** A typical breakdown time distribution.

## 4.3 Bath Tub Characteristics

The failure characteristics of a majority of the equipment follow the pattern shown therein, sometimes called like bath tub pattern, which has three distinct phases (Fig. 4.2).

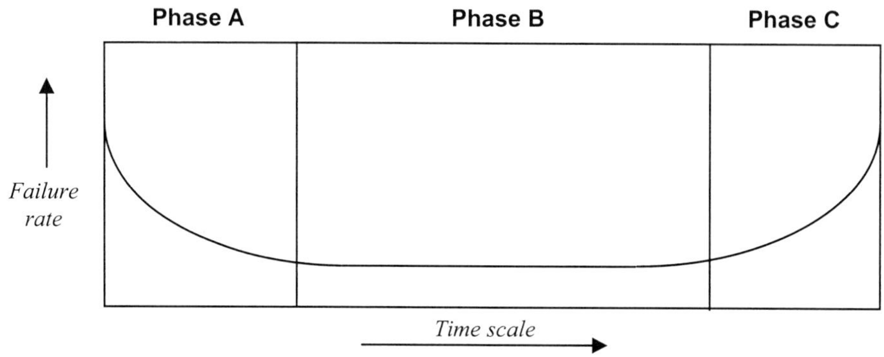


**Figure 4.2** Failure rate of an equipment.

*Phase A or the burning in period*: The major contributing factor for this failure is the poor component quality. When the equipment is given initial trials, there might be many initial failures due to poor design, workmanship, assembly errors etc. Damaged components and poor joints or connections also contribute to this failure. These are tested and replaced generally at the manufacturer's premises to improve the reliability.

*Phase B or the useful life period*: Here the failure rate is low, but may occur unexpectedly and at random intervals. They are known as random failures or normal failures. It is during this period, that all our availability or reliability analysis is based on. The major contributing factor is the stress to which the equipment or products subjected to and could be due to operating stresses, poor maintenance, operator abuse and accidents.

*Phase C or the wear out period*: Beyond the useful period, the wear rate is the major contributing factor because of ageing or wear of the components of the system and could be due to weak design, poor lubrication, wear, fatigue failure, corrosion and insulation breakdown. In short, Table 4.1 illustrates the contributing factors for each phase.

**Table 4.1** Contributing factors for failures.

| | Phase A | Phase B | Phase C |
|---|---|---|---|
| **Period** | Burning in | Useful life | Wear out |
| **Failure occurrence** | Trial | Random | Excessive |
| **Major contributing factor** | Low Quality | Stress | Wear |
| **Other contributing factors** | Weak design, assembly errors, damaged components, poor joints/connections | Operating stresses, poor maintenance, poor lubrication, operator abuse, accidents | Weak design, wear, insulation breakdown, fatigue, corrosion |

A statistical representation of the probability that a machine can have maintenance free performance for a given number of operating hours is given in Fig. 4.3.

Curve A shows a case where the machine is not subjected to severe conditions of services and tend to breakdown at nearly constant intervals following the last repair. The statistical variation of these intervals is given by a normally distributed curve with, its mean corresponding to the specified free run time ($T_a$).

Curve B shows a case of machine having more moving parts than in case A. As the failure of any of the moving part would result in the failures of the whole machine, the variations expected in the average free time are more than the case A, with the free time expected itself being lower. Thus the distribution curve for this will be a slanted type of normal curve.

Curve C shows wide variations in the free times, in the case of machines necessitating intricate and careful setting up for efficient performance.

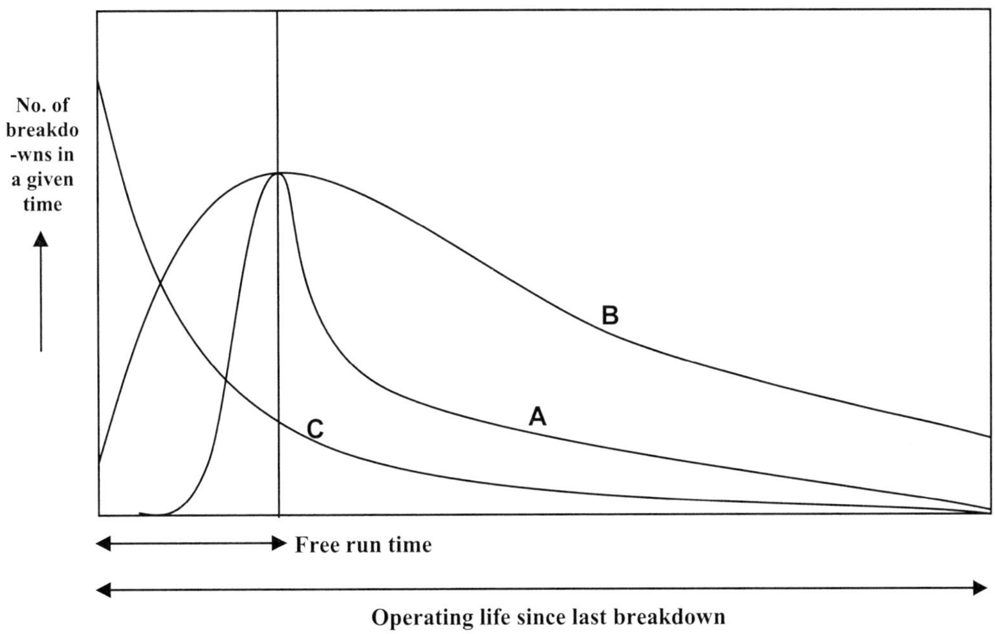


**Figure 4.3** Maintenance free performance curves.

## 4.4 Basic Definition of Availability

As illustrated in Paragraph 4.1, the availability can be defined as the ratio of the total working time to the total available time. In other words, the availability of a machine in specified time is given, by

$$A = \frac{T_{up}}{T_{up} + T_{down}}$$

where, $T_{up}$ = the cumulative working time of the machine and,

$T_{down}$ = the cumulative down time of the machine.

Referring to the Paragraph 4.5 it is clear that the system effectiveness can be increased by increasing the machine capacity and/or the machine availability. Nevertheless the increase of machine capacity involves high capital investments and other considerations that are beyond the control of the maintenance department. In fact, it is the availability factor that can be increased and contribute towards the common goal of increased output.

## 4.5 Factors of Availability

Availability of a machine is a function of how often a machine has a breakdown and how long it takes to repair it. These factors are defined, as follows.

(a) *Functional reliability:* It is the ability of an item to perform a required function under stated conditions for a stated period of time. It depends on how OFTEN the breakdown occurs and is expressed by the term MTBF, the Mean Time Between Failure. It is measured, in hours. Referring to Fig. 4.1 and also Fig. 4.5, it is also equal to the total time of working ($T_{up}$).

(b) *Maintainability:* It is a characteristic of the design and installation, and is given by the average expected time taken to restore an equipment to its specified condition in accordance with specified, procedures and resources. It is expressed by the term MTTR, the Mean Time to Repair, and its unit is hours.

(c) *Maintenance Efficiency:* It is an organisational factor influencing the promptness with which a maintenance job is attended to. It is expressed, by the term MWT, the Mean Waiting Time, and is measured in hours.

Thus the total availability of the machine can be expressed as in Fig. 4.4, while Fig. 4.5 illustrates some of the components of these factors,

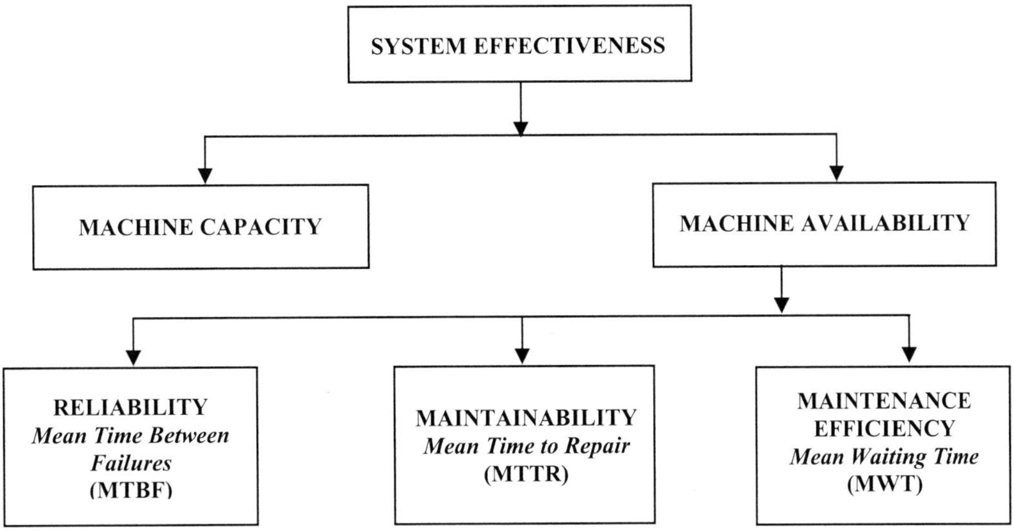


**Figure 4.4** Machine availability.

$$\text{Availability} = \frac{\text{Reliability}}{\text{Total Machine Available Time}}$$

$$= \frac{\text{MTBF}}{\text{MTBF} + \text{MTTR} + \text{MWT}}$$

**Figure 4.5** An expression for machine availability.

For example if a machine has an MTBF of 10 hours, MTTR of 2 hours and MWT of 0.5 hours, then availability is

$$\frac{10}{10+2+0.5} = 0.8$$

This relation between these factors of availability is further explained in Fig. 4.5. Some of the components of these factors are also explained in Fig. 4.6.

| | | | |
|---|---|---|---|
| Availability or Performance Reliability | Functional Reliability | Trouble Intensity | 1. Checking & Inspection<br>2. Dimensions<br>3. Design Principles<br>4. Supply |
| | | Redundancy | 5. Parallel<br>6. Standby<br>7. Bypass |
| | | Maintenance Intensity | 8. Preventive Maintenance<br>9. Opportunity maintenance<br>10. Replacement systems |
| | Maintainability | Trouble detectability | 11. Means of fault recording<br>12. Testing equipment<br>13. Condition checking |
| | | Spare Availability | 14. Standardization<br>15. Modularization<br>16. Transport & Storing |
| | | Repairability | 17. Kind of trouble<br>18. Standard tools<br>19. Accessibility<br>20. Working place environment<br>21. safety and protection |
| | Maintenance Efficiency | Maintenance Staff | 22. Quality<br>23. Skill<br>24. Contractors and consultants<br>25. Training needs<br>26. Wages and salaries |
| | | Repair equipment | 27. Machinery units and tools<br>28. Special equipment<br>29. Workshop location |
| | | Supply depot | 30. Replacement units<br>31. Spare parts<br>32. Consumables |
| | | Technical data | 33. Maintenance instructions<br>34. Operating instructions<br>35. Drawings<br>36. Index |
| | | Administration | 37. Organization<br>38. Reporting<br>39. Control systems |

**Figure 4.6** Factors influencing performance reliability.

The same can be represented by the Fig. 4.7 also.

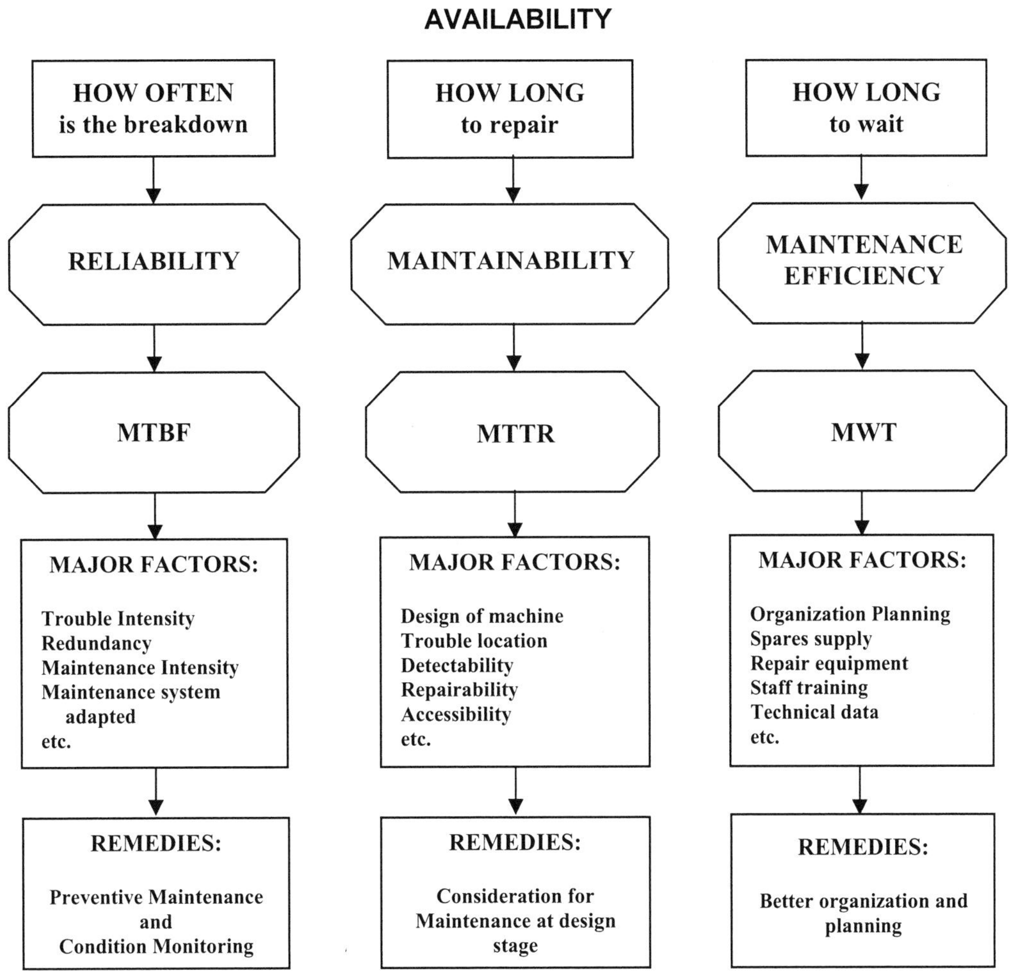


**Figure 4.7** Machine availability.

## 4.6 Failure Rate and Probability of Failures

Considering a system that has a large number of elements, the failure rate is the number of failures that occur in equipment, in a unit time say 100 hours.

or $$F = \frac{100}{MTBF}$$

Profitability of failure of an item in a given period of time is given by

$$P_f = \frac{nt}{T}$$

where

n = the number of elements in the system,

t = the length of the given period in hours,

T = the average MTBF of each element in hours and,

Reliability can also be expressed as

*Reliability* = 1 - probability of failure = 1 - $P_f$

*Reliability improvement factor* is the ratio of the probability of failure of the present system to that of the improved system.

Or in other words,

$$\textit{Reliability Improvement Factor} = \frac{P_{f_1}}{P_{f_2}}$$

where, $P_{f1}$ and $P_{f2}$ are the failure probabilities before and after the improvement respectively.

$$\text{R.I.F} = \frac{Pf_1}{Pf_2}$$

**Figure 4.8** Reliability improvement factor.

## 4.7 Other Indices of Maintenance Control

While reliability of a machine is an indirect but the most popular index, for the performance of the maintenance function, logically speaking the measurement of maintenance should be linked with productivity. The productivity of an equipment is reduced by,

(a) Scheduled stoppages,

(b) Breakdown stoppages,

(c) Reduced output due to defective plant,

(d) High production of scrap or waste products, etc.

To minimise the above causes, the maintenance has to put in its efforts in the following forms, which, no doubt cost money:

(a) Frequency of inspection and lubrication,

(b) Preventive checks and repairs,

(c) Scheduled, overhauls and major repairs,
(d) Predictive maintenance and use of condition monitoring equipment,
(e) Division of volume of work, and methods of work,
(f) Control on cost of spares and spares inventory,
(g) Use of proper tools, equipment and materials, standardisation of parts and equipment,
(h) Supervision and overheads.

With a consideration of above, the following indices can be developed to measure the maintenance performance.

(i) $\text{Maintenance Productivity Index} = \dfrac{\text{Production turnover (or output in Rs.)}}{\text{Cost of the maintenace effort}}$

(ii) $\text{Maintenance Cost Index} = \dfrac{\text{Maintenace cost} \times 100}{\text{The total capital cost}}$

(iii) $\text{Breakdown Maintenance Index} = \dfrac{\text{Total Man-hours Spent in Breakdown}}{\text{Total Man-hours Available}}$

(iv) $\text{Level of Maintenance} = \dfrac{\text{Total Hours spent on Scheduled Maintenace} \times 100}{\text{Total Man-hours Available}}$

## 4.8 Numeric Examples

### Example 1:

*A specific machine working in a factory, on a single shift basis, from 8 am to 12 noon and 1 pm to 5 pm, Monday through Friday, has the following working history during a particular week. Calculate MTBF, MTTR, MWT and Availability for that week. Presume that the maintenance workers follow the production timings only.*

| | | |
|---|---|---|
| Monday | 8.00 am | Start machine, |
| | 10.30 am | Breakdown, |
| | 11.00 am | Start repair, |
| | 3.30 pm | Finish repair and start machine, |
| Tuesday | 2.30 pm | Breakdown, |
| | 4.00 pm | Start repair |
| Wednesday | 10.30 pm | Finish repair and start machine |
| Thursday | | No breakdown |
| Friday | 2.30 pm | Breakdown, start repair, |
| | 3.30 pm | Finish repair and start machine |

## Solution:

Let us represent the down times on a time scale (Fig. 4.9).

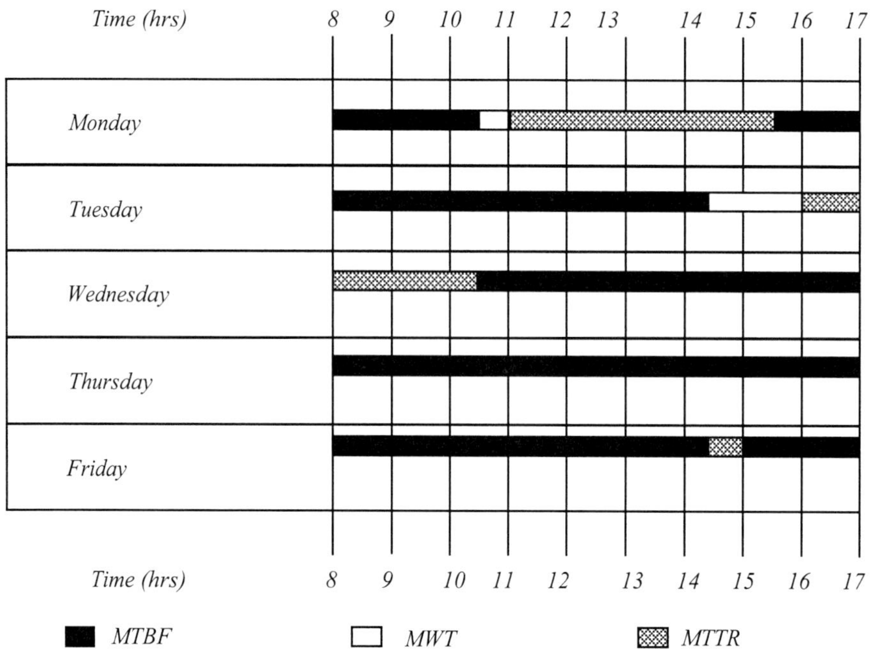


**Figure 4.9** Representation of waiting time on a time scale.

The working times, the waiting times and the repair times can be summarised in a table form.

| Period. No. | WORK DAY | WORKING PERIOD | WAITING PERIOD | REPAIR PERIOD |
|---|---|---|---|---|
| 1 | Monday | 08.00 to 10.30 | 10.30 to 11.00 | 11.00 to 15.30 |
| 2 | Monday | 15.30 to 17.00 | nil | - |
| 3 | Tuesday | 08.00 to 14.30 | 14.30 to 16.00 | 16.00 to 17.00 |
| 4 | Wednesday | | | 08.00 to 10.30 |
| 5 | Wednesday | 10.30 to 17.00 | | |
| 6 | Thursday | 08.00 to 17.00 | | |
| 7 | Friday | 08.00 to 14.30 | nil | 14.30 to 15.00 |
| 8 | Friday | 15.00 to 17.00 | | |

Therefore,

Total working time = 2.5 + 1.5 + 5.5 + 6.5 + 8.0 + 5.5 + 2.0 = 30.5 hours

Total waiting time = 0.5 + 1.5 + 0.0 = 2.0 hours

Total repair time = 3.5 + 1.0 + 2.5 + 0.5 = 7.5 hours

There are three failures during the week, and the periods 1 and 4 overlap with those of the adjoining weeks, and hence considered as ½ period as far as this week is concerned.

Reliability = MTBF = 30.5/3 = 10.16 hr

Maintainability = MTTR = 7.5/3 = 2.5 hr

Mean Waiting Time = MWT = 2.0/3 = 0.67 hr

$$\text{Availability} = \frac{\text{MTBF}}{\text{MTBF} + \text{MTTR} + \text{MWT}}$$

$$= \frac{1.16}{10.16 + 2.5 + 0.67} = \frac{10.16}{13.33}$$

$$= 0.76$$

**Example 2:**

*An electronic system is composed of 100 elements, each having an MTBF of 100,000 hours. Determine*

(a) *the probability of the failure of the system in an operating life of 100 hours*

(b) *the reliability of the system*

**Solution:**

(a) Probability of failure = $P_f = \frac{nt}{T}$

where,

n = number of the elements in the system = 100

t = the length of the given period in hours = 100

T = Average MTBF of each element = 100,000

Hence Probability of failure = 100 × 100/100,000=0.1

Reliability = 1 – $P_f$ = 1 – 0.1 = 0.9

**Example 3:**

*If the reliability of a system is increased from 0.9 to 0.95, what is the reliability improvement factor?*

**Solution:**

Probability of failure in first case

$$= 1 - \text{reliability} = 1\text{-}0.9 = 0.1$$

Probability of failure in second case

$$= 1 - \text{reliability} = 1\text{-}0.95 = 0.05$$

$$\text{Reliability improvement factor} \quad \frac{0.1}{0.05} = 2$$

Thus by increasing the reliability from 0.9 to 0.95 the probability of failure is reduced half or the reliability improvement factor is doubled.

**Example 4:**

A certain system can function alike, either with 5 vacuum tubes each with an MTBF of 10,000 hours or with 20 transistors each with an MTBF of 80,000 hours. Which system has the higher reliability?

**Solution:**

Probability of failure in the cases of the system with vacuum tubes

$$= P_{f1} = \frac{n_1 t}{T_1} = \frac{5t}{10000} = \frac{t}{2000}$$

Similarly for the second case with transistors,

$$= P_{f2} = \frac{n_2 t}{T_2} = \frac{20t}{80000} = \frac{t}{4000}$$

Obviously the second system is better because the probability of failure in the latter case is half of the former. The improvement in the reliability by switching over from system 1 to system 2 is hence.

$$= \frac{t/2000}{t/4000} = 2$$

In general, a system having elements each with reliability of $R_1$, $R_2$, $R_3$, ....$R_n$ etc., then the overall reliability of the system can be said to be the product of these reliabilities i.e., $R_1 \times R_2 \times R_3 \times \ldots R_n$. In other words if each of these elements have equal reliability of R the overall reliability would to be $R^n$.

## 4.9 Pareto Principle

The Paretto principle (also known as (i) the *80–20 rule*, (ii) the *law of the vital few*, and (iii) the *principle of factor sparsity*) states that, for many events, roughly 80% of the effect come from 20% of the causes. In other words, in any population, 20% of the people contribute to 80% of a parameter say the GDP. This is similar to the Principle of ABC analysis which states that in an engineering industry 10% of the production items contribute to 70% of the total annual consumption. When applied to the maintenance situation, we can say that as per Pareto principle, 20% of the machines cause 80% of the total machine down time.

This principle would help in identifying which of the machines that breakdown more often so as to concentrate the preventive maintenance efforts more on these machines.

## 4.10 Note on Reliability based Maintenance

In this chapter we have considered reliability as one of the physical factors affecting equipment availability. In another chapter on reliability based maintenance, we will see how reliability is a function of probability of failure occurrence and how simulation analysis based on this would help us plan preventive maintenance under the scenario of failures that occur at random.

## 4.11 Conclusion

Any manufacturing system's effectiveness not only in terms of the machine availability, but also in terms of machine reliability is wholly defendant on the machine parameters like MTBF, MTTR and MWT as explained and illustrated in this chapter. The performance indexes explained in paragraph 4.7 provide guidelines to the maintenance manager to increase the effectiveness of his department.

# Chapter 5

# Preventive Maintenance

*Chapter synopsis*: *Periodic inspection and service of plant facilities would expose conditions leading to breakdowns, thereby saving us lot of time, energy and money in repairing. This chapter emphasizes the need for the preventive maintenance and its several concepts. Some illustrations are provided how preventive maintenance is practiced in specific industries.*

***Keywords:*** *Preventive maintenance, optimal level of maintenance, history of preventive maintenance, maintenance costs, planned preventive maintenance, inspection and checkup, frequency of inspection, planning and analysis, TANITA, cashew shelling machine, TIPER, Hyundai Motor India.*

## 5.1 Introduction

'Prevention is better than cure'

'A stitch in time saves nine'

So far we have been discussing the fundamental principles behind maintenance activities. We have also seen that among the several categories of maintenance practice, preventive maintenance pays a very significant role in the upkeep of industrial equipment. This not only answers the *'where'* *'when'*, *'by whom'* and *'what'*, but also *'why'* of maintenance. It involves the policy of inspection, upkeep, replacement and/or modification rather than *operate to fail* or *repair after failure*. The preventive maintenance activities include partial or complete overhaul at predetermined periods. This involves stoppage of the machine before undertaking complete inspection. It is a systematic procedure where the condition of the equipment is constantly watched and maintained at a specific level. It hence lays emphasis on the prevention of breakdowns.

One major misconception about preventive maintenance is that it is unduly costlier than in letting the machine operate until repair is absolutely necessary. And that it would cost more for scheduled downtime and maintenance even before any breakdown occurs. But in fact it has been established that without preventive maintenance the cost of production loss when the machine breakdown unexpectedly. This can be illustrated by Fig. 5.1 which is more explained in chapter on maintenance costs.

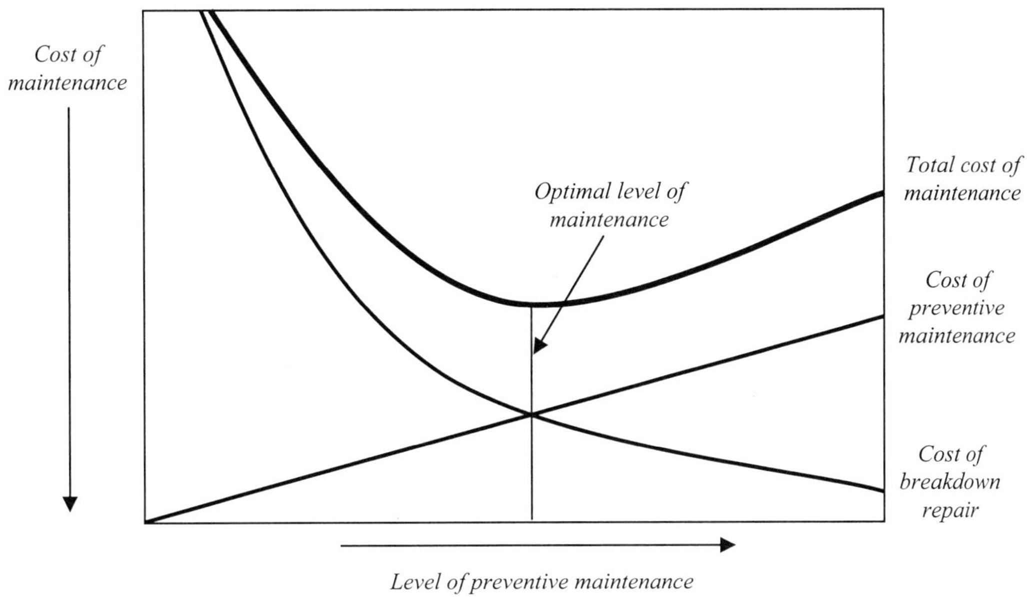


**Figure 5.1** Maintenance costs.

Inadequate maintenance would lead to both high cost of repair and loss of production time. Such losses can be minimized only by giving proper attention to the prevention of machine breakdowns. To avoid inconvenience to the production department with random and unannounced stoppage of machines, preventive maintenance is always preplanned, usually before the start of the calendar year or financial year, indicating which machine would be stopped when and for how long, so that the production personnel can plan their Production Planning and Control (PP&C) well with least production loss. This is also called as Planned Preventive Maintenance (PPM). Lubrication and oil change are the other major activities of preventive maintenance.

## 5.2 Definitions of Preventive Maintenance

Wikipedia defines preventive maintenance as

> *The care and servicing by personnel for the purpose of maintaining equipment and facilities in satisfactory condition by providing systematic inspection, fault detection and correction of incipient failures either before they occur or develop into major defects.*

Soviet Encyclopedia offers the following explanatory definition.

> *Preventive maintenance is a systematic and preventive operation or groups of operations for maintaining technical devices or items at a given level of reliability in a serviceable or active state. Preventive maintenance is generally performed at specified intervals that are determined in advance. The procedure consists of an inspection of the item and the replacement or repair of its parts and units. The item is also cleaned, lubricated and adjusted. Preventive maintenance reduces the likelihood of sudden loss of operating efficiency that can result from, for example, the deterioration of the parts of a device of the clogging of its contacts. Preventive maintenance may also be performed at regular intervals in conjunction with the restoration of the operating efficiency of an item after its failure.*

The British Standards Institute's definition is,

> *The combination of all technical and administrative actions, including supervision actions intended to retain an item in, or restore it to a state at which it can perform its required function.*

Other definitions offered on preventive maintenance by various authors are:

1. *The maintenance is a set of organized activities that are carried out in order to keep an item in its best operational condition with minimum cost acquired.*
2. *Activities of maintenance functions could either be repair or replacement activities, which are necessary for an item to reach its acceptable productivity condition and these activities should be carried out with a minimum possible cost.*
3. *Preventive maintenance is a procedure for inspecting, testing and reconditioning a system at regular intervals according to specific instructions intended to prevent failures in service or to retard deterioration.*

## 5.3 History of Preventive Maintenance

During the pre-war era, people were averse to stop the production machines just for doing routine checking. Maintenance was thought of adding to the cost of the plant without increasing the value of the product and was limited to the fixing of the unit only when it breaks down.

During second World War, the emphasis was for higher production, needing the machines to work non-stop for all three shifts. This resulted in increased occurrence of breakdowns of the machines. The production supervisors did feel the need for continuous on-line maintenance like lubrication and occasional quick off-line stoppages for repairs, but this feeling could not be sustained due to non-stop pressure from the army for higher production.

It is after the World War that this thinking deepened and mooted the idea of providing spare machines just to keep as stand-by when the main machine breakdown. As a compromise between this idea and the actual practice, the basic maintenance processes were developed. Subsequently increased awareness of such issues like environment safety, product quality and service made maintenance as one of the most important function that contributed for the success of the industry (Fig. 5.2). World class companies are in continuous need of a very well organized maintenance programme to compete globally. The advances made in engineering and sciences with advanced research specific to maintenance management like the wear debris analysis.

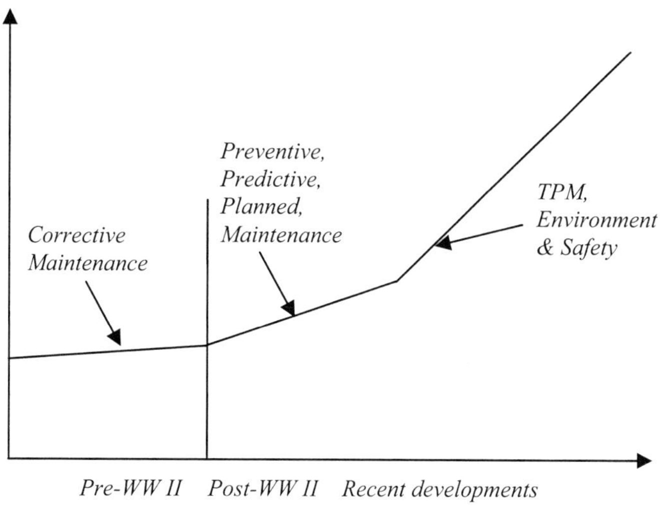


**Figure 5.2** History of maintenance (adapted from Wikipedia).

## 5.4 Objectives of Preventive Maintenance

The primary goal of preventive maintenance is to prevent failures before they actually occur. It helps in preserving and restoring equipment reliability by checking and replacing worn components before they fail. In general the primary objectives can be listed as under.

1. To obtain maximum availability of the plant by avoiding breakdowns thereby reducing the machine downtime to a minimum.
2. To keep the machines in good condition so as to maintain the quality product.
3. To optimise resources utilisation.
4. To achieve better spares stock control.
5. To preserve the value of the plant by minimizing wear and tear and reduce scrap rate.
6. To ensure the safety of the workers.
7. To keep the plant at the maximum production efficiency.
8. To achieve all the above objectives economically.
9. To provide reliable cost and budgetary control.
10. To minimise energy usage.

A good preventive maintenance program could achieve the following:

1. Identification of equipment or systems that may malfunction
2. Typical equipment to be inspected including switchgear, circuit breakers, transformers, switches, etc.
3. Establishment of schedules and procedures for routine inspections
4. Periodic testing of plant equipment for structural soundness
5. Prompt repair or replacement of defective equipment found during inspection and testing
6. Inventory of spare parts for equipment that needs frequent repairs
7. Use of an organized record-keeping system to schedule tests and document inspections
8. Commitment to ensure that records are complete and detailed, and that the record test results and follow-up actions
9. Preventive maintenance inspection records should be kept with other visual inspection records.

## 5.5 Elements of Preventive Maintenance

Preventive Maintenance activities include oil changes, lubrication and partial or complete replacement, overall maintenance or overhauls done at or specified periods. It also includes the recoding of the history of equipment deterioration so as to provide a knowledge on the precedence of the incidences and the best action to be taken. In general, the following are the elements that are generally included in a preventive maintenance program

1. Inspection and checkup
2. Lubrication
3. Planning and scheduling
4. Dispatching and execution
5. Records and analysis
6. Preparation of standard repair procedure
7. Training of maintenance staff
8. Planning and control of spare parts inventory.

These elements of preventive maintenance can be illustrated as per Table 5.1, further explained in the subsequent paragraphs.

**Table 5.1** Elements of preventive maintenance.

| No | Action | Details | Purpose |
|---|---|---|---|
| 1 | Lubricate | Friction points | Retard and slow down the process of wear and deterioration |
| 2 | Clean | Machine and surroundings | Remove dust and prevent corrosion of the external and internal parts. This step is also called 5S, further explained in the chapter 17. |
| 3 | Adjust | Nuts, loose belts, alignment etc. | Eliminate undue stresses and reduce vibration and noise |
| 4 | Paint | Protective coating | Protect against rusting and corrosion of internal or external parts |
| 5 | Inspect | Monitor condition | Apply condition monitoring principles and also by routine patrolling. |
| 6 | Analyze | History of behavior | Use past history to estimate and evaluate the cause and extent of the defect so as to set up an optimal procedure. This is also called the expert system. |
| 7 | Plan and schedule | PM procedure | As explained in subsequent chapters to adapt ideal maintenance procedure |
| 8 | Dispatch | Issue work orders and instructions | To ensure smooth and unambiguous procedure is followed by the maintenance workforce. |
| 9 | Replace | Worn out components | Mostly as per the PM schedules, but where necessary unscheduled components as indicated by condition monitoring also can be replaced. |

**Table 5.1** *Contd...*

| No | Action | Details | Purpose |
|---|---|---|---|
| 10 | Repair cracks etc. repairable damages | Repairable components. | Where spare is scarce or very expensive and if the repaired component can perform well. |
| 11 | Modify | Component design and assembly | This is also called Design Out Maintenance (DOM), as explained in chapter 3. |
| 12 | Replace | Equipment where needed | Considering the economics of replacement as further explained in chapter 23. |
| 13 | Record | All repair elements | To provide useful guidelines in future. |
| 14 | Establish | Standard repair procedures | For an efficient management of activities. |
| 15 | Train | Maintenance staff | In all the procedures described above |

These elements can also be illustrated as per Fig. 5.3.

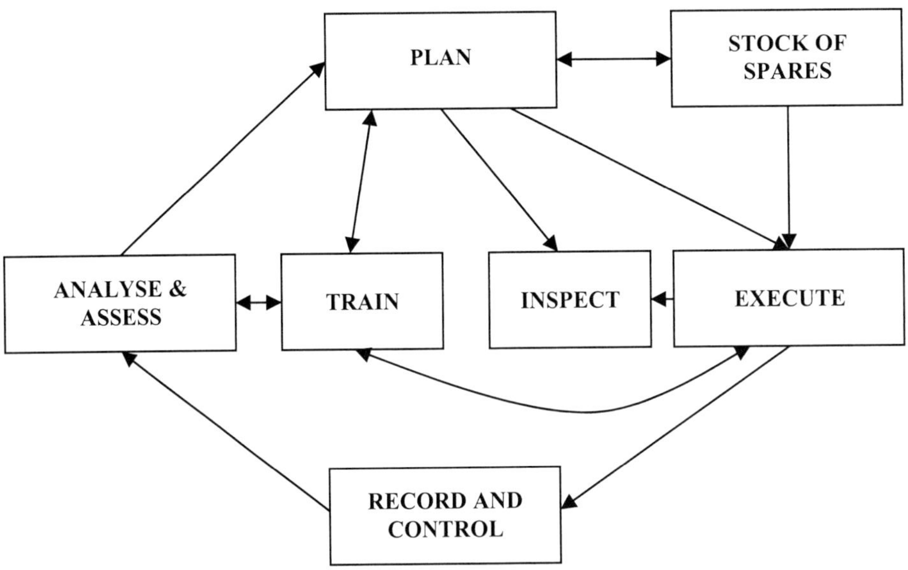


**Figure 5.3** Inter relationship between elements of preventive maintenance.

## 5.6 Inspection and Check Up

Inspection is the most essential element of the preventive maintenance program. The maintenance crew stops the equipment and carries out regular inspection of all the critical components of the equipment to check their working condition. Nevertheless this inspection can also be carried outside the preventive maintenance schedules, by constantly inspecting the critical parts while the machines are working. In other words, the inspection can either be *on-line inspection* or external inspection without stopping the machine or *off-line inspection* by stopping the machines, for inspecting the internal and normally inaccessible parts like gears, bushes and bearings. These types of inspections, more specifically the latter, are called condition monitoring, which is described more in detail in the next chapter. Frequency of inspection should be determined based on past experience since too much inspection may result in wastage of machine production time and too less inspection may lead to frequent breakdowns.

Some of the items that are normally subjected to routine inspection are the oil levels, commutator undercut, checking the functions of the relay, measuring insulation resistance of the motor, checking the foundation bolts of the equipment for tightness, checking the temperature of the motor, checking the slackness of the drive belt. It may be noted that most of these form part of condition monitoring except that the latter is performed by senior supervisors generally using simple instruments, as detailed in the next chapter.

Appendix I illustrates an inspection chart used for maintenance planning and scheduling in a cashew processing industry. The company has 168 shredding machines to break open the shells of partly roasted raw cashew nuts. These machines are laid in 28 rows with 6 machines per row. Each of the machines has a variety of moving components that need regular inspection. At the end of the day a maintenance inspection crew selects 25% of the machines in each row and inspects each and every component for their working condition and report in the records in the format as illustrated.

## 5.7 Frequency of Inspection

While the determination of the frequency depends on each specific machine, equipment or plant, the general guideline for the frequency is as follows:

- *Daily:* For vital and critical items and for machinery coming under purview of mine Rules.
- *Weekly:* For enclosed lubrication systems, force lubrication systems, coolant systems etc.
- Monthly: For equipment like compressors, air conditioners, filter and ore processing equipment. This inspection shall be more thorough and is generally off-line.
- *Quarterly*: For equipment of special review. Some equipment needing longer period of inspection could be inspected quarterly. This should be more

meticulous and any part that is suspected not to last next 4 months shall be replaced.

- *Annually*: This is the salient feature of preventive maintenance program. The group of machines that are interrelated in a production line, or large equipments like heavy vehicles, elevators, electroplating plant etc. could be stopped for period ranging from one day to fifteen days for performing the inspection together with part replacement and overhauling.
- *Opportunity preventive maintenance:* Certain periods like holidays, major power cuts, raw material shortages, certain factory closures due to bandh etc. can be used to rescheduled and conduct preventive maintenance of some major equipment wherever it is possible.

## 5.8 Lubrication

Mechanical components like bushes and bearings, give good performance for long periods when they are systematically lubricated by using the right type of oil at right time and with right method of application of oils. A large number of breakdowns can be prevented by proper lubrication, which hence forms one of the most important elements of preventive maintenance. This aspect is discussed in detail in chapter 10.

## 5.9 Planning and Scheduling

Every preventive maintenance job should be on the basis of the analysis of the past records. A scheduled program should be prepared giving the daily, weekly, monthly and annual attention for each and every machine and equipment. This should be followed strictly. Few examples of such programs are given at the end of this chapter.

## 5.10 Dispatching and Executing

The preventive maintenance schedules shall be updated every week as weekly planning instructions to each maintenance supervisor and/or maintenance worker followed by a work order for each maintenance job. This is discussed more in detail in chapter no. 21.

## 5.11 Records and Analysis

Good record keeping is essential for a good preventive maintenance and helps in forecasting the maintenance requirements. A typical record is the Equipment History Card in which all the basic details of the past maintenance history is recorded as illustrated in chapter 20. Nevertheless it should be noted that excessive paper work delays the adoption of preventive maintenance. To simplify paperwork minimize the number of forms and entries and integrate preventive maintenance paperwork with other maintenance records. The following basic reports are suggested to be prepared for each equipment and reviewed annually:

- Equipment record
- Check list
- Inspection schedules
- Equipment and maintenance cost record

It may also be noted that digitizing the record keeping would not only minimize difficulties related to excessive paperwork but also help in the analysis. This is illustrated more in chapters 20 and 22.

## 5.12 Training

For the success of any preventive maintenance program a sound training of the personnel is essential for carrying out the inspection and repair in a systematic manner.

## 5.13 Control of Spare Parts

In a majority of situations the machines remain idle for want of spares parts resulting in delays in the repair process which in turn result inconsiderable loss of production. At the same timeover stocking of less or unwanted parts would increase inventory costs. A well planned stock control is hence a must and is discussed in detail in chapters 15 & 16.

## 5.14 Reconditioning of Machine Tools

This is an extension of the annual overhauls, done at longer periods as necessary. This would be advantageous for large industrial units to avoid/defer capital expenditure in procuring additional equipment. In deciding to take up equipment for reconditioning, the following may be studied.

- The equipment history card having its complete maintenance history, would tell if the overhauling or the reconditioning is needed and if at all it is worthwhile to recondition it.
- The reports of the recent alignment tests.
- The nature of work performed and the extent of precision demanded from the machine tool.

## 5.15 Preventive Maintenance Program at Hyundai, Chennai

Hyundai Motor India at Chennai has outlined the following features of a well-planned preventive maintenance program and put them as posters as a part of employee awareness. *(courtesy- GM(Maint), HML)*

**A. Machine Down Time Reduction (MDTR)**

1. *Multi functioning*
   - (a) On the job training
   - (b) One point lesson
   - (c) Class room specialized training on PLC's, Hydraulics, pneumatics etc.
2. *Reliability centered maintenance*
   - (a) Robot operation check
   - (b) Shock pulse monitoring of bearing's condition
   - (c) Quality check on filling equipment
   - (d) Weld parameters check
3. *Implement computerization in CMMS*
   - (a) History of equipment
   - (b) Work out Mean Time Between Failures
   - (c) Check operation time and efficiency
   - (d) Follow preventive maintenance plan
   - (e) Prepare work standards for all activities
4. *Zero hour spares*
   - (a) Minimum stock on belts and bearings
   - (b) Develop original equipment suppliers
   - (c) Make spare supplying ancillaries to maintain some of their stock at HMI
5. *Heath check maintenance*
   - (a) Follow five sensory feed back
   - (b) Feedback from operator
   - (c) Morning greetings and status enquiry with operator
   - (d) Ready condition check

**B. Machine Time Availability (MTA)**

1. *Visual education*
   - (a) Prepare operation sheet

   (b) Check interlock sequence
   (c) Display kaizen, improvement themes
   (d) Mark TPM points
2. *Enhance quality of frequent failure parts*
   (a) Eliminate root cause of failure
   (b) Fix damping pad to prevent metal contact
   (c) Reduce speed wherever it is permitted
   (d) Do value engineering analysis to improve quality
3. Maintain 5S condition
   (a) Keep work space well arranged
   (b) Take responsibility for your action
   (c) Keep good attitude
   (d) Display neatness everywhere
4. *All round campaign*
   (a) Adhere to campaign plans
   (b) Practice change management
   (c) Circulate bench marking practice
   (d) Carry out self-audit for your performance

**C. Material Flow Management (MFM)**

1. *Stock computerization - AS 400*
   (a) Stock level planning
   (b) Re-order level planning
   (c) Standardization of spares
   (d) Easy search enquiry of history by members
2. *Continuous review of stock levels*
   (a) Rework of spare parts
   (b) Periodic planning and replacement of consumables
3. *Disposal of obsolete stock*
   (a) Fix color tags for items exceeding shelf life
   (b) Collect scrap copper, aluminum of high value

   (c) Send enquiries to vendor companies
   (d) Exchange obsolete devices for latest versions

4. *Water level management*
   (a) Fixing of automatic level controllers in cooling tower
   (b) Flushing of all lines to prevent scaling
   (c) Replacement of pumps to meet actual needs.

5. *Resourceful space utilization*
   (a) Rearrange work area
   (b) Minimize stocking of spares using Two Tier bins
   (c) Distribute racks and bins to all sections
   (d) Keep clean and ensure multipurpose usage of data room

**D. Machine Inspection Standards (MIS)**

1. *Visual marking*
   (a) Eye marking on all fasteners and gauges regarding opening, closing etc.
   (b) Fix values and visually communicate to all
   (c) Mark directions of flow of oil, air and gases
   (d) Follow uniform color coding of pipe lines.

2. *Performance checking*
   (a) Conduct audit on equipments
   (b) Reduce downtime and ensure equipment availability
   (c) Carry out root cause analysis on breakdowns
   (d) Follow annual safety inspection on equipment

**E. Machine Safety Awareness (MSA)**

1. *Safety interlocks*
   (a) Prepare try-out status reports on safety interlocks
   (b) Follow failure-safe mode in operation
   (c) Check and recheck interlocks in automatic equipments
   (d) Change wears parts in brake systems

2. *Two Button Operations*
   (a) Fix two button operations (TBO)s on all presses
   (b) Display TBO necessity in stations

   (c) Give training to operation on TBO
   (d) Standardize for wheel of progress

3. *By-pass status display*
   (a) Data liner for display of faults
   (b) Give alarm signal for by-pass condition
   (c) Install hooter in critical equipments
   (d) Encourage participation through shared information

4. *Safety guards*
   (a) Fixing of beam sensors in shuttles
   (b) Covert all rolling elements with guards
   (c) Prevent unauthorized entry into automatic lines
   (d) Check safety fences in all shops
   (e) Check functioning of all safety devices and alarms

5. *Safe work practices*
   (a) Wear helmets, goggles and safety shoes
   (b) Switch off power before starting work in electrical lines
   (c) Avoid/eliminate slippery surfaces
   (d) Follow hot work permits inside work places

**F. Machine Education and Training (MET)**

1. *On the job training*
   (a) Training - a new tool for communication and thought sharing
   (b) Clarify confusing terms
   (c) Teach elementary techniques
   (d) Update general knowledge of employees
   (e) Give examples from experience
   (f) Maintain technical reports, one point lesson etc.

2. Trouble shooting report
   (a) Identify the most common and all probable causes
   (b) Identify the methods to overcome them
   (c) Prepare trouble shooting charts.

3. Parameter display
   (a) Fix parameters to simplify the maintenance operation
   (b) Maintain visual documentations and manuals

## 5.16 Appendices

The following appendices I to IV are some of the forms studied and/or introduced by this author in Tanzania at TANITA Co. Ltd. (Cashew nut processing company), Tanzania Italy Oil Refineries (TIPER), Transport Depot (Shirika la Usafiri), etc. during his service at the University of Dar Es Salam, Tanzania.

## 5.17 How to Sell PM

It has been a common experience for maintenance engineers working in small or medium industries that the management is generally reluctant to stop the machines just for the sake of inspecting it, since it means production loss, and they give weightage value, the production manager with a line function more than the maintenance engineer with a staff function. Hence it is the duty of the maintenance engineer to sell the idea of preventive maintenance to the management, and for this he has to do his homework of proving how preventive maintenance saves the company a lot more than the production loss. The following steps are suggested:

1. Check over the previous records for the past one or more years for all machine breakdowns, the mean time to repair (MTTR), the number of hours lost, the number of persons working for the repair and the number of persons rendered idle because of the machine breakdown.
2. Workout the total repair cost in each case, involving the labor, material, overheads or any other charges.
3. Include the idle time rework etc. under specific categories.
4. Check if the breakdown resulted in any injury to the operatives, and include that also.
5. Make a detailed and clear cost comparison between the cost of production loss during the preventive maintenance period and the losses due to the breakdowns.
6. Add to this the detailed list of benefits as explained in this chapter.
7. Discuss this first with the production manager and the supervisor and win their appreciation. Finalize report and discuss with the top management in the presence of the production manager, who would have agreed with you to a great extent.

## 5.18 Conclusion

Having visited several industries especially in India and Tanzania, the author has the opportunity to study the preventive maintenance systems adapted universally. That could be explained in this charter. This not only answers the *'where' 'when'*, *'by whom'* and *'what'*, but also the *'why'* of maintenance.

## Appendix – I

| **TANITA CO. LTD. DAR ES SALAAM, TANZANIA**<br>**CASHEW PROCESSING UNIT**<br>**SHELLING MACHINE INSPECTION REPORT** | | | | |
|---|---|---|---|---|
| **Machine no.** | | | **Date** | |
| **Sl. No.** | **Part or operation** | **OK/ not OK** | **Inspector's remarks** | **Special remarks** |
| 1 | Blade Cams | | | |
| 2 | Blade cam followers | | | |
| 3 | Blade alignment | | | |
| 4 | Blade springs | | | |
| 5 | Blade beam guides | | | |
| 6 | Horizontality of top blade beam | | | |
| 7 | Horizontality of bottom blade beam | | | |
| 8 | Matrix springs | | | |
| 9 | Matrix cams | | | |
| 10 | Push pin carriage cam | | | |
| 11 | Setting of push pins | | | |
| 12 | Ratchet cam | | | |
| 13 | Ratchet roller guide block – 30mm | | | |
| 14 | Ratchet roller guide block – 22mm | | | |
| 15 | Shelling chain setting | | | |
| 16 | Condition of shell chain | | | |
| 17 | Tightness of shell chain | | | |
| 18 | Carriage slide | | | |
| 19 | Clutch | | | |
| 20 | Distance between matrix and blades | | | |
| 21 | Distance between top and bottom blades | | | |
| 22 | Setting of bottom blade lever | | | |
| 23 | Condition of felt | | | |
| | | | | |
| | | | | |
| *Inspected by:*<br>*Repairing checked by:*<br>*Final inspection by:* | | *Electrical inspection by:*<br>*Date of completion:*<br>*Plant manager:* | | |

**Figure 5.4** An inspection chart for a cashew processing unit.

## Appendix - II

*Form no. 2*

### TANITA CO. LTD. DAR ES SALAAM, TANZANIA

**ENGINEERING DEPARTMENT**

**ROUTINE INSPECTION REPORT**

| | |
|---|---|
| From: | To: |
| The chief Engineer | ................................ |
| Ref. No........................ date | ................................ |

1. Please make a careful inspection of the plant/machinery/Part as outlined below and enter the details of your inspection in the space provided below.
   ..........................................................................................................
   ..........................................................................................................
   ..........................................................................................................
   ..........................................................................................................
   Inspector's remarks
   ..........................................................................................................
   ..........................................................................................................
   ..........................................................................................................
   ..........................................................................................................
   Date ........................................Signature............................................

---

| | |
|---|---|
| From: | To: |
| The chief Engineer | ................................ |
| Ref. No........................ date | ................................ |

2. Please carry out the necessary repairs as detailed in 1 above and report to the inspector on the completion of the job.

   Repairs completed on ........................... Signature

3. Final inspection on .............................. Signature

   Inspector's remarks
   ..........................................................................................................
   ..........................................................................................................
   Chief Engineer's comments
   ..........................................................................................................
   Date.......................................... Signature..........................................

**Figure 5.5** An inspection report for a cashew processing unit.

## Appendix - III

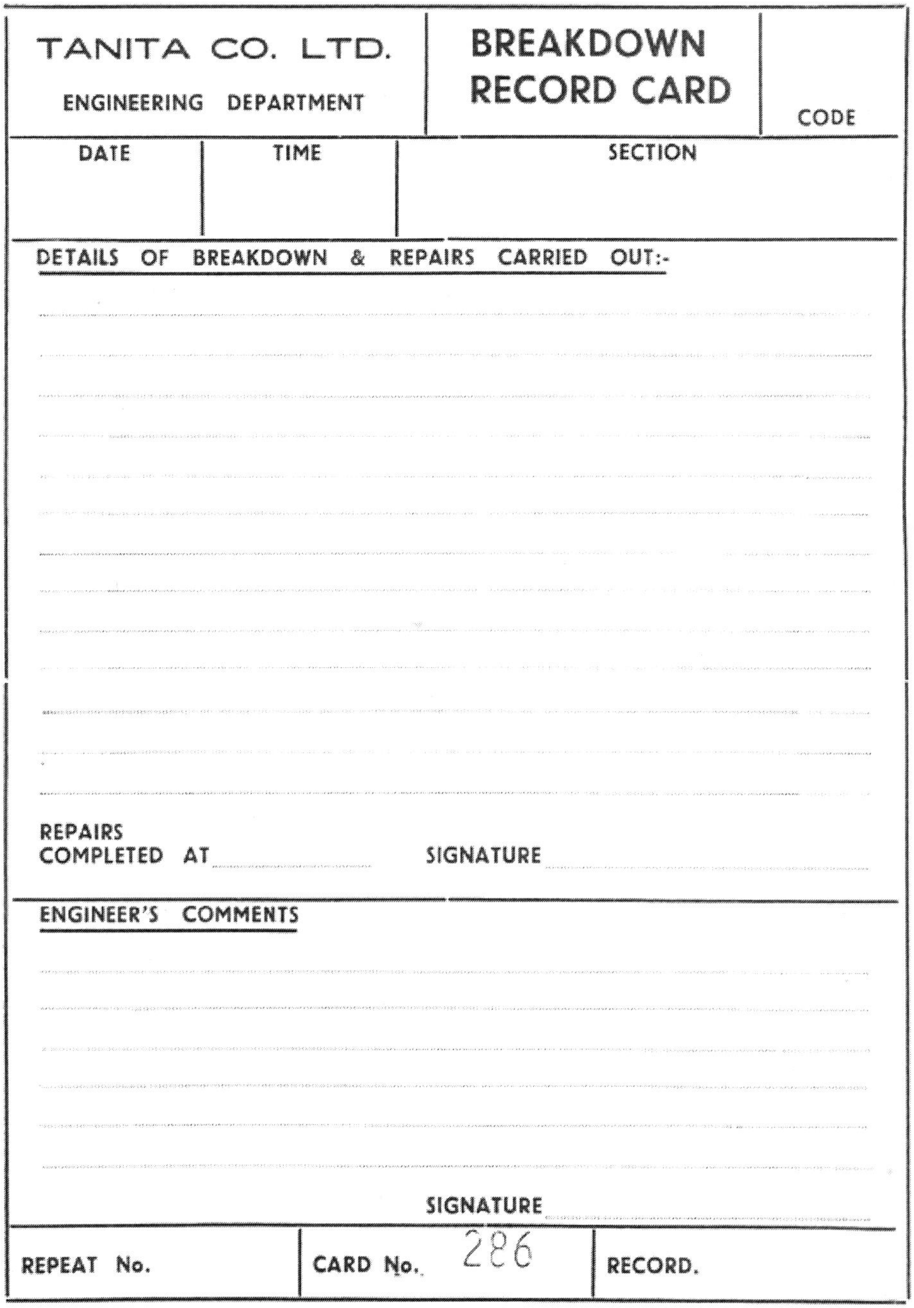

| TANITA CO. LTD.<br>ENGINEERING DEPARTMENT | | BREAKDOWN RECORD CARD | CODE |
|---|---|---|---|
| DATE | TIME | SECTION | |

DETAILS OF BREAKDOWN & REPAIRS CARRIED OUT:-

REPAIRS COMPLETED AT ............ SIGNATURE ............

ENGINEER'S COMMENTS

SIGNATURE ............

| REPEAT No. | CARD No. 286 | RECORD. |
|---|---|---|

**Figure 5.6** Breakdown Record Card at TANITA Co Ltd.

# Appendix - IV

**TIPER** OIL REFINERY

LUBRICATING SCHEDULE

Equipment.................................... Dept....................................

Insp. Card No. ........................

O = OPERATOR G = GREASER

| POINTS OF LUBRICATION | LUBRICANT | HOW APPLIED | | | | | | | FREQUENCY | | | | | | |
|---|---|---|---|---|---|---|---|---|---|---|---|---|---|---|---|
| | | Oil Can | Reservoir | Grease Cup | Grease Gun | Hand | | | Each Shift | Daily | Weekly | Fortnightly | Monthly | Quarterly | Semi-Annually |
| | | | | | | | | | | | | | | | |

**Figure 5.7** Lubrication Schedule Card of TIPER, Tanzania.

## Appendix - V(a)

SHIRIKA LA USAFIRI DAR ES SALAAM

CHECKLIST FOR PREVENTIVE MAINTENANCE

Depot .......... Fleet No. .......... Milage ..........

Mechanic Name/No. ..........

Date in .......... Date out .......... Time started .......

Time finished.......

Total hours.......

"A" SERVICE 4000 Km

- Check and top up radiator water level, clutch fluid level
- Check engine oil dilution
- If dilution is present locate and correct cause
- If dilution present complete "B"SERVICE

Check security, condition and adjustment of the following :

- Steering wheel, steering box and drop arm
- Steering arms, drag link and track rod
- Front spring blades, centre bolts, eye clamps and 'U'bolts
- Front spring hanger brackets, shackles and shackle pins
- Front shock absorbers and brackets
- Front brake shoes, back plates and brake chambers
- Exhaust system
- Air reservoirs-drain
- All oil, air, fuel-feed and fuel-return pipes
- Fuel tank, brackets, straps and caps
- Prop shaft flange, universal joints and centre bearings
- Rear spring blades, centre bolts, eye clamps and 'U'bolts
- Rear spring hanger brackets, shackles and shackle pins
- Rear brake shoes, back plates and brake chambers
- Rear axle oil leaks and breather
- All body bolts and chassis outriggers
- All wheel nuts and studs
- Engine moutings and brackets (injection pump)
- Engine oil leaks
- All engine bolts
- Radiators mountings and water leaks
- Gear box oil leaks and breather
- Clutch plate, hydraulic cylinders and pipes
- Check clutch pedal free play (1.5" required)

Check security, condition and adjustment of the body :

- Body panels, seat, windows and doors

Lubricate and top up the following :

- Engine oil
- Injection pump oil
- Fluid flywheel
- Gear box oil
- Rear axle oil
- King pin oil
- All grease nipples
- All accelerator, footbrake and handbrake joints

Electrical equipment - Repair or replace as required

- Check battery, fill distilled water
- Check, clean and tightne battery terminals
- Check self starter & tighten terminals
- Check charging system
- Check horn
- Check indicators

**Figure 5.8a** V(Page 1)

# Appendix – V(b)

Test Procedure

- Check air pressure built up time
- Check air pressure leakage (limit 5 lbs per minute) breakes applied

Road Test Vehicle

- Operation of steering
- Operation of electrical equipment
- Engine performance
- Engine speed and exhaust emission
- Efficiency of hand, foot and emergency brakes
- Check for fuel, oil, water and air leaks

ADDITIONAL WORK REQUIRED - "B" SERVICE 8000 Km

- Renew engine oil filter and change oil
- Check tire air pressure, bring up to prper value

ADDITIONAL WORK REQUIRED - "C" SERVICE 32,000 Km

- Renew chassis and engine fuel filters
- Adjust tappets

ADDITIONAL WORK REQUIRED - "D" SERVICE 64,000 Km

- Clean chassis
- Change gear box filters
- Clean gear box oil filter
- Drain gearbox air cylinders
- Check gearbox air pressure
- Check condition of piston seals and brake band adjustment
- Change rear axle oil
- Inspect and repack front wheel bearings
- Remove injection pump and injectors

Electrical Equipment

- Check dynamo output with volt ampmeter, adjust to proper valves

ADDITIONAL WORK REQUIRED

REMARKS

UNSCHEDULED MAINTENANCE

**Figure 5.8b** PM Checklist for vehicles at Usafiri Depot (Page 2)

# Chapter 6

# Condition Monitoring

*Chapter synopsis:* *The second major category of planned maintenance viz. condition monitoring or assessing the working condition of equipment is dealt with in detail in this chapter. It illustrates the various simple instruments used in industry to assess and monitor subjectively and objectively the different parameters of equipment's performance. This chapter also introduces signature analysis, the latest concept in vibration monitoring.*

*Keywords:* *Condition monitoring, human perceptions, deferring or advancing shut-down schedules, on-load and off-load monitoring, subjective or objective monitoring, primary and secondary signals, malfunctioning signals, simple instruments and gadgets used, inaccessible parts, dentist mirrors, borescopes, periscopes, endoscopes, thermal stickers, shock pulse meters, ultrasonic detectors, corrosion coupons, belt tension gages, V belt wear monitoring, shaft run-out, dial gages, optical sensors, hauling/hoisting wire ropes, wire rope monitoring, Intron Plus, lubrication monitoring, Internal component wear, NDT methods, PT, MT, ET, UT, RT, thermal radiography, thermographic inspection, computer tomography, signature analysis.*

## 6.1 What is Condition Monitoring?

Planned maintenance, as distinct from breakdown maintenance falls into two basic groups, preventive maintenance and condition-based maintenance, the main difference being the determination of moment when maintenance should be performed. While the previous chapter discusses preventive maintenance, this chapter discusses condition monitoring, also called predictive maintenance highlighting the difference between the two.

Condition monitoring is a system of measuring or assessing the service condition or change of the condition of any component or part of an equipment. In other words, condition monitoring is that type of planned maintenance wherein the actual working condition of major components of an equipment are constantly monitored to give an indication of which parts should be replaced and which need not be replaced in the near future. The preventive maintenance with condition monitoring is a system in which, in addition to the programmed stoppages of maintenance work, an assessment is made continually on the working condition of all the major parts of equipment even when a maintenance schedule is on. If by assessment, the condition of the part is satisfactory, the schedule of the programmed stoppage of the equipment is reviewed. That is, if the unit is working perfectly well just before the planned stoppage, then the schedule is reviewed and the planned stoppage is deferred for a later date. On the other hand if the condition of the unit is found deteriorating much before the planned stoppage, then the maintenance schedule is advanced and service undertaken to prevent unexpected breakdowns. This is also called condition based maintenance or predictive maintenance.

We have earlier seen that in breakdown maintenance the machine gets stopped when the repair would be undertaken. In preventive maintenance, the machine is stopped at predetermined intervals based on statistical failure data of the unit. Condition monitoring differs from both the above, that it requires continual monitoring of certain condition indicating parameters of the unit. Service/ replacement action would be undertaken only if the condition warrants it.

For example if a part is scheduled to be replaced in January ($T_2$ - 12 months after $T_1$) and in August ($T_3$) of the previous year the component is found to develop defects like cracking, and is not expected to last long, then the scheduled repair or replacement is advanced by 3 or 4 months to $B_1$ to avoid major breakdowns. If on the other hand, even in December ($T_4$), it is found to be in perfect order and can work for another 6 months , then its replacement decision is deferred to $B_2$ (Fig. 6.1). In general the condition monitoring results in several such instances of deferring stoppage schedules than advancing them. This means the machine downtime, whether planned or unplanned, would be less if the condition monitoring is perfectly conducted.

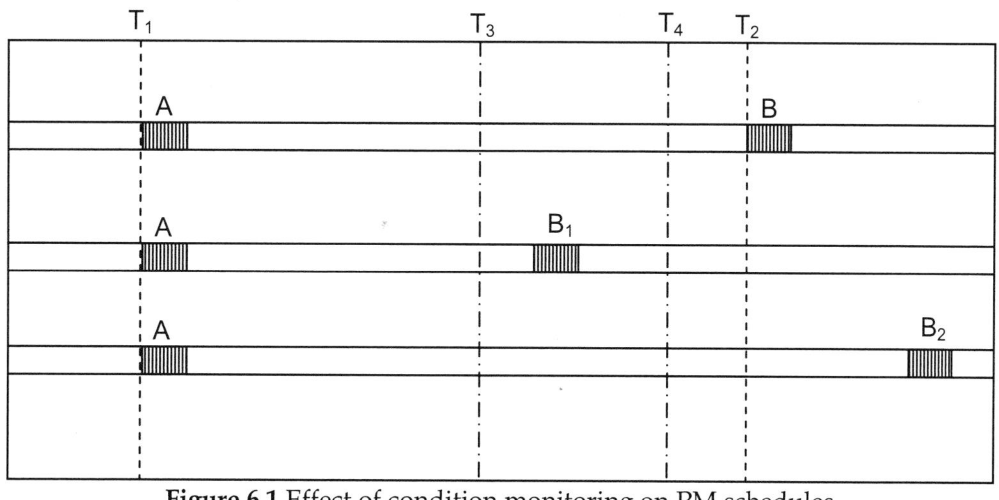


**Figure 6.1** Effect of condition monitoring on PM schedules.

Another illustration of condition monitoring can be cited here. In a unit, around 100 roller bearings run under similar conditions. After a certain time, say 6 months, 10 of these bearings fail, the next 50 bearings fail during the next 24 months and the last 40 bearings have a life of 48 months. In preventive maintenance all the 50 bearings have to be replaced during a scheduled stoppage after 6 months, and these 40 bearings too get replaced in a shorter span of life, though they could be used for 42 more months. On the other hand if condition monitoring is applied for each of these bearings, then most bearings could be used for their full life time.

## 6.2 Off-load vs On-load Condition Monitoring

*On-Load Monitoring* means, monitoring or adjusting the parameters while the machine or equipment is running. This is done for superficial, easily accessible & non-interfering parts of the equipment which can be carried out without any interruption to the machine operation.

*Off-Load Monitoring* means, monitoring or adjusting the parameters while the machine or equipment is stopped temporarily to check the condition and make the needed adjustments. This is done for interior and inaccessible parts. However there may be several situations like the two-shift working or the plants' temporary shut down for other reasons, when this type can be conducted without production loss.

## 6.3 Subjective or Objective Condition Monitoring

In subjective condition monitoring, the personnel use their perception of senses and judgment to note any change in the condition. The four senses a man is bestowed with perceptions are illustrated below

| Sense | Organ | Perception | Contribution to fault identification |
|---|---|---|---|
| Seeing |  | *Emanation of smoke* | 80% |
| Hearing | 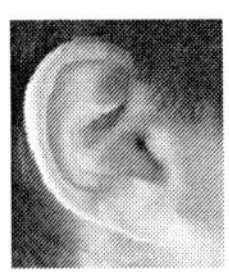 | *Hearing noise* | 14% |
| Touching |  | *Touching the motor or shaft to feel the excessive heating or touching the bearing housing extent of vibrations* | 2% |

*Contd...*

| Sense | Organ | Perception | Contribution to fault identification |
|---|---|---|---|
| Smelling | 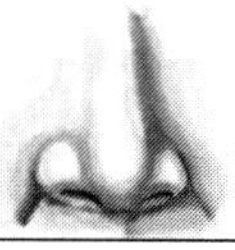 | *Burning of rubber components due to excessive heat* | 2% |

It is important that the person carrying out subjective condition monitoring has the adequate qualifications, experience and skill as the result depends on their judgment. In subjective condition monitoring, it is also possible to use instructions, guidelines or hints where to look for leakage, bearing play etc. Posters or figures illustrating different conditions of components, mounted on walls, may be helpful

In aided Subjective Condition Monitoring, the monitoring personnel use simple gadgets to add to their ability to perceive the conditions better, like bore scopes, periscopes, endoscopes etc. or the use of screw drivers or stethoscopes to feel the vibrations and slight noise.

In objective condition monitoring, different instruments and facilities are used for obtaining the direct or indirect measure of the parametric condition of the component either on-line or off-line. These mechanical or electronic instruments can be located at convenient locations for better monitoring and control

We can illustrate these three situations as:

- A doctor feeling the pulse (*Subjective*)
- A doctor using a stethoscope to feel any abnormal heart beat (*Aided Subjective*)
- A doctor looking into the ECG monitor (*Objective*).

## 6.4 Signals Indicating Malfunctioning of Equipment

*Primary signals* are those parameters that are designed to emanate in cases of malfunctioning. Examples are high pressure relief valves, pressure sensors, dial gages and those that are designed to emanate identifiable sounds in case of malfunctioning. They sometimes are designed to indicate the quantum of deviation from normal. Another example is the automatic lifting of a signboard whenever a thread breaks off in a weaving loom.

*Secondary signals* are that which indicate changes in parameters like vibrations, temperatures, chemical or physical parameters, leakage sounds. They are identified as a loss output rather than direct measurement. Since the failure of equipment generally occurs due to factors that might not have been foreseen during the design stage, it is always inevitable to monitor these secondary signals for maintaining equipment's health.

## 6.5 Condition Monitoring Techniques and Gadgets used

*(figures with courtesy of Edward Arnold Publishers)*

Based in these emanated secondary signals several parameters of the functioning of the equipment can be monitored.

- Monitoring of visual condition of inaccessible parts and locations
- Monitoring of inside condition of tubes
- Temperature monitoring
- Vibration monitoring
- Leak Monitoring
- Lubricant Monitoring
- Corrosion Monitoring
- Belt Tension Monitoring
- Shaft Run out Monitoring
- Thickness Monitoring
- Crack formation Monitoring
- Internal components' wear monitoring.

**(i)** ***Monitoring of Visual Condition of Inaccessible Parts and Locations***

Quite often the visual condition of the components located at inaccessible locations are to be monitored, may be with respect to corrosion or fatigue cracking or even dust/sediment collection that is deleterious to the functioning of the component. Depending upon the extent of the change as perceived, the equipment can be stopped for detailed check and/or rectification. Several gadgets are used for this purpose as illustrated in the figures.

*Inspection mirrors* used to inspect crevices, dark spots, inside of narrow mouthed cylinders, or the heat exchanger pipes or boiler pipes. These mirrors can be simple like those of the dentist or developed for special applications as below.

(a) With adjustable mirror angle controlled by a cable running along the holding rod,

(b) Illuminated with a light bulb near the mirror of special use in dark spaces, inside the cylinders or piping, or

(c) With telescopic rods for adjusting the length of the gadget,

(d) These borescopes are extensively used by the NDE practitioners Co in view of their versatality to inspect the inside of the boiler pipes and tell a lot about

about their inside condition regarding rusting, cracking, sedimetation etc., enabling early preventive action. (Fig. 6.3). Here mirrors are fitted inside a small tube rather than on a rod to prevent damage to the mirror during insertion or other operation.

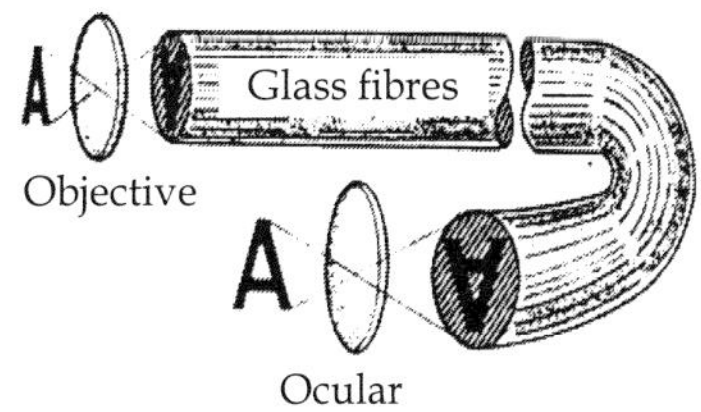


**Figure 6.2** Principle of endoscope.

(e) Periscopes for specialized applications or for assisting the viewer.

(f) Endoscopes, or optical fibro scopes which work like periscopes but flexible. They work on the principle of image transfer by total internal reflection in a bundle of glass fibers as illustrated by Fig. 6.2. In medical field they are popular in diagnosing ulcers in the stomach. Sometimes they are permanently fixed for constant monitoring of critical components or inserted when needed into pre-fixed tubular guides leading to the location that is to be monitored

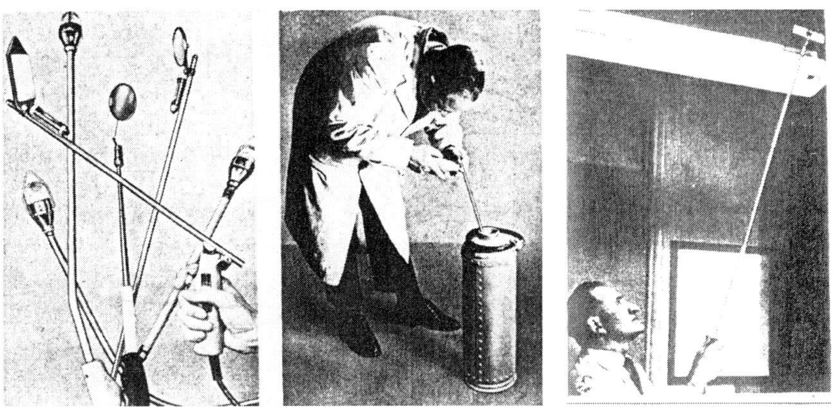

**Figure 6.3** **(a)** a collection of inspection mirrors **(b)** Inspecting inside a cylinder **(c)** inspecting the inside of crevices.

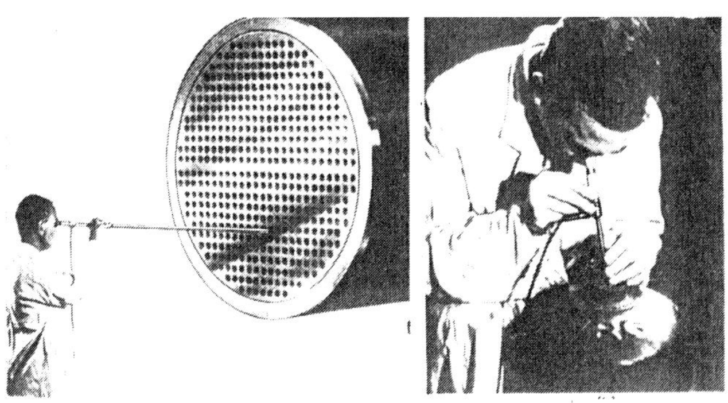

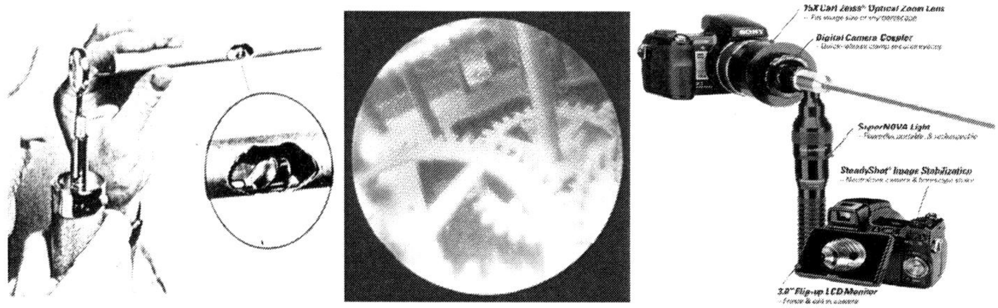


**Figure 6.4** Use of borescopes to inspect heat exchanger pipes.

(g) Having illustrated the gadgets for visual inspection as above, some of other gadgets used in other aided subjective monitoring systems may be cited as below, classified into the three categories of subjective, aided subjective and objective monitoring systems.

*Subjective Condition Monitoring gadgets,* where the monitoring personnel use their perception of senses and judgment to note any change of the condition, as explained before.

*Aided Subjective Condition Monitoring gadgets* where the monitoring personnel use simple gadgets to add to their ability to perceive the conditions better, like use of screw drivers or stethoscopes to feel the vibrations and slight noise or bore scopes, periscopes, endoscopes etc. explained before.

*Objective Condition Monitoring gadgets,* where different instruments and facilities are used for obtaining the direct or indirect measure of the parametric condition of the component either on-line or off-line. These mechanical or electronic instruments can be located at convenient locations for better monitoring and control.

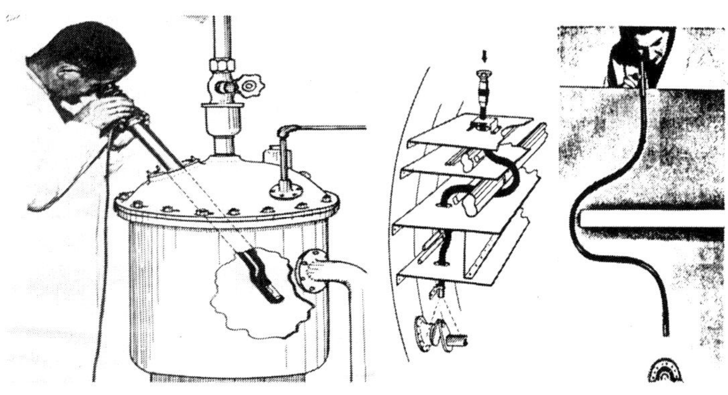

**Figure 6.5 (a)** Principle of endoscope **(b)** inspecting a cylinder through a periscope **(c)** using permanently fixed endoscope and **(d)** inserting an endoscope through tubular guides

**(ii)** ***Temperature Monitoring***

*Subjective:* Touching the motor etc. and assessing if overheated.

*Aided subjective:* Temperature sensitive stickers are the most common and cost effective. A sticker having four or five 20 mm diameter dots of special paints, each of which changes its color at a particular temperature, varying from 35°C to 75°C, is stuck on to the heat prone equipment.

They are based on the principle that when a certain chemical is mixed, the paint changes its colour from say pink to black at a definite temperature. By using different compositions of such chemicals a spectrum of temperature sensitive paints are developed to indicate the temperature of the body as illustrated. The operatives or the supervisor can identify the temperature range by looking at these stickers during their periodic patrol rounds. In the illustration alongside, temperature of the body would be between 55°C and 65°C, if the middle sticker has changed its colour or the intensity. This category includes temperature sensitive chalks, thermal paints with which larger parts of the heat prone body can be painted for identification from a farther distance. Other self acting gadgets like thermocouples and thermal plugs may be grouped under this category.

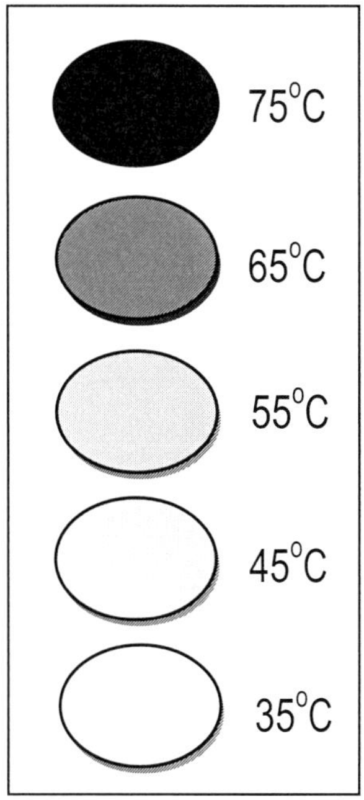


**Figure 6.6** Thermal stickers.

*Objective:* Pyrometers, thermometers, pistol thermometers, infrared radiation scanners etc., fall under this category. Depending upon the convenience and need, these instruments can be of thermostat type connected to some sort of warning system in case of overheating.

**(iii)** ***Vibration Monitoring***

*Subjective:* By touching and feeling the vibrations in a rotating body or by listening to the vibration sounds.

*Aided subjective:* Use of stethoscopes or screw driver ends by expert foremen.

*Objective:* Shock Pulse Meters (SPM), vibration measuring instruments etc. can be used. These instruments generate the characteristic sine curve. Its amplitude indicates how good or bad the situation is, while the frequency indicates the source of the problem, especially when the machine has multiple rotating parts each with a different frequency. The vibration measurements made near the source of vibration might also indicate the following:

- Imbalance,
- Shaft misalignment,
- Damaged bearings (worn out or chipped balls or races),
- Damaged gears and other power transmission components,
- Mechanical looseness,
- Cavitation, stalling etc.

**(iv)** *Leak Monitoring*

Leakage can either be internal and external. In the former, the gas or liquid leaks from inside to outside, while in the latter, the gas leaks from outside to inside

*Subjective:* Listening to and identifying any hissing sound near hydraulic or pneumatic circuits. In case of carbon based gases like cooking gas, a smell indicates leakage.

*Aided subjective:* A number of detecting techniques are available, including

- The soap and water method, where bubbles coming out indicates an external leakage.
- Use of proprietary preparations can make this method more effective, capable of detecting leaks as low as 1 ml/sec.
- Thin paper or fiber strips placed near leakage points gets either sucked in case of vacuum pipe leakage or gets vibrated in case of pressure pipe leakage.
- Sensors reacting chemically with such gasses can be placed at strategic points, so that they can give visual or sound indication of any leakage.

*Objective:* One powerful technique used is ultrasonic detection. When a fluid is forced through an orifice, under internal or external pressure, sound is generated in the frequency range of 40 to 80 kHz. The ultrasonic leak detector identifies this high frequency which can easily be separated from lower frequencies of ordinary machine noises.

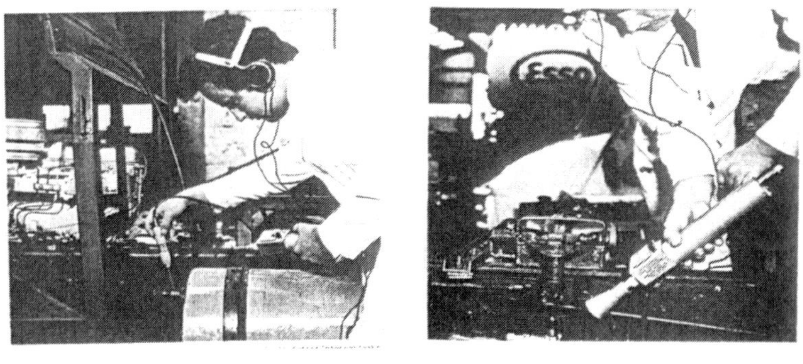


**Figure 6.7** Illustration of leak monitoring. **(a)** Internal and **(b)** External

### (v) *Corrosion Monitoring*

*Subjective:* In general, corrosion cannot be judged visually and is not as effective as those discussed below.

*Aided subjective:* Corrosion coupons, small pieces of 20 × 200 mm, of the same material as the base of the vessel are stuck at locations suspected to be highly corrosive prone, and analyzed once in a month or as required. The corrosion characteristics of the coupon are expected to be the same as the base material to which they are stuck and hence give an indirect measurement of the corrosion of the body.

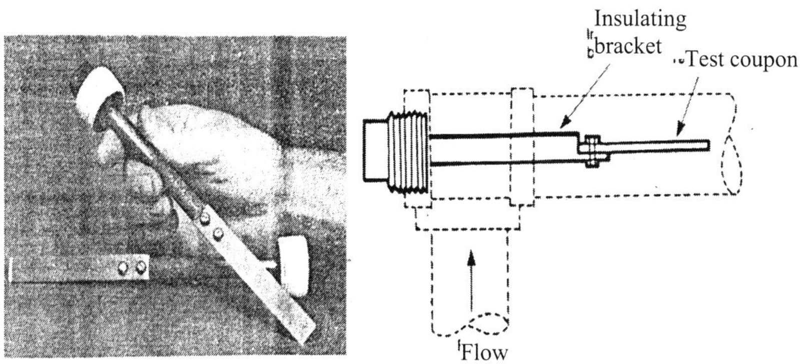


**Figure 6.8** Corrosion test coupons.

*Objective:* Electrical and electronic devices can be developed to measure the corrosion characteristics based on the following principles, the details of which however are beyond the scope of this book.

- The resistance of special electric elements changes in corrosive environment.
- The potential difference between the reference electrode and the system will indicate the presence and extent of corrosion.
- The polarization resistance of specially designed probe indicate the rate of corrosion.
- Hydrogen gas formation during corrosion can be detected with a thin walled blind tube which can be analyzed to determine the corrosion.

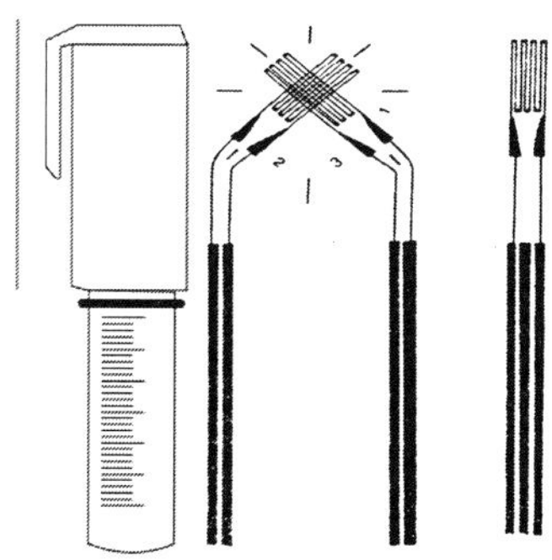

**Figure 6.9 (a)** Belt tension gage and **(b)** Strain gage.

### (vi) *Belt Tension Monitoring*

*Subjective*: Slightly pressing V-belt down.

*Aided subjective*: A pen type tension gage as illustrated is a simple and handy gadget. Depending upon the belt looseness, the lower part of the gage telescopes into the upper part, leaving the rubber ring stuck on to the lower part at the lowest point of insertion. On removing the gage, the position of the ring on the graduated scale of the lower part indicates the belt looseness that can be corrected if necessary.

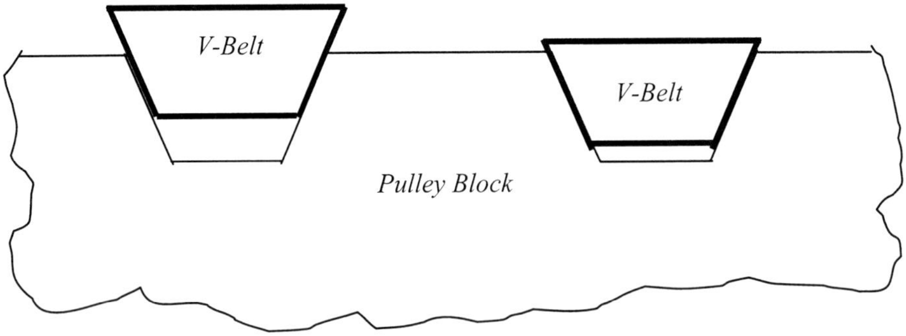


**Figure 6.10** V belt wear monitoring.

*Objective:* Strain gages or other electronic instruments as illustrated can be used to measure the stresses in metallic parts or components.

### (vii) *Shaft Run-out Monitoring*

When a lathe is being tested for accuracy, a mandrel is fixed in the spindle and rotated and checked for the run-out.

*Subjective:* If the run-out is high, it can visually be seen.

*Aided subjective:* By use of dial gages fixed on the body, the amount of run-out can be seen by turning the spindle by hand or at lower speeds.

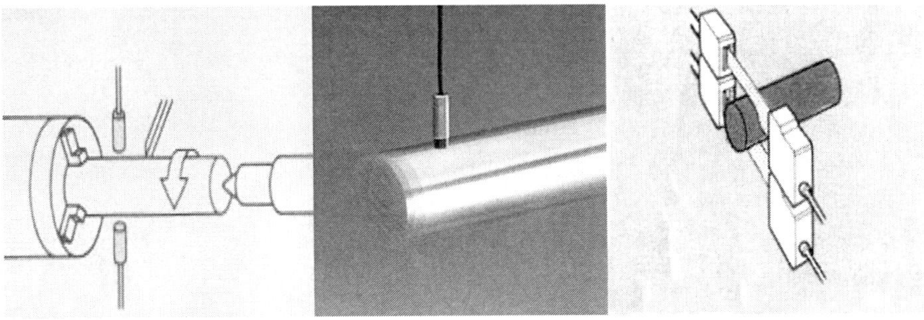

**Figure 6.11** Run-out checking by optical sensors.

*Objective:* Optical sensors, as illustrated below can be used to check the run-out even at high speeds. They are mounted on the body (e.g. a tool post of a CNC machine).They work on the principle that, when a sheet of light is passed around the shaft as shown, the interruption of the light beam can be sensed and measured to an accuracy of 10 microns. This can be an on-line check for the run-out. In fact some CNC machines have such gadgets mounted on them so that the sensors can automatically adjust the tooling parameters to maintain the accuracy.

***(viii) Wire rope monitoring***

A steel wire rope comprises of a considerable number of simultaneously working wires stranded together. These are used extensively for hauling and/or hoisting in mining, port handling and building construction. If one wire breaks during operation the remaining wires take over its share of load due to which this breakage, especially of the interior wires cannot be observed visually. Nevertheless the wire rope's haulage strength gets lowered. If more than one wire breaks, the rope's strength is largely diminished leading to major accidents while hauling or lifting weights.

The condition monitoring of such ropes can be done by using a simple gadget consisting of the DC coil and permanent magnets as illustrated in Fig. 6.12. It can be mounted on the cable rope initially before slinging it over the pulleys and the gadget is periodically moved along the rope, either manually or by motorised friction gear system.

A strong D.C. magnetic field is generated by the moving coil and along the rope's length. When discontinuities occur such as broken wires, a radial field is produced by leakage across the separated ends of the break which can be identified and measured as the discontinuity is passed through the coil system. This identification and analyzing can be done off-load ether in the ground control room or in the gantry control cabin.

Identification of the fringe fields near the surface of the rope can detect local defects like broken wires and corrosion pitting, while measurements of changes in magnetic flux passing through a short length of rope may quantify changes in metallic cross section mostly caused due to corrosion and pitting.

Recent developments have seen the use of custom designed electronics and custom developed software for use on proprietary computing equipment. *Intron Plus* from Russia is versatile software that can be used with this gadget to analyse the internal structure of the wire ropes.

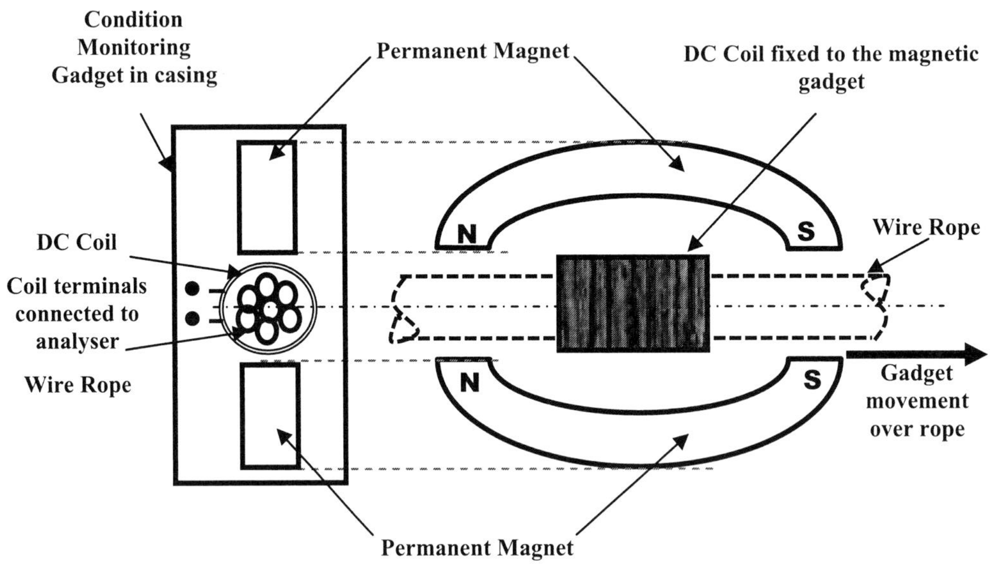


**Figure 6.12** Wire rope condition monitoring gadget.

### *(ix) Lubrication monitoring*

Lubrication monitoring has two components

(a) Lubrication monitoring, which is to check if the quality and quantity of the lubricant is as per norms, to ensure that the lubricant used serves its purpose. This is like using a dipstick in a car.

(b) Lubricant monitoring, which is basically to identify if any internal component is wearing. This is as explained in the next paragraph and also as detailed further in a later chapter on wear debris analysis.

### (x) *Internal Component Wear Monitoring*

In industrial machinery for metal cutting, like the lathes, there are several internal rotating components like bearings and gears that keep wearing out. Because they are inaccessible, their wear cannot be noticed until they fail or the machine is dismantled due to the failure of some other part. Hence it is essential to monitor the wear out of the internal rotating components and this is especially true in case of CNC machines. A recent advancement in this direction is wear debris analysis, by which the wear debris collected from the lubricant oils is analyzed to identify the exact component and the extent to which it is wearing out. Just as the doctor analyses the blood sample to diagnose the internal problems of a human body, the analysis of the wear debris among the oil samples drawn from the machines can help in pinpointing the impending internal problems. In view of the significance this is discussed in detail in a chapter 8.

### (xi) *Crack Monitoring*

This is mostly an off-line monitoring test and most tests under this category form the non-destructive evaluation (NDE), the primary function of NDE specialists, trained extensively by the Indian Society of Non-destructive Testing (ISNT). Most serious failures especially in pressure vessels are preceded by a crack growth at the point of stress concentration or form a material defect on or under the surface of the component. It was widely believed that failure due to sub-surface cracks occur without warning. This is mostly untrue, though most cracks are not normally visible on casual inspection. A number of crack detection tests as cited below are available to overcome this difficulty. While the first two can be classified as aided subjective, the latter can be categorized as objective.

*Dye penetrant test (PT):*

A dye is applied to penetrate into the cracks and by spreading a white absorbent powder on this, when cracks as small as 0.25 mm can be revealed to naked eye. This is done in 6 steps as illustrate alongside.

1. Wipe the surface of the suspected crack
2. Remove dirt inside crack by thinners
3. Apply penetrant
4. Wipe excess penetrant
5. Apply Developer
6. Inspect the crack clearly shown now.

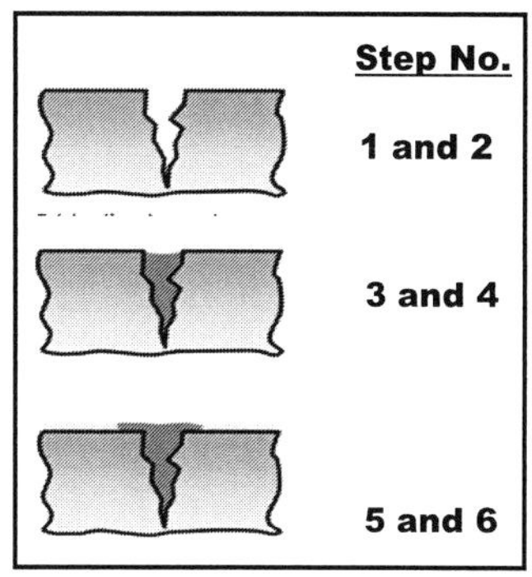


**Figure 6.13** Dye penetrant test.

*Magnetic particle test (MT):*

Here a magnetic field is applied on ferrous materials. When a crack or other defect crosses the path of a magnetic field, it causes the magnetic flux to spread round the crack into the space above the surface. The existence of this field and hence the crack is revealed using a magnetic powder.

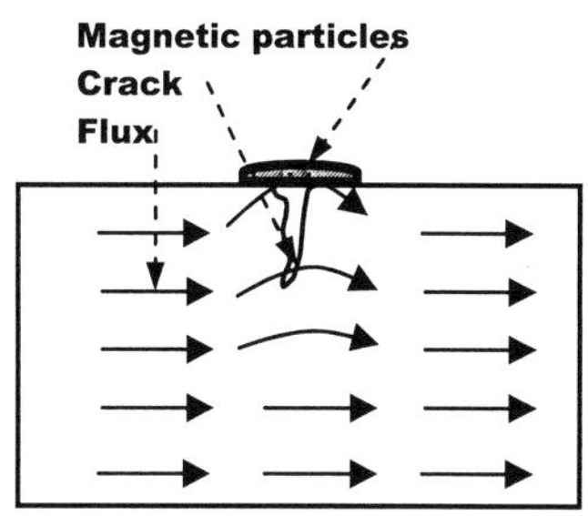


**Figure 6.14** MP test.

*Electrical resistance test or Eddy current test (ET):*

The presence of a crack at the surface will increase the resistive path measured between two probes in contact with the surface. In spite of difficulties in ensuring a perfect surface contact, this method can be used to detect the depth of the crack.

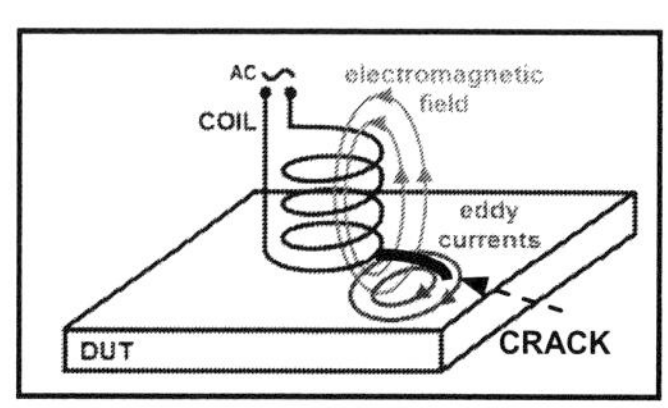


**Figure 6.15** Eddy current test.

*Ultrasonic test (UT):*

Ultrasonic waves generated at the surface of the component will be reflected by any surface in the path of the wave. The time delay between the generation and detection of the sound pulses is displayed to give a measure of the distance of the crack from the surface.

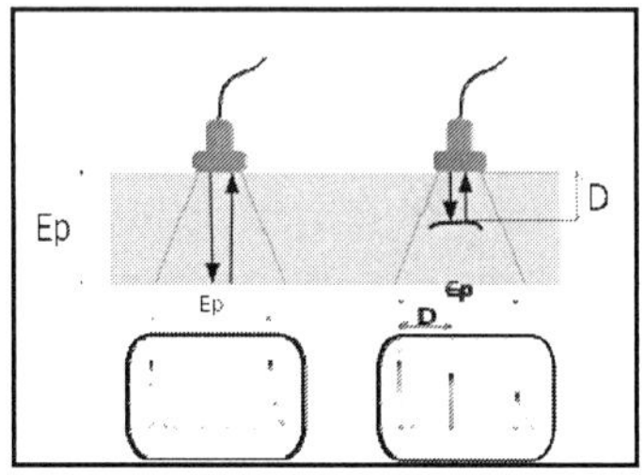


**Figure 6.16** Ultrasonic

*Radiographic test (RT):*

Imperfections can be photographed using either X-rays or gamma rays from a radioactive source and by special photographic techniques, a change in thickness upto 2% can be detected in steel plates/parts as thick as 50 mm.

The method usually requires dismantling of the unit being examined and also raises the problem of personnel protection. Fig. 6.12 alongside is a representation of a radiograph taken on engine manifold showing development of cracks that are not normally visible from outside or by any other NDT technique.

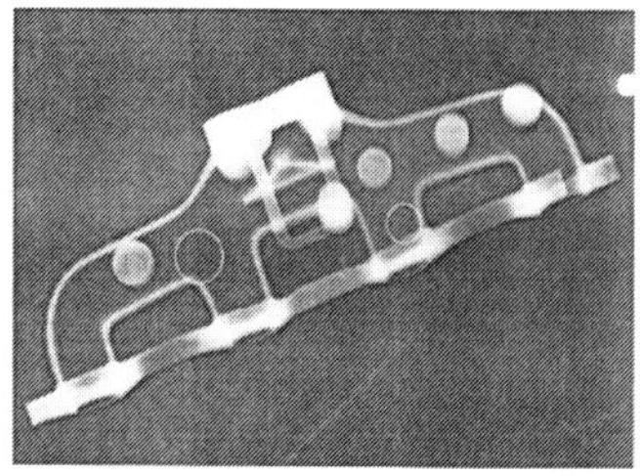

**Figure 6.17** Radiograph of an engine manifold.

*Thermographic inspection(TM):*

Thermographic inspection is the mapping of the thermal patterns or *thermograms,* on the surface of objects of the thermal patterns at the object's surface by use of infrared radiation. Unlike the radiographic testing, thermographic inspection is safe, nonintrusive and noncontact, allowing the detection of relatively shallow subsurface defects. Thermal measurement methods have a wide range of uses for preventative maintenance, processes control and energy audits besides NDT. They are used by the police and military for night vision, surveillance, and navigation aid as well as by firemen and emergency rescue personnel for fire assessment, for search and rescue. They are also used by the medical profession as a diagnostic tool.

Illustrated alongside are thermograms of a motor, indicating the hot points as white patches. TM can be done on objects as large as an aircraft as illustrated in Fig. 6.18(b).

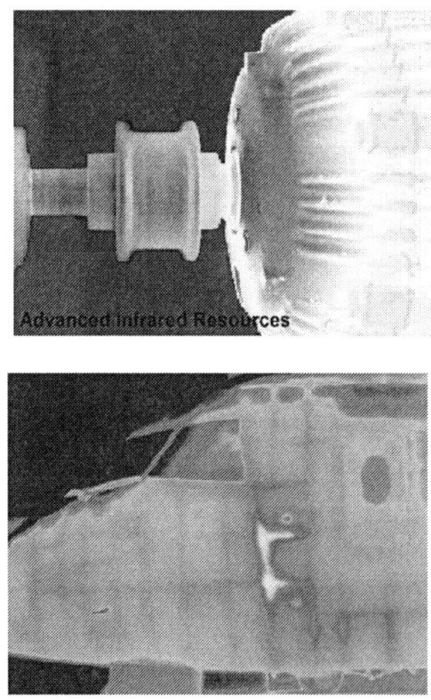


**Figure 6.18** Thermograms of **(a)** motor and **(b)** aircraft

*Computer Tomography (CT):*
Computer tomography is a powerful nondestructive evaluation technique for condition monitoring and identification of the internal defects by producing 2-D and 3-D cross-sectional images of an object from flat X-ray images.

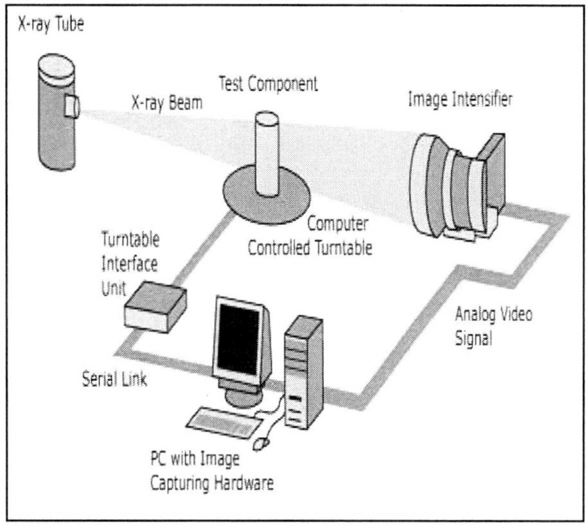


**Figure 6.19** Principle of CT scan.

The object is irradiated by an X-ray cone beam, while a detector collects radiation passing through the object. During the inspection, the part is rotated 360° under controlled conditions, and producing several projections for each angular position and finally producing computerized reconstruction of images of any desired part of the object. This CT scan is done not only on industrial components as a part of NDT, but also on humans as a part of diagnosis.

## 6.6 Signature Analysis

Forensic experts analyze the human signature to evaluate not only the contours, but the movements originally made to create, a signature. Microscopic examination of a person's signature indicates a sort of vibrations in the ink lines, caused by the vibrations of the hand while signing. These vibrations are peculiar to a person. Even though a forged signature looks 100% identical to the original signature, forensic experts can identify the difference by studying the vibrations. In a similar manner the vibrations of a faulty machine can be distinguished from the vibrations of a healthy machine by analyzing the sinusoidal curve. This is called the signature analysis and has recently become an effective conditioning monitoring tool for vibration engineers.

In analog electronics, signature analysis is a troubleshooting technique in which an AC signal with a specific waveform (usually a sine wave) is applied across a component. The resulting current-versus-voltage characteristic, called the analog signature, is rendered on an oscilloscope, with current displayed along the vertical axis and voltage displayed along the horizontal axis. The resulting analog signature is unique for each particular type of "healthy" component (resistor, capacitor, or diode, for example). Any deviation from the normal signature strongly suggests the existence of a fault in that component (adapted from http://whatis.techtarget.com).

## 6.7 Conclusion

While only a few types of monitoring techniques have been discussed, a host of simple gadgets are being developed to suit the specific use for many other monitoring techniques that are not discussed here. Also regarding to the crack detection by NDE techniques, several other methods have been developed for NDE and the reader is advised to refer to books on this subject for further information.

The different condition monitoring techniques are summarized in Table 6.1, while a short comparison of the three common techniques of temperature, lubricant and vibration monitoring is given in Table 6.2.

**Table 6.1** Summary of the condition monitoring techniques.

| Sl. No. | Method | On/off load | Location of fault | Skill required | Remarks |
|---|---|---|---|---|---|
| 1 | **Visual** | On | Surface only | Experience | Covers a wide range of ad-hoc methods |
| | | Off | Accessible components | No special skill needed | Turn round inspection mostly in aero industry |
| 2 | **Temperature** | On | Surface or internal | Little skill for most methods | Subjective, aided subjective and objective methods |
| 3 | **Vibration** | On or Off | By analysis of total signal, frequency, amplitude, peak level, phase etc. | Some skill needed | Subjective, aided subjective and objective methods. |
| 4 | **Leak** | On or Off | Any joint or weak point subjected to pressure environment | No skill for subjective, but skill needed for using instruments . | Subjective, aided subjective and objective methods |
| 5 | **Corrosion** | | | | |
| 5a | **Weight loss coupons** | Off | | In coupon preparation | Monitoring when plant is shut down. |
| 5b | **Ultrasonic** | Off | Can detect 0.5 mm thinning | Skill needed both in operation and interpretation. | |

**Table 6.1** *Contd...*

| Sl. No. | Method | On/off load | Location of fault | Skill required | Remarks |
|---|---|---|---|---|---|
| 5c | **Corrosometer** | On | Can detect 1 mm thinning | Skill for interpretation | |
| 5d | **Polarization resistance** | On | | Medium skill | Only indicates that excessive corrosion is occurring |
| 5e | **Hydrogen probe** | On | | Skill needed to drill exact depth for probe | Hydrogen evolved diffuses into probe with a rise in pressure |
| 5f | **Ultrasonic** | Off | Near to surface only, depending on closeness of probe to surface | Skill needed | Can detect cracks, inclusions, hardness etc. |
| 6 | **Belt tension** | Off and on | Looseness of belt | Medium skill | Mostly aided subjective. Strain gages in specific cases |
| 7 | **V- belt wear** | Off and on | Wear in the pulley groove and belt bottom | Medium skill | Constant vigilance |
| 8 | **Shaft run out** | Slow speed or off | Misalignment or improper spindle tightening | Medium skill | Mostly dial gages and optical sensors while production is on. |
| 9 | **Lubrication** | Off and on | Lubricate all required points as and when needed | No special skill but constant vigilance | Normally done by patrol maintenance gang or the foreman. |
| 10 | **Internal component wear** | Off and on | Identification often component and the type of wear/chipping | High skill needed in microscopic analysis | Analysis of the wear debris from the lubricants |

**Table 6.1** *Contd...*

| Sl. No. | Method | On/off load | Location of fault | Skill required | Remarks |
|---|---|---|---|---|---|
| 11 | Crack monitoring | | | | |
| 11a | **Dye penetrant test** | Off | Invisible cracks on the surface | Medium skill needed | Detects only surface cracks |
| 11b | **Magnetic particle test** | Off | On surface and in surface | Easy to miss cracks. Some skill needed | Only for magnetic materials, but can estimate the depth of crack from surface |
| 11c | **Electrical resistance test** | On or off | At clean, and smooth surface | Some skill needed | Can indicate crack orientation and crack depth |
| 11d | **Eddy current test** | On or off | Any crack inclusion or hardness near surface. | Skill needed | Can detect defects close to surface. Clearance between surface and probe matters |
| 11e | **Ultrasonic test** | On or off | The depth, size and orientation of any subsurface defect, (pore or inclusion) | High skill essential | Can be used both for leak testing as well fault identification by signature analysis |
| 11f | **Radiographic test** | Off | The best method to indicate the photograph like picture of the internal defects | Very high skill needed | It can cover a larger area at one time. Access to both sides of component is necessary and max. part size is 50mm. Has radiation hazard to inspectors. |
| 11g | **Thermographic inspection** | On or Off | Shallow sub-surface | Normal skill | Can scan large objects as large as airplanes |
| 11h | **Computed Tomo graphy** | Off | Internal defects | Very high skill | Reproduces computerized reconstruction of images of all internal defects |

**Table 6.2** Comparison of temperature, vibration and lubricant monitoring.

| Sl. No. | Parameter | Temperature Monitoring | Vibration Monitoring | Lubricant Monitoring |
|---|---|---|---|---|
| 1 | Medium of transfer of information | - Solid, casting, shaft body etc.<br>- Fluids, lubricants, coolants or air<br>- Depends on thermal conductivity | - Any solid part of the machine<br>- Depends on elastic and mass characteristics of solids | - Lubricant oils and coolants<br>- Depends on the lubricant properties |
| 2 | Components monitored | - Any heat generating device<br>- Bearings. Gear boxes, and any rotating parts | - Any component that moves<br>- Surfaces between components with relative motion<br>- Clearances | - Any lubricated component like bearings,<br>- Transmission components like gears, couplings, cams, etc.<br>- Oil pumps. |
| 3 | Faults detected | - Failures of drives,<br>- Blockages of ducts,<br>- Loss of cooling, fouling of coolers,<br>- Overheating of motors | - Change in any moving parts,<br>- Wear and failure of bearings,<br>- Misbalance,<br>- Change in clearances etc. | - Any form of wear or failure of rotating parts,<br>- Failure of lubrication systems,<br>- Leakage of contaminants into the system, etc. |
| 4 | Monitoring equipment | - Thermometer<br>- Fluid or bimetallic, thermocouples,<br>- Pistol (infra-red) thermometers, pyrometers etc.<br>- Thermal stickers & paints | - SPM's,<br>- Accelerometers.<br>- Electronic processing equipment to display time values,<br>- Frequency filters and recorders. | - On-load removable filters,<br>- Magnetic plugs for visual examination,<br>- Ferroscope for separating debris,<br>- Microscope with debris analysis software, |
| 5 | Frequency | Continuous and periodic | Continuous and periodic | Continuous and periodic |

# Chapter 7

# Logical Fault Location

***Chapter synopsis:*** *Quite often it is more difficult to identify the defective component than to repair it, even after splitting open the machine. This chapter emphasizes the need for locating the faults logically rather than by trial and error method after splitting open the equipment. The several types of faults and their development are discussed and a logical procedure is proposed based on past experience. The series, parallel and compound circuits and functional blocks are illustrated.*

***Keywords:*** *Symptom, fault, cause, fault location, time dependent, random, observable and non-observable faults, six step procedure for logical fault location, sequential elimination , Left-right right-left and midway check methods, half split rule, functional block diagram, series, parallel and compound circuits of functional blocks.*

## 7.1 Introduction

Despite the best condition monitoring systems and equipment a company may have in taking necessary steps to prevent a breakdown, and despite the optimal preventive maintenance schedules and procedures the company may meticulously practice, there is every chance of a breakdown during the service of some component or other of the machine tool or other equipment, forcing the adoption of breakdown maintenance practices.

Most of the machine tools or the production equipment comprise of a series of working units. Even if one of them develops a defect and fails, the whole system breaks down. More difficult than the repair of this component is the process of identifying which particular component is defective. The skill of the maintenance engineer lies in his quick and accurate identification of the source of the fault. This is analogous to the person becoming sick and the doctors' skill being reflected more by his diagnosis than by his treatment.

The concept of logical fault location has attained significant development over the years, more specifically in the electronic industry. An electronic circuit generally consists of a large number of components in series. Failure of a single component leads to the failure of the whole circuit and normally 80% of the service time is spent in identifying the faulty item. Hence much research has been done in developing logical fault location procedures in electronics industry as evidenced by the availability of a hoard of text books and references.

On the other hand, even though mechanical systems like machine tools and other service equipment especially the power transmission systems, contain a number of individual units running in series, and even though they can be likened to the electronic circuits, with reference to the fault location, not much work has been done in this direction and surprisingly hardly any text book or paper on mechanical maintenance deal with this concept. This chapter discusses these concepts and procedures and also explains and illustrates how this logic can be applied in the case of engineering industry's machine tools and other production or service equipment.

## 7.2 Logical Fault Location

The skill in the diagnosis involves the location of the faults which arises in a machine or system, and the quick and accurate identification of the cause of the fault. To develop this skill we should, adopt a logical approach to the diagnosis and to follow it till the cause is found. Past experience of a machine must be taken into account, by consulting the machine history records, as this would save much of the time consuming diagnosis. Fig. 7.1 illustrates how to recognise the symptoms that can indicate a fault and its cause.

An illustration from a turret lathe can be as follows

**Figure 7.1** An example of diagnosing a cause from a symptom.

## 7.3 Basic Definitions

*Symptom* : An indication of the presence of a disturbance or a deviation from normal.

*Fault* : The physical abnormality, whose correction will allow the system to return to normal.

*Cause* : The reason for the occurrence of the fault.

*Fault Location* : The process of locating the fault by systematic methods on the basic of the symptoms.

## 7.4 Faults and their Development

In general various faults can be classified into the following groups, which are further illustrated in Fig. 7.2.

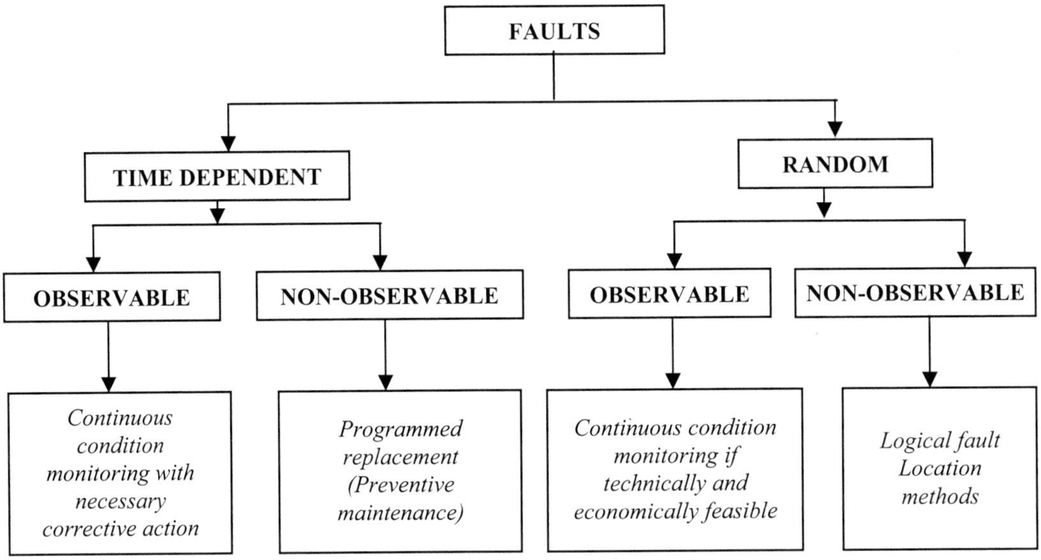


**Figure 7.2** Classification of faults.

*Time dependent failures:* These generally show a known or discoverable pattern of failures, for example, worn out shafts, tyres etc.

*Random faults:* These faults are distributed in time and occur at random, so that they cannot be predicted. For example, breaking of tension springs, blowing of electric bulbs etc.

*Observable time dependent faults*: They can be detected by condition monitoring since the time between the fault appearance and failure can be estimated. Nevertheless since even the process of time dependent failures are affected by environmental factors like lubricating oils, handling, fuel quality, dirt etc., we cannot say with 100%

confidence that the process of failure is deterministic. Fig. 7.3 illustrates the development of time dependent faults.

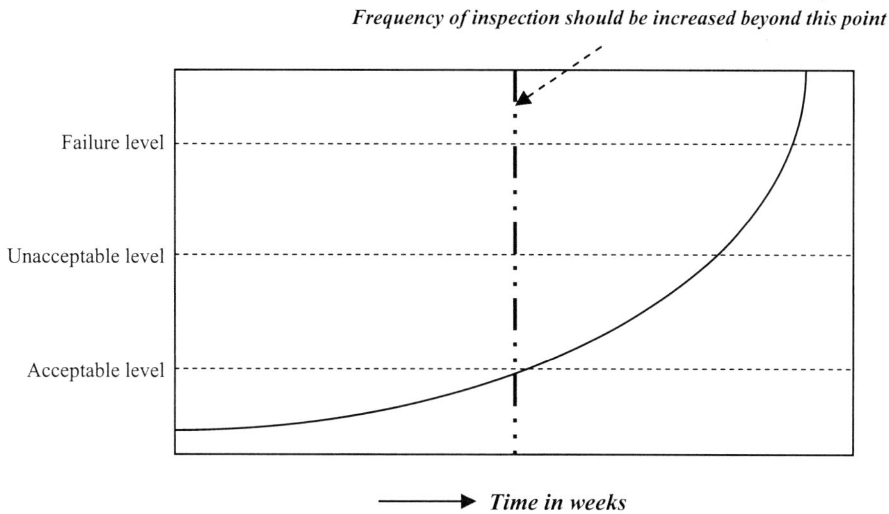


**Figure 7.3** Development of time dependent faults.

*Non-observable time dependent faults:* These can be prevented only by programmed replacement policy or repair after logical fault location. Programmed replacement and/or repair is often used in the case of equipment that is run continuously for some time and then taken out of service for some period. For example, the equipment used in process or continuous industries.

*Observable random faults:* They develop over a period of time. If the inspection periods are shorter than the period of fault development, these can often be detected by the inspection or the condition monitoring. If the development period is shorter, condition monitoring period/frequency should also be increased, if economically viable.

*Non-observable random faults:* They have no development period, that is the failure occurs as soon as the fault appears. Therefore they cannot be forestalled by condition monitoring. Neither can such fault be prevented by programmed replacements before they occur since they are randomly distributed in time. This according to our earlier definitions is Breakdown Maintenance. The faults do occur and it is the skill of locating the fault that is the essence of this chapter. The only thing we can do to prevent such faults, is to redesign, if economical, as a result of recording and analysis of the fault pattern.

## 7.5 Six Step Procedure for Logical Fault Location

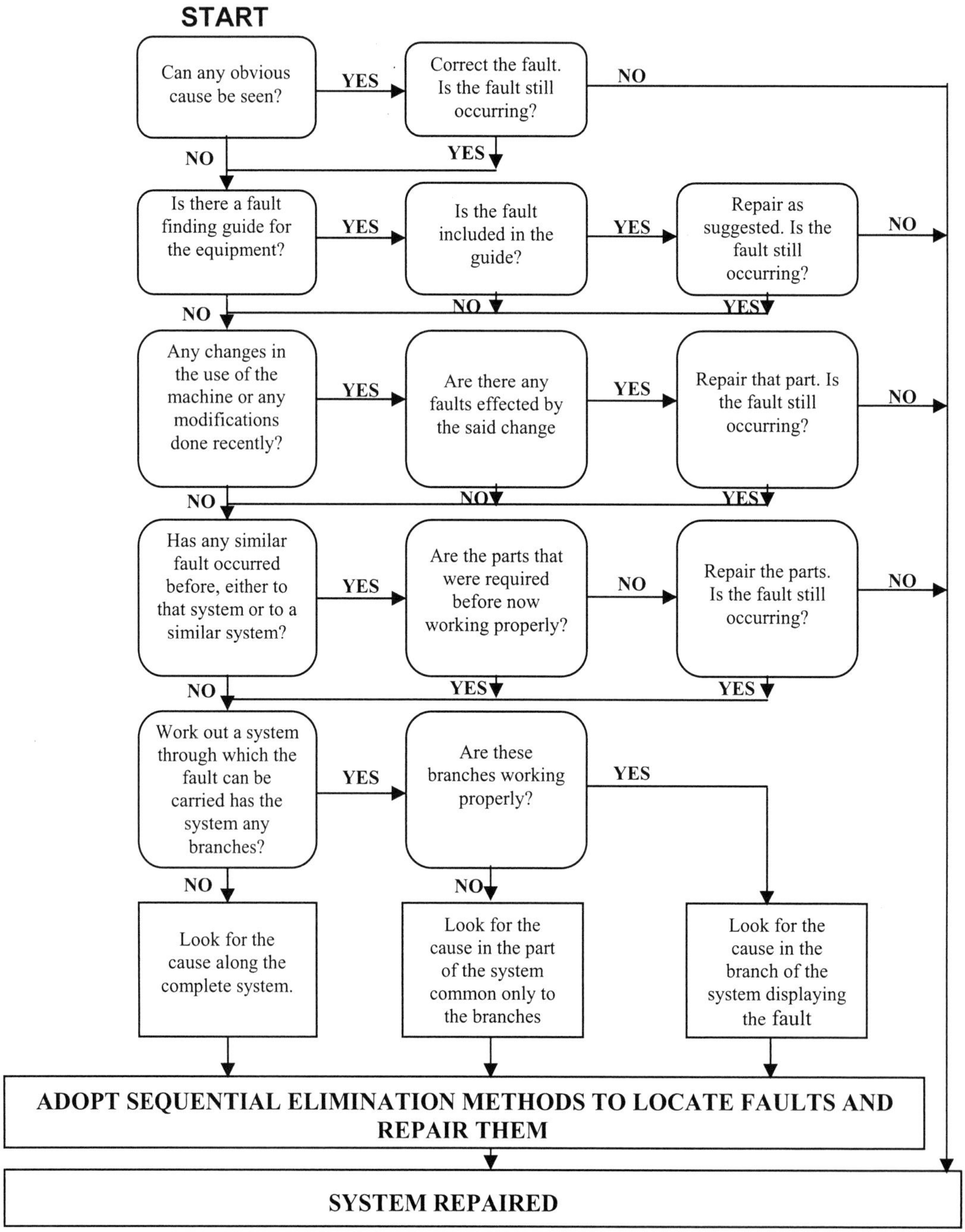


**Figure 7.4** A Sequential guide for logical fault location based on equipment history.

Repair time includes location or detection time as well as the rectification time. In many situations, especially those where there might lie a substantial chain of consequent effects, the diagnosis time can be significant.

The fundamental philosophy of logical fault finding is to begin with the most general and proceed to the more specific. From this emerges the following six step procedure, which is illustrated in Fig. 7.4.

1. Symptom analysis-determination and analysis of the effects of the fault,
2. Equipment inspection,
3. Location of the fault area,
4. Location and elimination of fault,
5. Repairing or replacing, and
6. Performance testing and forwarding, information for the cause diagnosis.

A systematic procedure for identification and rectification of the faults can also be indicated by the sequential guide as illustrated below.

If this is to be effectively implemented the fault finder must be well trained and provided with documentation, and text facilities. The logical identification of faults should be introduced at an early stage in the training.

## 7.6 Types of Logical Fault Location Systems

The six stage approach should form the basis of the methods adapted for fault location which includes understanding of such fundamental techniques, as classified in Fig. 7.5.

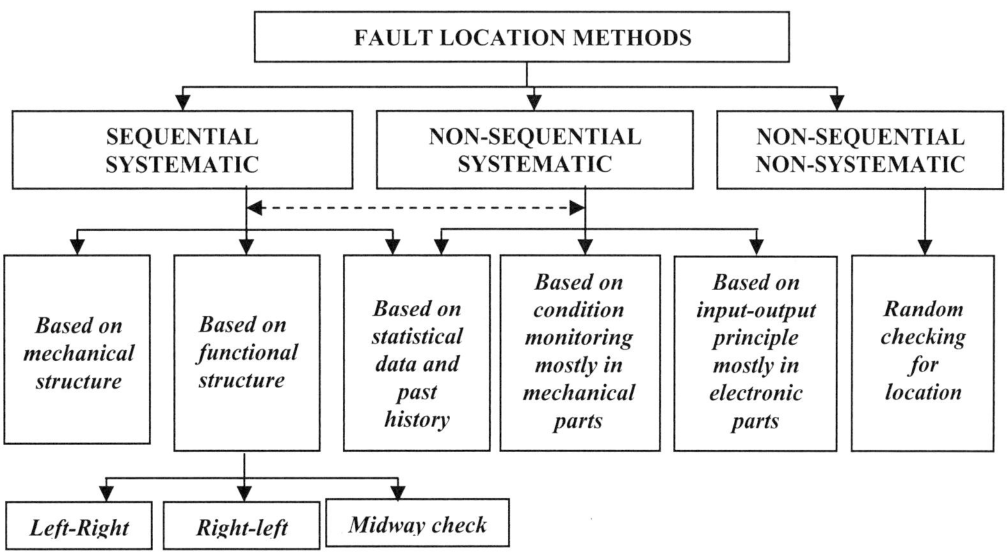


**Figure 7.5** Classification of logical fault location systems.

*Non sequential methods:* The fault is located by one measurement or a number of measurements made rapidly and automatically. This is mostly adopted in electronic circuits or machines controlled with electronic circuits where the determination of faults is by the measurement of the transfer function which is the relationship between the output and input signals. A set of signals of varying frequencies is applied to the input circuit, and the output characteristic curve is measured. This output is compared with a set of pre-determined or pre-established characteristics obtained with specific components' failure in the circuit. For example in a circuit tuned to a particular frequency, a fault in any of the timing equipment will change the resonant frequency. The nature of the detected change can indicate which component is faulty. This technique has limited application, but it has been known to be feasible for small linear electronic circuits, where it is possible to a large extent to incorporate such fault location provision at the design stage.

A similar principle, which compares the voltage instead of the frequency characteristics can be designed and adopted, in certain electrical circuits and machines, whereby the voltage at a number of key points in the circuit is measured from which the fault can be located. If such different voltages at different points are continually and permanently indicated in a central control board, any fault can easily be identified, as soon as it occurs and before a damage occurs.

Both these methods involve a large number of measurements and comparisons with established norms. Hence use of data processing system aided by a computer is useful. One of such systems of using data processing in this case is called the Expert system.

*Sequential and systematic method*: Here, the fault location is made in a sequence, by narrowing down the fault finding area. A set of pre-determined rules is used to decide the method for each check, as well as the procedure. These decisions can be based on the reliability data, or the functional structure of the equipment or its mechanical structure. This is illustrated in Fig. 7.1.

*Decision based on reliability data or equipment history:* The reliabilities of different components are fairly well established and if the equipment is similar to one in which, the components are of known reliability, checks and procedures can be made on the basis of these component reliabilities. For example a decision may be to change all bearings at an early check in fault location procedure.

*Decision based on the mechanical structure of the equipment*: Looking for the fault purely in the visual or audio senses come under this category, provided it is done systematically. For example, a search can be made by looking for the fault, at one side of the chassis, and looking for each component and connection in turn, working, towards the other side of the chassis. A fault may either be seen directly like broken wire or shorted terminal points or cracked component, or can be detected by secondary effect like melting wax or burnt paint. The fault can also give audio indications like body rattling or peculiar noise in the exhaust tube, that indicates a hole in the exhaust pipe of an automobile.

*Decision based on the functional structure equipment:* Functional structure means the functional relationship between components as far as the operation of the equipment is concerned. This is a very important aspect of the fault location and hence discussed more in detail in the subsequent paragraphs.

## 7.7 Sequential Elimination Methods

The common fault location methods are the following sequential methods, based on the functional structure:

- The left-right check
- The right-left check and
- The midway check

(i) *The left-right check* commences with a check near the input of an equipment and eliminates fault location areas by a series of checks, that move gradually to line output. This method is also called the left to right sequential elimination method. This same is derived from the fact that in a lathe which is the most common machine handled by the maintenance engineers, the driving motor is located on the left spindle of the machine, while the power flows towards the right to drive the head-stock or the carriage. Even in a majority of the functional diagrams of several process equipment the power flow or the product flow is from left to right. Consideration of non-functioning unit consisting of a driver, a clutch and a driven component as in Figure 7.6 to determine exactly which portion is non-functional, disengage the clutch and see if the driver side is functioning well, in which case the cause of the fault is more likely to be on the right of the clutch. On the other hand, if the driver still does not work, then the fault is more likely to be on the left of the clutch.

(ii) *The right to left check* is the reverse of the above, that is commencing from the output and checking towards the input.

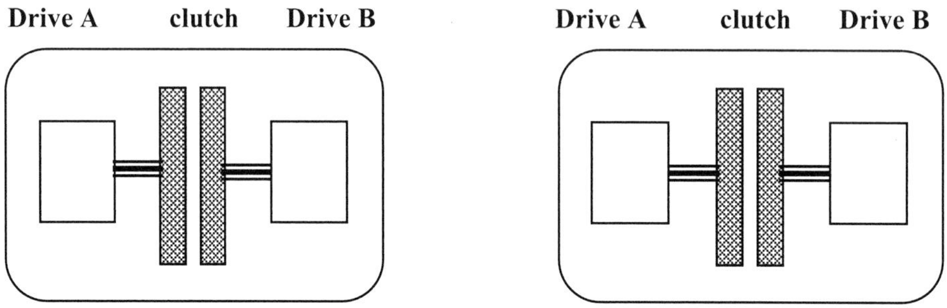


*Left drive works - fault is on the right* *Left drive does not work - fault is on the left*

**Figure 7.6** An illustration of the left to right rule.

(iii) *The half split rule or the midway check* method states that to locate a fault in a group of units or components, one of which is known to contain a fault, each check must be made at a midway point of the group that is having almost equal number of components on either side. This will enable maximum number of component to be eliminated from the fault checking list.

## 7.8 Functional Block Diagram

For logical diagnosis, we should consider an equipment as a series of components lined to each other is a way to form a chain, along which the power and movement are transmitted. This can be represented graphically and is called a Functional Block Diagram as illustrated in Fig. 7.7.

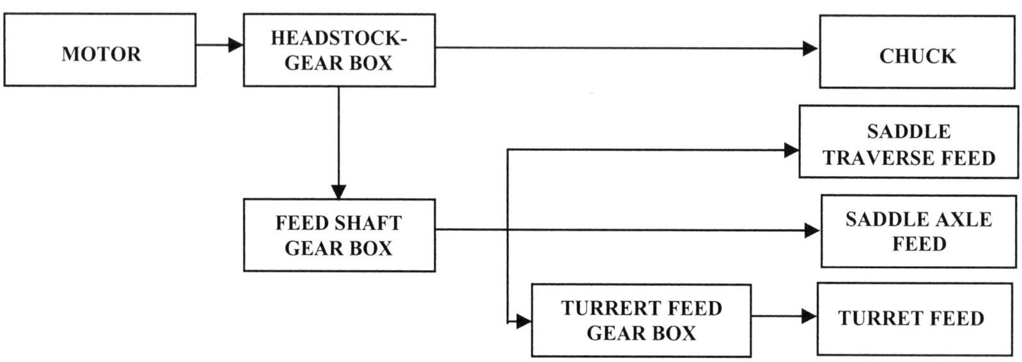


**Figure 7.7** An example of a functional block diagram.

Split the system into blocks, each of which represents a definite function and has measurable input and output. The following are some of the general rules in creating the block diagrams.

1. While the block represents a function, it is reasonable to try to coincide with the physical boundaries of the equipment.
2. It is essential that each functional block should have measurable input and output.
3. As far as possible in the initial step, the blocks should be visualised in such a way to include the major functions only. That is the whole system may be represented by a maximum of 6 blocks. When the faulty block is identified, it may itself be further broken down into sub-functions, each with its own block diagram and analysed as above.

## 7.9 The Midway Check Method or the Half Split Method

The half split rule states that to locate a fault in a group of units or components, one of which is known to contain at fault, each check must be made at a midway point of the group that is having almost equal number of components on either side. This will enable maximum number of components to be eliminated from the fault checking list irrespective of the result of the check, whether the left side is faulty or the right side is faulty.

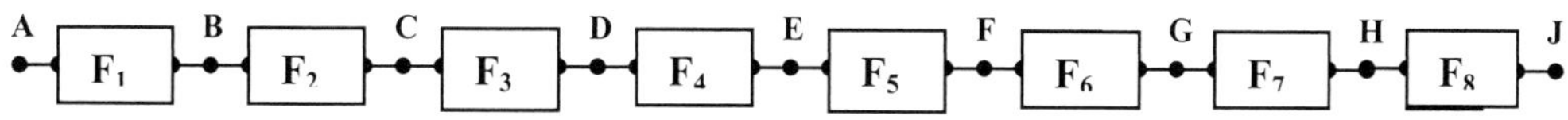


**Figure 7.8** A simple series circuit.

Consider the above system of 8 units of which one is known to contain a fault. The first check is made at a test point E. If the result is still not O.K., the fault is in any of the components $F_1$, $F_2$, $F_3$ or $F_4$ but if the result is O.K., then the fault is in any of the units $F_5$, $F_6$, $F_7$ or $F_8$. Thus we can see that atleast four units are eliminated straightaway with a single test, whatever may be the result of this test. For example, a first check at point, C for instance would have eliminated only two units if an O.K result is obtained, with 6 more units yet to be checked.

If the result of the test point E is O.K., the fault must be in right units of the system. The next test must therefore be made at the halfway point of those remaining units e.g. at test point G. If the result is not O.K, for example, the fault must be in $F_5$

or $F_6$ and one more test at point F will determine exactly which component is faulty, The decision tree as per Fig. 7.9 illustrates the above procedure.

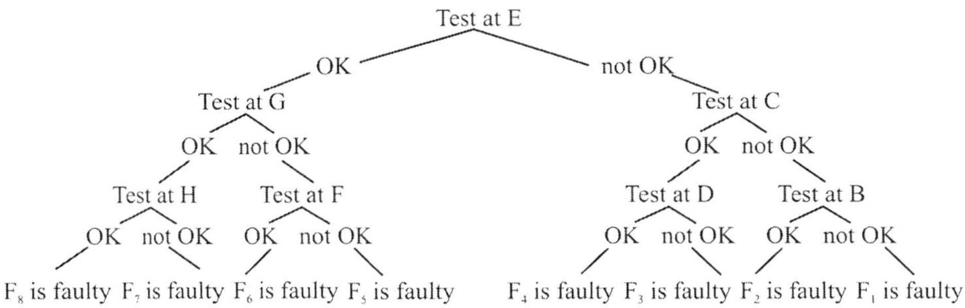


**Figure 7.9** Decision tree for the midway check method.

In case the probabilities of failure of certain specific components of the system is known to be more than the others, then each of the test point shall be assigned with a probability weightage as indicated in Fig. 7.10 and the procedure for the test would be as indicated in Fig. 7.11.

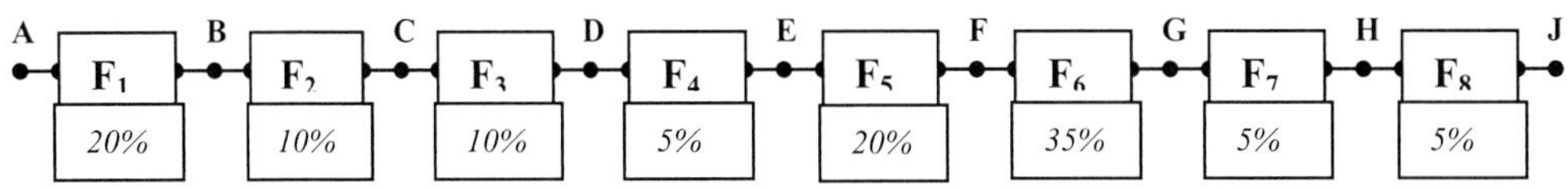


**Figure 7.10** A simple series circuit with know failure probabilities.

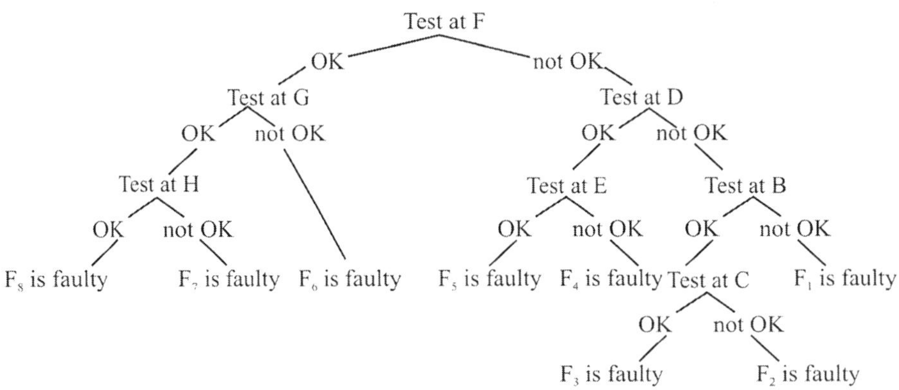


**Figure 7.11** Decision tree for the midway check method for the above situation.

## 7.10 Assumptions made in the Half Split Rule or Midway Check Method

The following assumptions are made in application of half split rule.

(a) Only one component is faulty

(b) All components are equally reliable

(c) The test signal tolerances are known.

(d) All checks are of the same difficulties and take the same time to complete.

## 7.11 Choosing the Best Method

Fault location is carried out using a combination of the above methods and the particular combination used depends on a number of factors such as:

- Whether reliability information is available or not,
- Whether units can be easily moved or not, and
- The availability of test points.

A method based on reliability is clearly useless when there in a lack of failure data, and also a method based on unit replacement is not possible for an equipment that is not constructed with easily removable units.

Whatever may be the method used, most maintenance tasks require a series of checks in order to reduce progressively the faulty area. If maintenance efficiency is related to the number of checks made to locate the fault, then the best maintenance performance must be the one which requires the lowest mean number of checks. It is important that the average must be taken from a large number of maintenance tasks. The reason for this is that for one specific task, a particular method could be brightly successful but for some other tasks, it may not be so.

## 7.12 Case Study

A bicycle manufacturing company in Tanzania, National Bicycle Company, has many automatic plants each of which are made up of a series of individual units. These plants have had a history of high rate of breakdowns. An external consultancy study revealed that whenever a plant broke down, line engineers spent a lot of time in identifying the location. As a matter of fact their log books showed that 60% of their maintenance time was spent on wasteful dismantling, testing and reassembling of units. This sparked the critical thinking and development of logical fault location methods by the author as explained in this paper. The maintenance engineers were given guidance in adapting this procedures to most of their automatic machinery like the automatic electroplating plant.

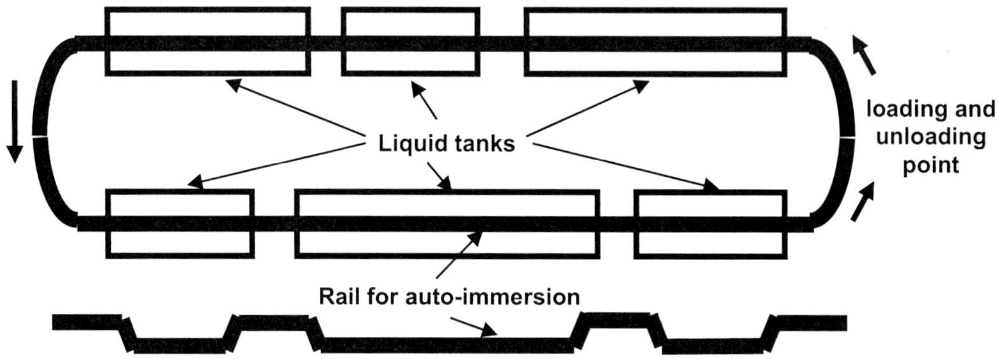


**Figure 7.12** Schematic diagram of the railing and the tanks of the automatic plating plant.

## 7.13 Conclusion

While the rules and the illustrations indicated in this chapter for the logical fault location are universally applicable in general, custom designed procedures can be developed for each machine and equipment, reducing a lot of tension filled moments for the maintenance engineer. It is the author's experience that though many engineers apply these short cut methods knowingly or unknowingly, it is the lack of a streamlined concept based on logic, which has prompted this chapter. Practicing engineers are earnestly requested, to write to the author either directly or through professional journals, their experiences in this direction so that more case studies can be examined in developing this into a systematic procedure aiding all mechanical maintenance engineers.

# Chapter 8

# Wear Debris Analysis

*Chapter synopsis:* *Among the several condition monitoring techniques, wear debris analysis has emerged as a significant on-line NDE technique for assessing the wear condition. It forms the latest and very accurate technique among the World class Maintenance Practices. This technique involves in collection of wear debris that flows along with the lubricant and analyzing their shape, chemistry, size, and density to determine the wear mechanism, the wear mode and wear phases undergoing in the equipment. This chapter details this technique with reference to the principles and methodology highlighting some of the softwares available in this field.*

*Keywords:* *Condition monitoring, wear debris analysis, history of WDA, ferrography, abrasive wear particles, adhesive wear particles, fatigue wear particles, corrosive wear particles, type of material, shape of the particle, size of the particle, numerical density, factors for excessive wear, material identification, platelets and flakes, spherical debris, pebbles and smooth ovoids, chunks and slabs, curls, spirals or slivers, strands and fibers, methodology, plain glass slide, magnetic slide, rotary particle depositor, binocular microscope, trifocal microscope, magnetic plugs, computer aided systematic particle analysis, computer aided visual engineering, wear particle atlas, 5200tv trivector analyzer, metallurgy plus, pyro science plus, national conference on condition monitoring, Nitte, Karnataka.*

## 8.1 Introduction

Condition monitoring is that type of planned maintenance wherein the actual working condition of major components of an equipment are constantly monitored to give an indication of which parts should be replaced and which need not be replaced in the near future. Several conditional monitoring techniques are available to monitor this process and identify the source and type of the problem.

For industrial machinery and equipment, the following are generally monitored

- Monitoring of visual condition of inaccessible locations

- Monitoring of inside condition of tubes
- Temperature monitoring
- Vibration monitoring
- Leak monitoring
- Corrosion monitoring
- Belt tension monitoring
- Shaft run out monitoring
- Crack formation monitoring
- Wear debris analysis

Among the above cited condition monitoring methods for industrial machinery and equipment like temperature monitoring, and vibration monitoring, wear debris monitoring is the latest and very accurate technique of identifying the source and type of wear and forms the foremost technique among the World class maintenance Practices. This becomes more significant since the interior and inaccessible parts can be monitored with relative accuracy before the failure occurs.

*Just as the doctor analyses the blood sample to diagnose the internal problems of a human body, the analysis of the wear debris among the oil samples drawn from the machines can help in pinpointing the impending internal problems.*

## 8.2 History of Wear Debris Analysis

The condition of the lubricant has always been an indication of the condition of the machinery and other industrial equipment involving eroding elements. The early debris observation was in the defense equipment operating in dusty atmosphere. The main cause of concern then has been the contamination of the oil by dust particles which gradually add to the abrasion of the bearing components. Subsequently other metallic fragments other than dust had been noticed which led to concept of conducting wear debris analysis that could identify the bearing components that get subjected to wear and tear even before any malfunction could be identified by other condition monitoring methods.

In this connection let us distinguish between contamination and wear debris. Contamination is the turbidity created by the atmospheric dust that leaks into the oil and gets settled there. Wear debris is the fragmented particles chipped off or otherwise created due to the wear and tear of the components.

It is the identification of the presence of the latter that distinguishes the WDA of machineries and plant equipment from IC engines or generator sets.

This analysis, also called Ferrography, can determine

- The start of the abnormal wear,
- Root cause of the failure,
- The components that are wearing,
- Usability of the lubricant beyond its rated use.

## 8.3 Achievements of WDA

Since early 1986 WDA became more wide spread and accepted as an effective condition monitoring technique. G. Fogel, in his book *Wear Debris Analysis* cites some of the successes achieved by the application of WDA as:

1. Failure prevention in diesel engines used in haul trucks.
2. Failure prevention in large mining especially in miner gear boxes.
3. Failure prediction on a large selection of both surface and underground mining equipment in gold mining industry in South Africa.
4. Contamination control in surface gold mining process equipment.
5. Grease and oil selection through comparative lubricant trials.
6. Condition monitoring of critical coal mining gearboxes in the power generation industry.
7. Failure investigation of critical bearings in ball mills used in the gold mining industry.
8. Identification of wear on the bearings inside the CNC machines before appearance of any other symptom, in an automobile industry in India during the late 1990's and early 2000's.

## 8.4 The Four Basic Types of Wear Particles

1. *Abrasive wear particles,* which normally indicate excess dirt and other hard particles that are cutting away at the load bearing surfaces.
2. *Adhesive wear particles* that reveal problems of lubricant starvation, either due to low viscosity, high load, high temperature or slow speed.
3. *Fatigue wear particles* that are due to mechanical criteria like improper assembly, misalignment, imbalance etc.
4. *Corrosive wear particles* that indicate leaking of water or other corrosive liquids getting into contact with metal surfaces.

## 8.5 Types of Debris

WDA involves the analysis of the four basic parameters which are

- *The type of material* of the debris, indicating the possible element of the machine tool,
- *The shape of the particle*, indicating the type of wear, like gear pitting or slide-way chipping,
- *The size of the particle* indicating the extent of damage,
- *The numerical density* of the particle indicating the extent of wear and tear.

## 8.6 Factors for Excessive Wear

- Excessive load,
- Onset of fatigue,
- Corrosion,
- Misalignment,
- Imbalance,
- Lack of lubrication,
- Lack of monitoring,
- Total acid content (TAN) in oil,
- Total base content (TBN) in oil,
- Period of run-in.

## 8.7 Wear Indicated by Debris Material Identified in the Oil Samples

Analysis and identification of abrasion material by metallurgical analysis generally indicates the following:

1. Presence of iron
    - Wear of the cylinder liner
    - Wear of the piston rings
    - Wear of crank-shaft
    - Wear of oil pump

2. Presence of aluminum
   - Wear of pistons
   - Babbitt white bearing metal
3. Presence of copper
   - Copper bearing metals and/or thrust bearings
4. Presence of silicon
   - Generally through sucked in air
5. Presence of lead, silver or tin
   - Wear of bearings that contain the respective metals.
6. Presence of nickel
   - Wear of bearings or valves
7. Presence of magnesium
   - Wear of crank covers
8. Presence of zinc, calcium, phosphorous etc.
   - Contamination from oil additives.

## 8.8 Wear Indicated by the Shape of the Debris Particles *(figures from Wikipedia)*

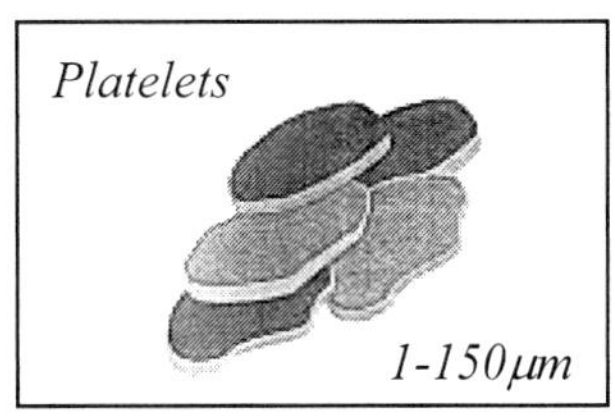


1. *Platelets and flakes*: These are two-dimensional particles caused by metal to metal sliding, generally in absence of lubrication, mostly indicating running in metal wear. It can also be from paint or rust.

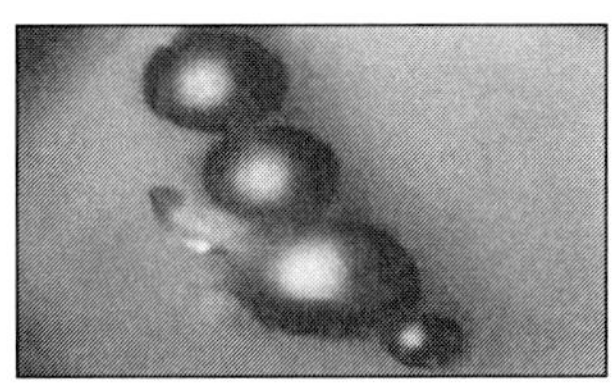

2. *Spherical debris*: These are mostly from the antifriction bearings. Due to the continued rolling of the other balls, the fragmented debris get rolled (like the marbles) and become smooth balls till they fall off the bearings and enter the oil flow. These particles of size about 50 microns indicate the severity of rolling fatigue wear.

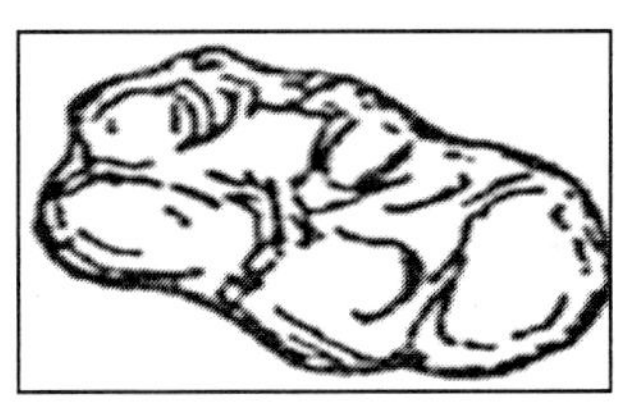

3. *Pebbles and smooth ovoids*: In most cases the material is silicon from atmospheric dust. However if the material is steel or other bearing material, they can indicate rolling debris, similar to the above.

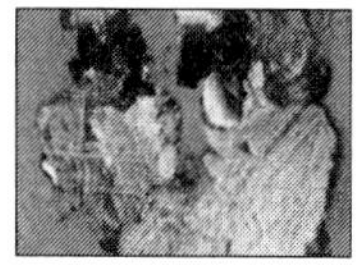

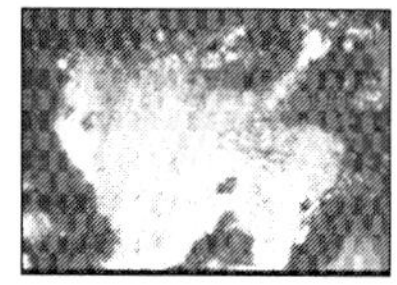

4. *Chunks and slabs*: They indicate metal fatigue or bearing pitting
5. *Curls, spirals or slivers:* They are similar to the machine swarf, caused by a hard metal particle abrading into a softer metal. Sometimes machining debris produced at high temperature take this form.

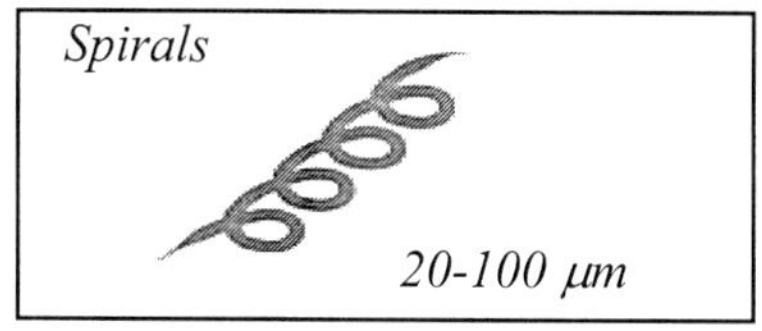


6. *Strands and fibers:* They are mostly of polymers, cotton and wood which are more from work pieces are otherwise than the equipment components. However if the machine or equipment is known to contain such polymer components, they may indicate the wear debris.

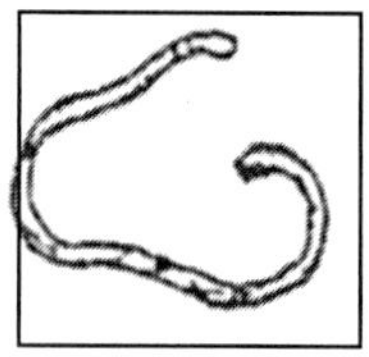

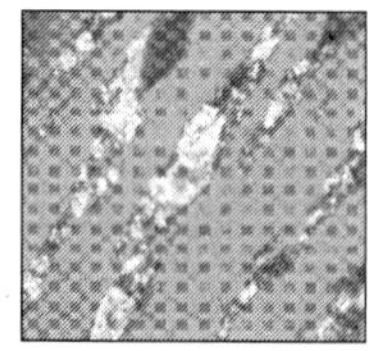

7. *Miscellaneous debris:* Occasionally metallic strands or fibers too can be found.

## 8.9 Wear Indicated by Debris Size

In general the size of the debris particles indicate the seriousness of the wear mechanism. As the wear worsens, the average size of the particle increases. For the purpose of WDA the size can be classified as follows:

| | |
|---|---|
| Fine | Less than 5 microns |
| Small | 5 to 40 microns |
| Medium | 40 to 100 microns |
| Large | above 100 microns |

## 8.10 Methodology for WDA

1. Collect oil sample with debris,
2. Mount on slide,
3. View through the image analyzer,
4. Analyze the debris morphology, using a magnification of atleast 100x,
5. Compare with previous samples,
6. Identify the source and location of the wear debris,
7. Dissemble the suspected component,
8. Inspect and repair or replace as needed.

## 8.11 Normal Method of Collecting Samples

1. *Use of plain glass slide:* Normal method is by collecting the particles on a slide and viewing under microscope to the adhesion, the debris particles would stick on to the oil film and remain on the slide. By this both the magnetic and non-magnetic debris can be collected and analyzed
2. *Use of magnetic slide:* If the machine under investigation is either machining cast iron or grinding, then it is very difficult to segregate the swarf from the wear debris. In this case a magnet is located under the upper part of the slide so that only the non-ferrous debris slides to the bottom for easy analysis.

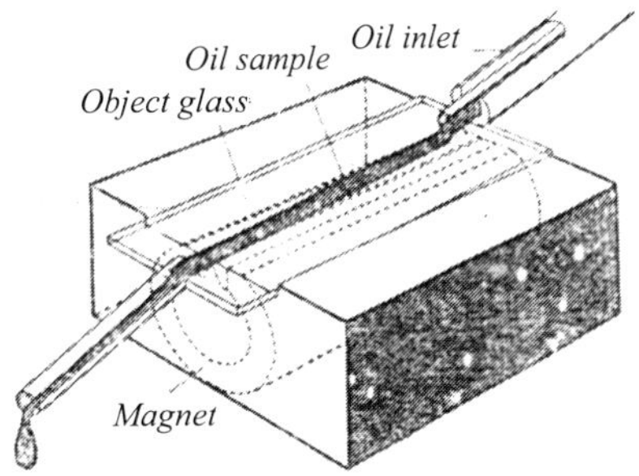


3. *Rotary particle depositor*: This has three concentric rotating rings. When the contaminated oil flows along the axis, they get deposited into the three rings graded by the size, using a combination of static and dynamic magnetic rings. Here after collecting the sample from the machine, it is first diluted in solvent and then passed through a membrane filter or magnetic separator, before loading into the rotary depositor. The Tribology and Condition Monitoring Group and Swansea Tribology Services were credited to have developed this.

## 8.12 Instruments for WDA *(figures from Wikipedia)*

- The binocular microscope
- Trifocal microscope with a web cam or digital camera fixed on to the third eye piece, that can be made to take photographs for future analysis
- Similar trifocal arrangement with the camera being connected to a computer to do on-line analysis
- Magnetic plugs.

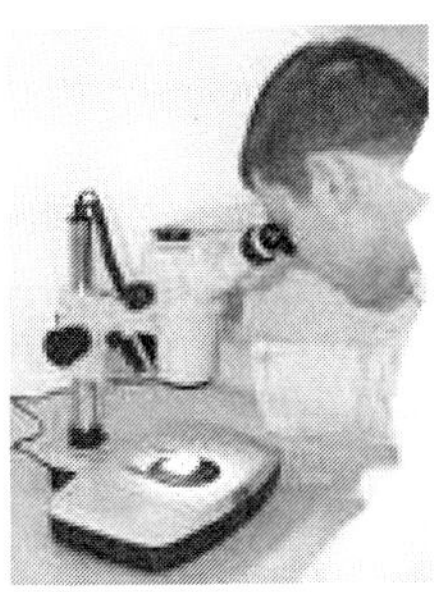

*The binocular microscope*

*Trifocal microscope*

*Magnetic plug*

## 8.13 Software for Computer Aided WDA

1. CASPA (Computer Aided systematic Particle Analysis), developed by RAF's Early Fatigue Detection Centre.
2. CAVE (Computer Aided Visual Engineering) CAVE is an automated system to classify and identify wear particles in terms of their morphological and textural features and by comparing with a library of expert data on available predetermined samples. This software uses video graphic generation of the images of the wear debris.
3. Wear Particle Atlas and Classification Scheme by Defense Evaluation and Research Agency (DERA) of UK. This provides data on critical components of defense equipment which can now be extended to industrial equipment.
4. The 5200TV Trivector Analyzer from Minilab measures three basic parameters of wear, contamination and chemistry, besides magnetic differentiation of wear particles as well as laser particle counting. This kit has a filter patch maker to aid the preparation of debris particles to be viewed through the microscope.
5. Metallurgy Plus, is the basic, user friendly software to determine metallurgical parameters as well as the particle size and count.
6. Pyro Science Plus is similar to above but also can help in acquiring, organizing, storing, retrieving, color adjustment and editing of an image and supports formats like BMP, TIFF, IMG, JPG, and GIF. Comparison of the image with standard specimens is possible with instantaneous report generation. We can also print the images at 6 images per A4 page.

   In fact, the last two software were used in the current study conducted by the author on wear debris analysis.

## 8.14 Side Benefits of WDA

Regular oil sampling ensures

- The presence of enough oil in required quality.
- Less breakdowns allows better use of manpower.
- Extension of the life of the machines.
- Assessing the condition of the machines so that they can either be replaced or allowed to run till major rectification.
- Reduction in inventory cost for the spares.
- Grease and oil selection through comparative lubricant selection.
- Failure investigation of critical bearings in ball mills etc.

- Contamination control in gold mining process equipment.
- Condition monitoring of critical gearboxes, other components of heavy vehicle and also in power generation industry.

## 8.15 Conclusion

Despite all these developments abroad, not much research has been done in India and no technical paper in this field could be located in any of the Indian Journals. During the National Conference on Condition Monitoring organized at Nitte, Karnataka, in 2002, around 32 papers were presented but none were on Wear Debris Analysis. Visits by the author to some maintenance departments of some leading industries in Chennai showed the debris analysis is generally limited to generators, motors and other service facility equipment, only to test the turbidity or the quantum of dust in the lubricant oil. However it is strongly advised that this wear debris analysis is done for the production machines and machine tools.

## 8.16 References


1. Brian J. Roylance and Trevor M. Hunt, *Increasing Reliability through Oil Analysis*
2. Kiran D.R., Condition Monitoring and Wear Debris Analysis, a PP presentation.
3. Kiran D.R., Wear Debris Analysis (thesis topic).
4. Coxmoor, *Wear Debris Analysis Handbook*
5. Alan Davies, Handbook of Condition Monitoring: Techniques and Methodology.
6. IIPM, Wear Debris Analysis: *www.iipm.ac.in/wear_debris_analysis.html.*


## 8.17 Relevant Websites for Further Reference


1. www.analystsinc.com/marksmith_poa.html - *Oil Analysis vs. Microscopic Debris Analysis*
2. www.stinet.dtic.mil/oai - *A Phenomenological Approach to Wear Debris Analysis*
3. www.stinet.dtic.mil/oai - Wear Debris Analysis of Grease Lubricated Ball Bearings
4. www.intota.com/multisearch. - *Wear-Debris Analysis Expert*
5. www.linkinghub.elsevier.com/retrieve/pii - *Wear : The use of color in wear debris analysis*

6. www.aggman.com/articles/feb04d.htm - *A Quick Guide to Oil Analysis Tests*
7. www.ingentaconnect.com – Ingenta Connect - *Wear-Debris Analysis in Expert Systems*
8. www.inderscience.com - *Computer vision techniques for wear debris analysis (CVTWDA)*
9. www.practicingoilanalysis.com - Fundamental Principles in Setting Alarms and Limits in Wear Debris
10. www.atypon-link.com - Academic and commercial efforts in Wear Debris Analysis Automation
11. www.polarislabs1.com/testing - Preventative Maintenance Programs
12. www.testoil.com/filter-debris- Filter Debris Analysis
13. www.practicingoilanalysis.com - *X-Ray Fluorescence Spectroscopy - The Next Generation of Wear*
14. www.kellysearch.com/us-product - Oil or Lubricant Wear Debris Analysis
15. www.aspexcorp.com/html - *Wear Debris Analysis by ASPEX*

# Appendix

## Illustration for Interpretation of the analysis

(Courtesy: www.analystinc.com)

| | Interpretive Method | Particle Size Range | Type of Debris Detected | How Debris is Collected |
|---|---|---|---|---|
| **Instrument-based Magnetic Attraction/Flux Techniques** | | | | |
| Direct Reading Ferrography | Trend and/or fixed limits depending on application. | Two ranges:<br>(1) DS: 1 μm to 3 μm<br>(2) DL: 5 μm to 100 μm | Ferromagnetic particles of iron and steel | Oil sample |
| PQA Ferrous Debris Monitor | Trend and /or fixed limits depending on application. | 1 μm to 1,000 μm (Slight efficiency fall off below 5 μm). | Ferromagnetic particles of iron and steel | Oil Sample |
| **Microscopic Analysis-based Magnetic Attraction Techniques** | | | | |
| Magnetic Filter Plug | Visual/microscopic inspection for density, size, shape, etc. | Depends on magnification, particle type, viscosity and flow rate. Typically 25 μm to > 1,000 μm. | Ferromagnetic particles of iron and steel | Removable magnetic plug |
| Magnetic Chip Detector | Visual/microscopic inspection for density, size, shape, etc. | Depends on magnification, particle type, viscosity and flow rate. Typically 25 μm to 1000 μm | Ferromagnetic particles of iron and steel | Removable magnetic cartridge |
| **Microscopic Analysis-based Filter Debris Techniques** | | | | |
| Patch Test | Visual/microscopic inspection for density, size, shape, etc. | Depends on magnification power and membrane pore size. Typically 3 μm to 1,000 μm. | All types of filterable debris | Oil sample |
| Rotrode Filter Spectroscopy (RFS) | Trended multi-element measurements reported in ppm. No standerdized measurments due to proprietary test methodology. | 5 μm to 150 μm | Wear debris and contaminate particles | Oil sample |
| Acid Digestion | Typically run as an exception test. Provides absolute quantification of the concentration soluble and insoluble elemental constituents in the oil, sludge, sediment, etc. | Generally not limited by particle size, i.e., from soluble to insoluble suspensions > 1,000 μm. | Wear debris and contaminate particles | Samples of oil, sludge, sediment and filter residue |
| Single Particle Analysis such as SEM-EDX (also referred to as chip analysis) | Typically run as an exception test to identify elemental constituents, including alloying elements, of specific particles in a failure investigation. | From submicron to macroscopic size particles (32 mm in some instruments). | Wear debris and contaminate particles | Samples of oil, sludge, sediment and filter residue |

# Chapter 9

# Reliability Based Maintenance

***Chapter synopsis:*** *As seen in the previous chapter, the reliability of a system is the ultimate hallmark of all maintenance activities. This chapter defines and discusses the concepts of reliability and availability of a system and how they are related to the breakdown history. To make the concept more clear, illustrations and worked out examples are provided.*

***Key words:*** *Reliability centered maintenance, House of RBM, induced, intermittent and wear out type of failures, severity of failures, SAE JA1011, functional failures, failure modes, failure effects, failure consequences, tasks and task intervals, on-condition tasks, scheduled discard tasks, scheduled restoration tasks, failure finding tasks, one-time changes, run-to-failure, soft foot condition, DOD-STD-2101, statistical distribution curves of failures – normal, Poisson and Weibull, probability density function (PDF), priority allocation of maintenance activity, Monte Carlo simulation, Markov Analysis.*

## 9.1 Functional Reliability

In Chapter 8 we have seen with several illustrations how the functional reliability affects the availability of equipment for manufacturing activities. We defined reliability as the mean time between failures (MTBF). We have also defined it as the probability of failure free operation of the product or system for a given period of time under specific conditions and the ability of equipment to perform a required function under stated conditions for a stated period of time and in other words how often the breakdown occurs as measured in time units say, hours.

Reliability centered maintenance is an engineering framework that enables the definition of a complete maintenance regime. It regards maintenance as the means to maintain the functions a user may require of machinery in a defined operating context. Reliability engineering deals with the study of the ability of the equipment to perform its required functions under stated conditions for a specified period of time. It characterizes, measures and analyses the failure and repair of the systems to improve their use by increasing their design life, mitigating defect risks and reducing the likelihood of failures.

## 9.2 Quality vs Reliability

While both the terms Quality and Reliability are generally used to represent the engineering disciplines to assure that the product performs its intended function as the customer wants it, the later implies a sustained quality performance over a period of time. This time factor marks the difference between traditional quality control, and the approach to reliability. We can further distinguish between them as

**Table 9.1** Quality vs reliability.

| Quality | Reliability |
|---|---|
| It is everything until put into operation | It is everything happening after 0-hours |
| It is independent of time | It is time dependent |
| Latent failures are removed by quality control methods. | Latent failures can be detected. |
| It is lot dependent | Numerical estimates like MTBF and failure rates are possible |

In other words quality is today and reliability is tomorrow.

## 9.3 What is RBM?

Reliability Based Maintenance, originally called by its synonym *reliability centered maintenance* (RCM) incorporates sound guidance for managers who wish to attain high standards of maintenance at their operating plants. The amount and type of maintenance which is applied depend strongly on

- The age of the machine or components
- Its replacement cost and
- The cost and safety consequences of system failure

Thus a careful analysis of the system components based on their reliability data should be done in order to optimize the maintenance program.

This chapter views reliability as influenced by the random failures of equipment and analyses the several types of failures that affect the equipment reliability. A maintenance system planned based on this statistical analysis is thus called Reliability Based Maintenance. This concept was first developed by the joint studies of US Department of Defense and United Airlines in the 1980's as DOD/UAL's Reliability Centered maintenance (RCM). This emphasizes the use of Predictive

Maintenance techniques in addition to traditional preventive measures by attempting to predict the random failures more accurately than leaving them to providence as GOK (God only knows). This concept was first applied in the 60's and 70's for the new Boeing 747 Jet.

Since then the concept of RCM had been applied at several industries for their physical asset management strategies. However the widespread use of this concept resulted in different interpretations of RCM by different practitioners leading to the emergence of processes that differ somewhat from the original process. The Society of Automobile Engineers (SAE) responding to this growing demand for a standardized concept initiated JA 1011 in 1999 a standard that sets up criteria that a process must be complied with, in order to be called RCM. This standard is discussed more in detail in paragraph 9.4.

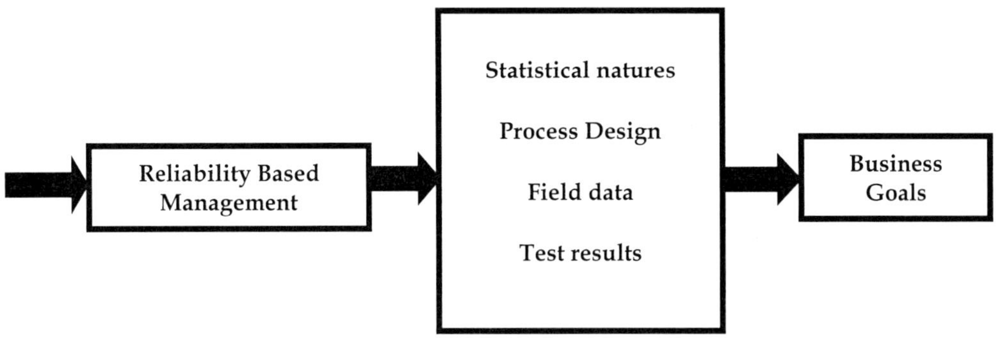


**Figure 9.1** Factors for RBM.

This author nostalgically remembers that during the eighties, Idhammer Konsult of Sweden conducted a training programme at Dar Es Salaam on "Results Oriented Reliability Maintenance Management" when "Christer Idhammar gave the "Introduction to Root Cause Problem Elimination" while Ian McKinnon's gave "Precision Maintenance Overview".

## 9.4 Nomenclature of RBM processes as used in SAE JA1011

SAE JA1011 [3], *Evaluation Criteria for RBM Processes*, sets out the minimum criteria that any process should meet, but uses a certain nomenclature for these processes like on-condition tasks, scheduled discard tasks, failure finding tasks etc, which can be equated to the processes described in chapter 3 of this book as illustrated in Table 9.2 below.

**Table 9.2** Nomenclature used in SAE Standard vis-a-vis that used in this text book.

| no | JA1011 Section | Nomenclature used in SAE-JA1011 | Nomenclature used in this text book |
|---|---|---|---|
| 1 | 5 | Reliability centered maintenance (RCM) | Reliability based maintenance (RBM) |
| 2 | 5.6 | Failure management strategy | Maintenance management strategy |
| 3 | 5.7.2 | On-condition tasks | Condition monitoring, Predictive maintenance |
| 4 | 5.7.3 | Scheduled discard tasks | Preventive maintenance, Fixed time replacement |
| 5 | 5.7.4 | Scheduled restoration tasks | Corrective maintenance, Remedial maintenance |
| 6 | 5.7.5 | Failure finding tasks | Logical fault location |
| 7 | 5.8.1 | One-time changes | Design out maintenance, |
| 8 | 5.8.2 | Run-to-failure | Breakdown maintenance, Operate-to-failure |

## 9.5 Principles of Reliability Based Maintenance

The EBME website (http://www.ebme.co.uk) lists the following principles upon which RBM is based

1. RBM is function oriented and seeks to preserve system or equipment function.
2. It is group focused and is concerned with maintaining the overall functionality of a group of devices rather than an individual device
3. It uses failure statistics in an actuarial manner to look at the relationship between operating age and the failures. However, RBM is not overly concerned with simple failure rate; it seeks to know the probability of failure at specific ages.
4. Acknowledges design limitations recognizing that changes in reliability are the province of design rather than maintenance. Maintenance can only achieve and maintain the level provided for by design.
5. RBM is driven by safety and economics. Safety must be ensured at any cost; thereafter, cost-effectiveness becomes the criterion.
6. It defines failure as any unsatisfactory condition as either a loss of function (operation ceases) or a loss of acceptable quality (operation continues).
7. It uses a logic tree to screen maintenance tasks to provide a consistent approach to the maintenance of all kinds of equipment.
8. Its tasks must address the failure mode and consider the failure mode characteristics.

9. RBM must reduce the probability of failure and be cost effective.
10. RBM tasks are interval (time or cycle) based and condition-based. Here run-to-failure, is a conscious decision and is acceptable for some equipment.
11. It is dynamic and gathers data from the results achieved which is fed back to improve future maintenance. This feedback is an important part of the Proactive Maintenance element of the RBM program.

## 9.6 House of RBM

The strength provided by RBM to an organization can be illustrated in the following Figure 9.2.

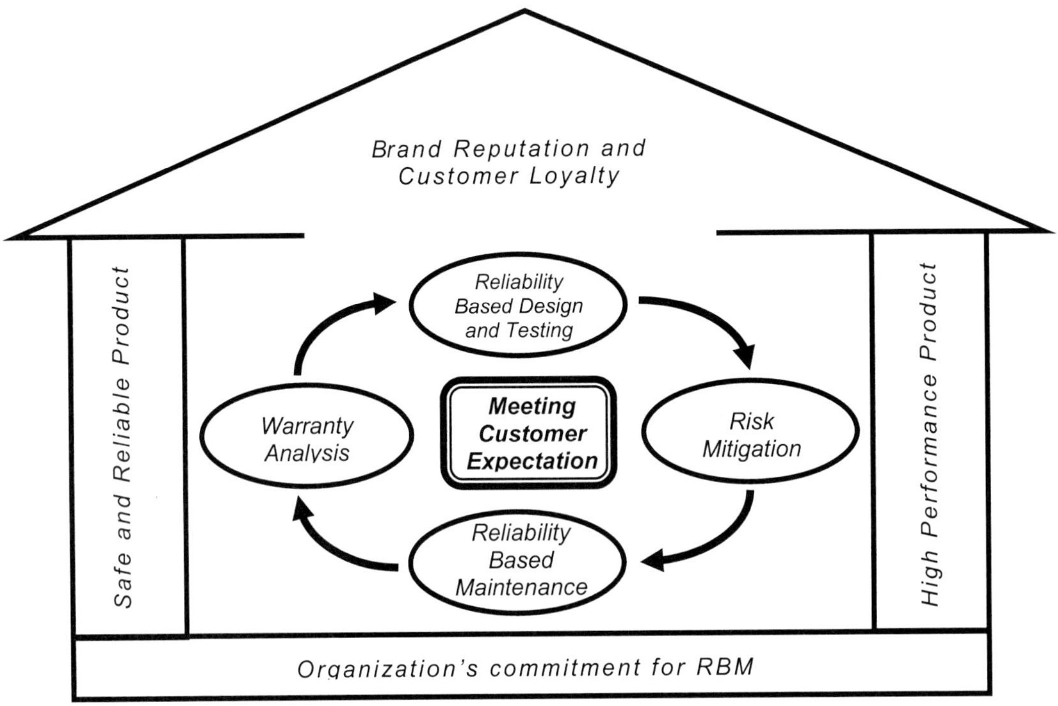


**Figure 9.2** House of reliability.

## 9.7 When do you apply RBM?

- High Cost of unscheduled breakdowns
- Lack of alignment between design and manufacturing capability on required reliability and maintainability.
- High warranty costs resulting in direct cut in operating profits.

## 9.8 Types of Failures

Failures can be grouped into the following three categories. Understanding these categories is critical when assigning maintenance tasks.

- Induced
- Intermittent
- Wear out

*Induced failures* are a result of an outside force causing the failure mode, like a soft foot condition on an equipment train causing coupling misalignment eventually leading to an inboard bearing failure. In soft foot condition imperfect surface contact between the underside of the machine's feet and the surface of the base plate, or frame causes misalignment of rotating equipment leading to the vibration.

It is important to understand that the induced failure must be recognized and analysis performed to determine the root cause, as explained further by the FMEA concept.

*Intermittent failures* can happen at any time at random and the mean time between failures (MTBF) cannot be pre-determined and the repair cannot be effectively planned and scheduled. A plant can best detect these failure modes through process monitoring and predictive maintenance to some extent.

*Wear-out failures* have a known MTBF and they occur when the useful life of a component is expanded. These types of failure modes are often detectable through process monitoring and predictive maintenance. However, time-based refurbishment or preventive maintenance sometimes could prove to be an effective maintenance strategy.

## 9.9 Severity of Failures

DOD-STD-2101 defines the characteristics of a component and system defects as

- Critical, if the failures will have adverse impact on safety,
- Major, if the defective characteristics will degrade with age, and
- Minor, if the defects donot have significant impact on the performance

## 9.10 Statistical Distribution Curves of Failures

While chapter 4 gives physical illustrations and computed the failure rate etc., by simple arithmetic computations, this chapter briefs the several distribution curves that the failures conform to. While the detailed statistical explanation is beyond the scope of this book, the basic explanation of their concepts to the extent a maintenance engineer should know is discussed in this chapter.

(a) *The normal distribution* is a probability distribution that associates the normal random variable around central value called the mean. This is generally applied in the analysis of variations around a nominal fixed value, like the variations in the machining of a bar to say 50 mm dia. It is also called a bell curve, since it looks like a bell with a central peak as in the figure 9.3.

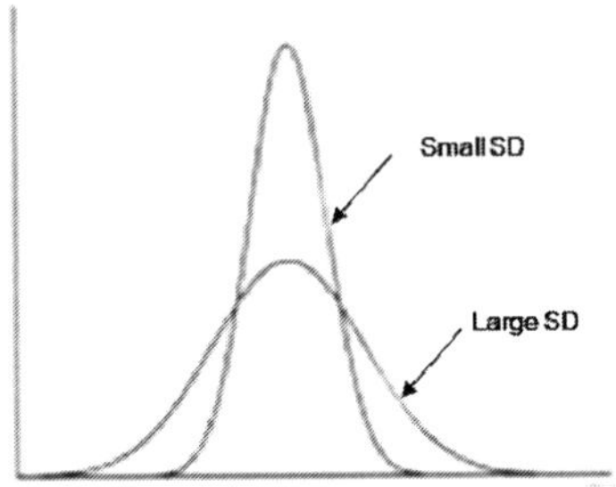


**Figure 9.3** Normal distribution.

(b) *The Poisson distribution* is a curve expressing the probability of a given discreet number of events occurring in a fixed interval of time or certain sample sizes like the number of defects found in each of the several samples picked up from a lot. This figure can be 0 or 1 or 2 etc., unlike the normal distribution which the variations cluster around a central figure. Here like the s of normal distribution, the variation factor is represented by λ (lambda). As λ = 1, it is a simple c shaped curve, reaching a maximum when the occurrence is zero and smoothening out to a normal curve when the occurrences are high.

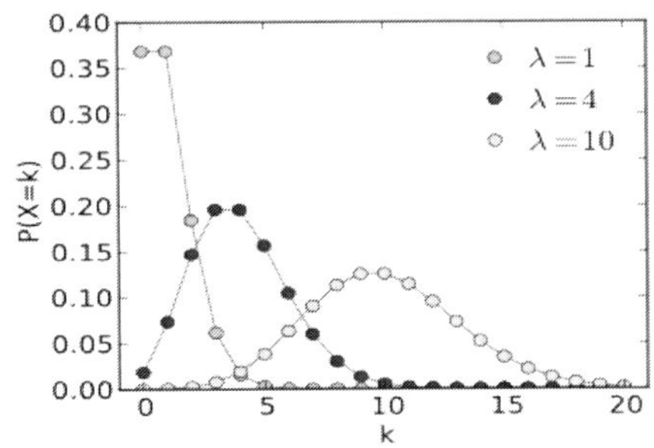


**Figure 9.4** Poisson distribution.

If X has a Poisson distribution with a mean of n (say 2) failures per year, then the probability that no more than r (say 1) failures occur per year is given by

$$\sum P(X = r) = \frac{e^{-2} n^{x}}{X!} \Sigma$$

If n = 2 and r = 1, the n (X = 1) would be 0.406 or 40.6% possible.

(c) *The Weibull distribution* is similar to Poisson, but uses three parameters

- The shape parameter (β), also known as the Weibull slope
- The scale parameter (η)

- The location parameter (γ)
- The most general expression of the Weibull *pdf (probability density function)* is given by the three-parameter Weibull distribution expression, or:

$$f(T) = \frac{\beta}{\eta}\left(\frac{T-\gamma}{\eta}\right)^{\beta-1} e^{-\left(\frac{T-\gamma}{\eta}\right)^{\beta}}$$

or if $t = (T - \lambda)/\eta$, *then* $f(T) = \beta t^{\beta-1} \exp^{(-t^{\beta})}$

You can notice that when β = 0, it is more or less similar to Poisson distribution, but as β increases, the curve assumes normal shape and becomes steeper showing that the determination of failure would be more precise. The Weibull distribution is very useful not only for the failure analysis in determining the equipment reliability but also in In survival analysis in insurance industry, in industrial engineering to represent manufacturing and delivery times, and also in weather forecasting.

More statistical curves can be deduced and represented as follows indicating the different conditions of failures

- When a fraction of items are expected to fail by a time t that is the probability of failure f(t)
- When a fraction of items are expected to survive by a time t that is the probability of survival f(t)

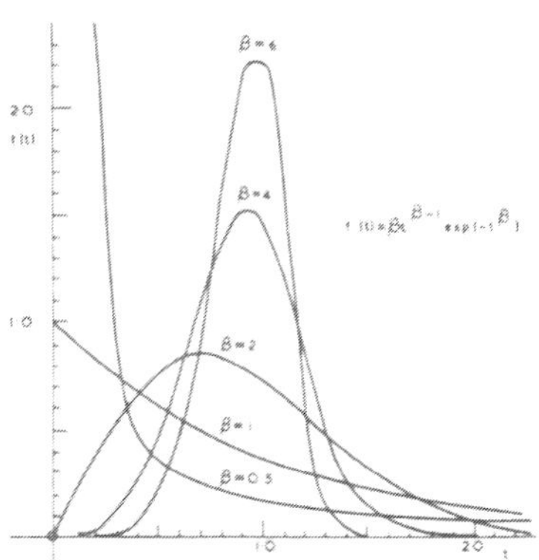

**Figure 9.5** Weibull distribution.

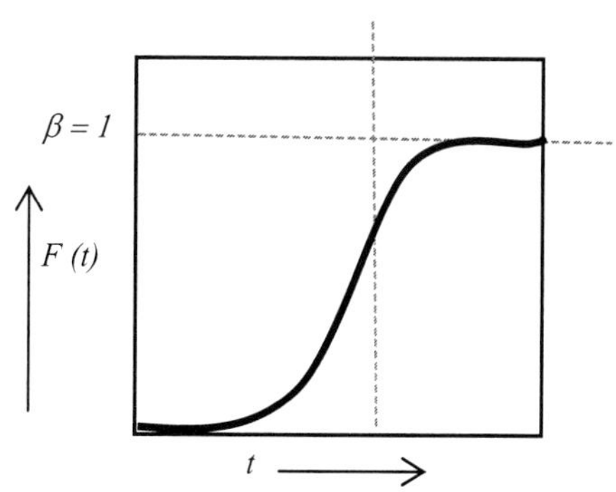


**Figure 9.6** Probability of failure.

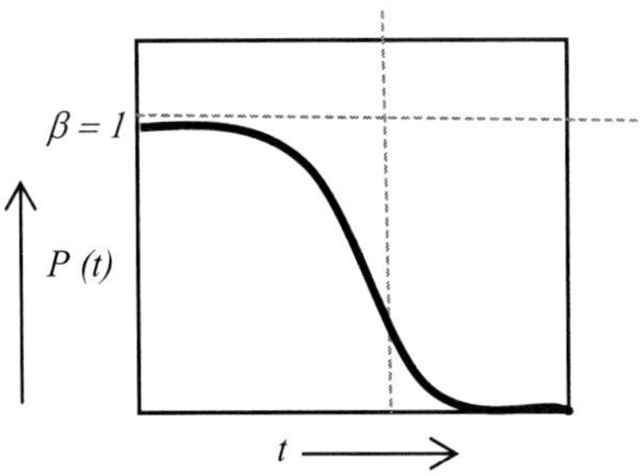


**Figure 9.7** Probability of survival.

## 9.11 Probability Density Function

The probability density function (PDF), or density of a continuous random variable, is a function that describes the relative likelihood for this random variable to take on a given value. This probability density function is most commonly associated with absolutely continuous univariate distributions and for the random variable to fall within a particular region is given by the integral of this variable's density over the region. In the illustration given alongside, the probability density for the median to fall within the limits of $Q_1$ and $Q_2$ is given by the darker shaded area (*source – Wikipedia*).

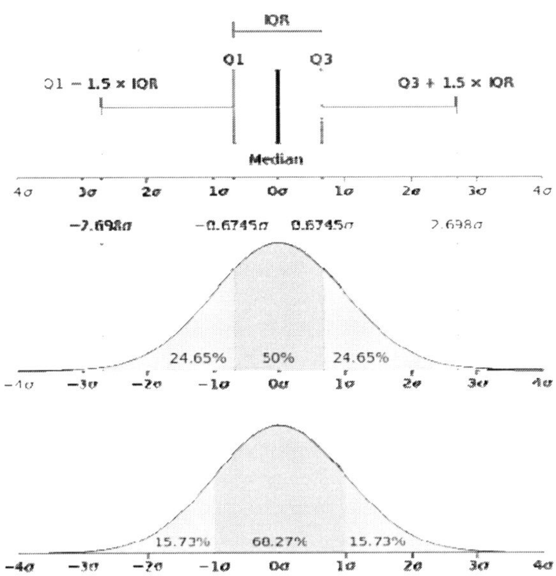


**Figure 9.8** Probability density function.

## 9.12 Procedure of RBM

1. Quantify reliability requirements as design goals or specifications.
2. Allocate and apportion the reliability requirements to specific system components and parts.
3. Apply reliability design methods during the equipment design and development. Perform reliability and maintainability analysis like block diagrams, stress-strength analysis, redundancy etc.
4. Conduct FMEA and criticality analysis.
5. Participate in design reviews.
6. Establish test procedures and conduct reliability testing.
7. Perform reliability prediction and demonstration.
8. Develop a reliability plan.

## 9.13 Reliability Prediction

Reliability Prediction, the process of forecasting the probability of success from available data is one of the important techniques in knowing the reliability of an equipment or system involves estimating the reliability (i.e. Performance of the system over a period of time) based on the failure rate of the components. It thus helps in identifying weak areas in a design and also in choosing the best design from among alternate configurations.

### 9.13.1 Elements of Reliability Prediction

- Reliability relationships
- Reliability concepts
- Constant failure rate
- The "Bathtub" Failure Rate curve
- System redundancy
- Fault tolerance
- Functional redundancy
- Fault avoidance.

### 9.13.2 Purposes of Reliability Prediction

1. Assuring the feasibility of reliability requirements (down time etc.) for the design proposed,
2. Comparing competing designs,
3. Identifying potential reliability problems,

4. Planning maintenance and logistic support strategies,
5. Using reliability predictions to assess the effect of product,
6. Predicting reliability of the maintenance activity and on the quantity of spare,
7. Estimating units required for acceptable field performance of any particular system. For example, predictions of the frequency of unit level,
8. Estimating unit and system life cycle costs.
9. Providing necessary input to system level reliability models.
10. Assisting in decision making which product to purchase from a list of competing products.
11. Setting standards for factory reliability tests and field performance.

## 9.14 Monte Carlo Simulation

**Monte Carlo method** is a broad class of computational algorithms that relies on repeated random sampling to obtain numerical results letting us to account for risk in the quantitative analysis and decision making. Just as we keep playing and recording our results in a real casino situation, the simulation is run many times over in order to calculate the probabilities heuristically. Hence the name Monte Carlo simulation. It is specifically useful in systems with several degrees of freedom with uncertainty in the inputs, like the calculation of risk in purchasing shares.

This author, who studied about Monte Carlo simulation in 1968, made it a point to visit the Monte Carlo Casino at Monaco during his visit to Europe in 1983. He had the first hand information of how each player keeps recording results of every move made by him as well as the other players on the roulette tables, thereby performing better in the subsequent moves.

When Monte Carlo simulations have been applied in space exploration and oil exploration, their predictions of failures, cost overruns and schedule overruns are routinely better than human intuition or alternative "soft" methods

Since testing of statistical significance of any activity or parameter is time consuming, simulation techniques can effectively be adapted for known probabilistic models to predict the outcome. Monte Carlo simulation is one such technique. The steps involved are:

1. Define a domain of possible inputs.
2. Generate random numbers between 0 and 1 to represent the reliability of a component.
3. Replace the reliability as a function of time into the model and calculate the corresponding time to failure by deterministic computation.

4. Generate another random number and repeat the process till sufficient number of trials made.'
5. Summaries the results.

Monte Carlo simulation can be successfully applied not only in engineering situations like microelectronics engineering, geometallurgy, aerospace engineering etc., but also in medical situations like Biological systems such as proteins, membranes, images of cancer etc.

## 9.15 Markov Analysis

Markov Analysis is an analytical method to determine the reliability and availability of a system whose components exhibit strong dependencies or constraints, similar to the tree analysis discussed elsewhere in this book.

This method proposed by Russian mathematician Andrei Andreyevich Markov used to forecast the value of a variable whose future value is independent of its past history and is a sure method for forecasting random variables.

Some typical constraints that can be considered for using Markov models are

- Components in cold or warm standby
- Common maintenance personnel
- Common spares win limited in-site stock.

## 9.16 Conclusion

As stated in paragraph 9.1, reliability based maintenance engineering deals with the study of the ability of the equipment to perform its required functions under stated conditions for a specified period of time. The maintenance engineer should have a full grasp of the principles of FMEA to be able to effectively adapt reliability based maintenance.

# Chapter 10

# Lubrication

---

*Chapter synopsis:* *One of the most important elements of preventive maintenance is lubrication. Proper lubrication is essential to obtain dependable performance of today's modern machinery and equipment. This chapter thus deals with various aspects of lubrication and describes in detail the different characteristics and varieties of lubricants and greases used in industry. A glossary of terms used in relation to lubrication is appended.*

**Keywords:** *Leonardo da Vinci, tribology, terotechnology, surface engineering, animal fats, key functions, planned lubrication, characteristics, ABCD PEN $F^2$ PVC LOSS, viscosity, flash point, pour point, carbon residue, emulsification, neutralization number, specification number, sulphur content, film strength, lubricity, adhesion, drip resistance, brittle resistance, oxidation resistance, greases, calcium soap grease, lithium soap grease, aluminum soap grease, lanolin, mineral oils, vegetable oils, SAE viscosity index, lubricant Base, slax wax, glaze formation, lubricant additives, NEWS FOR CD POP TV, lubrication feeding, charts, oil bath lubrication, splash lubrication, circulatory system, chain lubrication, drip feed, grease cups, grease guns, standard coding, lubrication terminology, GM lubricant coding.*

---

## 10.1 Introduction

The root cause of a majority of the plant breakdowns is due to the incorrect lubrication. In fact the British standard definition for lubrication is the work which is directed to the smooth working of components and prevention of their failures. It is hence, much essential to a company as the preventive maintenance program.

When machines were no more than simple aids to human muscles, like the wheelbarrow or the wood planer, the problems posed by these machines were very few and the maintenance could normally be done by the operators themselves. But modern machines or equipment that have become complex in operation with automatic controls and call for more precise and detailed attention. The problems are numerous and beyond the capacity of the operator who is normally expected to possess expertise and concentrate on the production. This necessitated an exclusive assistant or a gang to apply lubricants to all the machines and equipment of the work place. The reliability of these machines depends on the correct type of lubricant in the right quantity applied at certain predetermined intervals.

A lubricant is a substance, often a liquid, introduced between two moving surfaces to reduce the friction between them, improving efficiency and reducing wear. The earliest known form of lubrication may perhaps be by Romans when they used rags dipped in animal fat to lubricate wagon wheels. The industrial revolution during the nineteenth century made lubrication more popular with several applications. Subsequent researches in lubrications opened up a new science called tribology, which, as a branch of mechanical engineering, is the science of interacting surfaces in relative motion and includes the study and application of the principles of friction, lubrication and wear. This word is derived from the Greek expression *tribo* meaning 'I rub'. The other terminologies involved are *Surface Engineering*, meaning modification of the surface properties of solids by use of lubricants, and *Terotechnology* where multidisciplinary engineering and management techniques are used to protect equipment and machinery from degradation.

Historically, Leonardo da Vinci (1452–1519), the famous classical painter and creator of *Mona Lisa*, was the first to enunciate two laws of friction.

(a) The frictional resistance is the same for two different objects of the same weight but making contacts over different widths and lengths. That is, friction depends more on the weight of the object and less on the area of contact.

(b) The force needed to overcome friction is doubled when the weight is doubled.

He drew several sketches of his experiments in his notebooks, but his findings remained unpublished (Fig. 10.1).

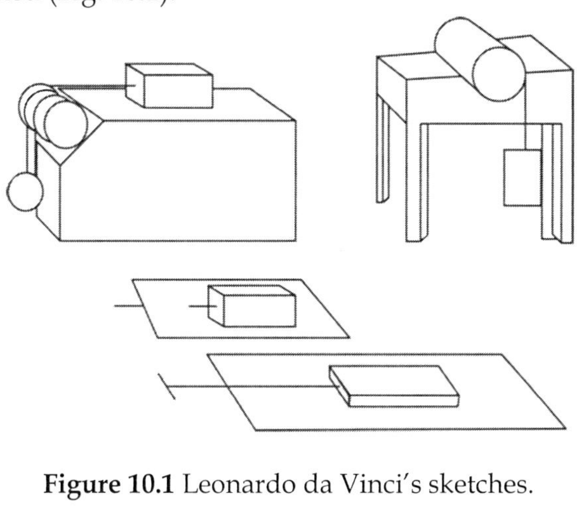

**Figure 10.1** Leonardo da Vinci's sketches.

## 10.2 Key Functions of Lubricants

- *Keep moving parts apart:* By forming a thin layer of lubricant between the moving parts.
- *Reduce friction:* Which has the benefit of reducing heat generation and reduced formation of wear particles as well as improved efficiency.
- *Transfer heat:* Typically the liquid lubricant is constantly circulated to and from a cooler part of the system. This circulating flow also determines the amount of heat that is carried away in any given unit of time.
- *Carry away contaminants & debris:* Lubricants in circulation can carry away internally generated debris and external contaminants that get introduced into the system to a filter where they can be removed. In closed systems such as gear boxes the filter may be supplemented by a magnet to attract any iron debris that gets created especially in grinders. Chapter 8 on wear debris analysis can be referred for further information.
- *Transmit power*: Some of the lubricating oils can also perform the function of power transmission. When they are used for such power transmission applications they are called hydraulic fluids.
- *Protect against wear:* Lubricants prevent wear by keeping the moving parts apart. Lubricants may also contain anti-wear or extreme pressure additives to boost their performance against wear and fatigue.
- *Prevent corrosion*: Sometimes the lubricants are provided with additives that form chemical bonds with surfaces to prevent corrosion and rust.
- *Seal for gasses*: The lubricating oils will occupy the clearance between moving parts through the capillary force and seal up any clearance. The most common application is the sealing of pistons and shafts.

## 10.3 Aims of Planned Lubrication

1. Every Machine or component must be correctly lubricated as specified.
2. The correct type, grade and quality of the lubricant must be applied at regular, specified periods.
3. The lubrication history should be recorded to indicate its effectiveness.

## 10.4 Characteristics of Lubricants

### (a) For Oils

The lubricants in the liquid state have as many as 16 characteristics that determine the choice of the right lubricant. Since it is difficult to remember so many, the author developed the following acronym formed by their initial letters to aid remembrance.

**A B C D P E N $F^2$ P V C L O S S**

1. **A** *Adhesion.* Helps prevent loss of lubrication due to throw-off associated with gravity or centrifugal force especially at high speeds.
2. **B** *Brittle resistance* so that the lubricant is capable of resisting embrittlement, especially at very low temperatures
3. **C** *Carbon residue*, as given out by the engine oil used in two-stroke petrol engines
4. **D** *Drip resistance* to prevent loss of lubricant, especially at high temperature which reduces its viscosity.
5. **P** *Pour point,* the lowest temperature at which the oil can flow.
6. **E** *Emulsification* which influences its effectiveness when it comes into contact with water as in water pumps.
7. **N** *Neutralization number,* which is the number of milligrams of potassium hydroxide needed to neutralize one gram of oil
8. **F** *Flash point,* at which the vapours given off at high temperature will burn when ignited.
9. **F** *Film strength*: Good film strength helps prevent metal contact and scouring *between the gear teeth.*
10. **P** *Pressure protection under extreme conditions* provides additional galling and welding protection for heavily loaded gears when the lubricant film thickness fails. Extreme pressure lubricants are available for mild and severe (hypoid) lubricant applications.
11. **V** *Viscosity,* an index of its thickness and fluidity. Good viscosity is essential to ensure cushioning and quiet operation. An oil viscosity that is too low would result in the oil film which is too thin to provide effective lubrication, whereas too high viscosity will result in excessive stickiness increasing the friction and also degradation of oil properties associated with high oil operating temperature.
12. **C** *Color,* which is significant in case of transparent cutting oils.
13. **L** *Lubricity (oiliness):*Lubricity is necessary to reduce friction
14. **O** *Oxidation resistance* and *chemical stability* prevents thickening and formation of varnish or sludge. This requirement is especially significant in high-

speed gears because the oil is subjected to high operating oil and air temperatures.

15. **S** *Specification number,* which is the number of milligrams of caustic potash required for neutralizing one gram of oil.

16. **S** *Sulphur content* that causes corrosion to the containers or metal components that are lubricated

**(b) For Greases**

1. *Base,* which is an indication of the type of soap (calcium, sodium, lithium barium etc.) used in its manufacture.
2. *Penetration,* the ability to extend its texture and enter the narrow gaps between the components to be lubricated.
3. *Melting point.*

## 10.5 Classification of Hydraulic Fluids

*General classification of lubricants* is based on their use like:

(a) Circulating oils,
(b) Gear oils,
(c) Brake oils,
(d) Machine or engine oils,
(e) Spindle oils,
(f) Steam cylinder lubricants,
(g) Wire rope lubricants,
(h) Hydraulic oils,
(i) Refrigeration grade oils,
(j) Lubricant oils,
(k) Solid or synthetic lubricants and
(l) Greases.

This classification by use is further detailed in paragraph 10.8.

*Classification of lubricants is based on their form:*

1. **Gaseous lubricants** like air, steam or liquid metal vapors,
2. **Liquid lubricants** like

   (a) *Lanolin,* which is a form of grease derived from sheep wool, and is an alternative to the more common petro-chemical based lubricants. Being water repellant it is a corrosion inhibitor, protecting against rust, salts, and acids.

(b) *Water,* which is sometimes used in engineering processes like hacksaw cutting.

(c) *Mineral oils,* which are described and classified further in the following paragraph.

(d) *Vegetable oils* like canola oil, castor oil, palm oil, sunflower seed oil and for ages rapeseed oil have been in use historically. During the latter part of the $20^{th}$ century these are used as friction modifier additives for automatic transmission fluids.

(e) *Synthetic oils* like:

- (i) Polyalphaolefin (PAO)
- (ii) Synthetic esters
- (iii) Polyalkyleneglycols (PAG)
- (iv) Phosphate esters
- (v) Alkylated naphthalene (AN)
- (vi) Silicate esters
- (vii) Ionic fluids

3. **Solid lubricants**, which are dry like

(a) Teflon, which is used as a coating layer on cooking vessels to provide a non-stick surface.

(b) Non-metals like graphite, boron nitride, molybdenum disulphide, which are effective at very high temperatures, but have poor resistance to oxidation.

(c) Metal alloys like that of lead, tin, zinc and bronze offer good corrosion resistance and sliding properties. They are also used as additives to grease.

4. **Greases:** These are oils or other fluid lubricants mixed with thickening substances like soaps to form solids or pastes. Unlike oils, grease does not flow or thrown out of the bearings, thus making it most suitable for lubricating ball or roller bearings. It is particularly useful when the contamination of components, by dripping or splashing oils is not acceptable. The property of the grease may vary with the type of soap used in the manufacture.

(a) *Calcium soap grease,* which is highly resistant to water but has a low melting point. It can be used for medium duty, plain and roller bearings.

(b) *Sodium soap grease,* which is poorly resistant to water but has a high melting point. It can be used for normal duty plain and roller bearings.

(c) *Lithium soap grease,* which is highly resistant to water and at the same time, has a high melting point. It can be used for rolling contact bearings. With special additives, they can be used for extremely low temperature applications.

(d) *Aluminum soap grease,* which is highly resistant to and insoluble in water. It has good adhesive property and so can be used for fast moving cams and reciprocating components.

## 10.6 Mineral Oils

These oils are derived from crude petroleum and almost all the lubricants used in industry are of this category. The American Petroleum Institute (API) further classifies the mineral oils into the following groups basically based on the viscosity index. The viscosity index is defined and assigned to each oil by Society of Automotive Engineers (SAE). It may be noted that there is a certain amount of overlapping between mineral oils and the synthetic oils as discussed below.

- Group I - SAE viscosity index of 80 to 120. These contain saturates < 90% and/or sulfur > 0.03%. These are manufactured by solvent extraction, solvent or catalytic dewaxing, and hydro-finishing processes. Common Group I base oil are 150SN (solvent neutral), 500SN, and 150BS (bright stock)
- Group II - SAE viscosity index of 80 to 120. These contain saturates > 90% and sulfur < 0.03%, and are manufactured by hydrocracking and solvent or catalytic dewaxing processes. These oils have superior anti-oxidation properties since virtually all hydrocarbon molecules are saturated. They are clear transparent like water.
- Group III - SAE viscosity index over 120, saturates > 90%, sulfur <0.03%, and manufactured by special processes such as isohydromerization. They also can be manufactured from base oil or slax wax by dewaxing process. Slax wax is a liquid form wax emulsified for ease of use. It gives a high gloss, long lasting shine and protective wax finish on used cars.
- Group IV - Polyalphaolefins (PAO)
- Group V - All others not included above, such as naphthenic, PAG, esters, etc.

*Lubricant Base:* Yet another classification may be by the base on which the lubricant belongs to. In North America, Group III is frequently described as synthesized hydrocarbons, or SHCs, while IV and V are described as other synthetic lubricants. However in Europe, only Groups IV and V are classed as synthetics.

## 10.7 Glaze Formation

Glaze formation is the generation of a compacted oxide layer which sinters together to form a crystalline 'glaze' generally at high temperatures due to metallic surfaces sliding against each other (or a metallic surface against a ceramic surface).

*Characteristics of glaze surface:*

1. Friction and wear is reduced due to the elimination of metallic contact and adhesion by the generation of oxides.
2. The surface hence becomes self-lubricating.
3. It can survive very high temperatures in air or oxidizing environments as the 'glaze' is already an oxide.
4. However, its disadvantage is that the base metal first undergoes some wear to generate sufficient oxide debris.

## 10.8 Fluid Types by Application

- Automotive
  - Engine oils
    - Petrol (Gasoline) engine oils
    - Diesel engine oils
  - Automatic transmission fluid
  - Gearbox fluids
  - Brake fluids
  - Hydraulic fluids
- Tractor (one lubricant for all systems)
  - Universal Tractor Transmission Oil - UTTO
  - Super Tractor Oil Universal - STOU - including the engine
- Other engines
  - 2-stroke engine oils
- Industrial
  - Hydraulic oils
  - Air compressor oils
  - Gas compressor oils
  - Gear oils
  - Bearing and circulating system oils
  - Refrigerator compressor oils
  - Steam and gas turbine oils

- Aviation
    - Gas turbine engine oils
    - Piston engine oils
- Marine
    - Crosshead cylinder oils
    - Crosshead crankcase oils
    - Trunk piston engine oils
    - Stern tube lubricants

## 10.9 Lubricant Additives

Lubricants are generally added with one or more additives which enhance the performance characteristics of the base oil. Some additives only affect one aspect of performance. For example viscosity index improvers allow the finished lubricant to attain different viscosities, depending on temperature. Other additives may have multiple effects. For example zinc dithiophosphate provides wear protection and helps inhibit oxidation.

The lubricant additives are added to a variety of lubricants used in:

- Heavy duty diesel engine oils
- Passenger car engine oils
- Four-stroke cycle small engine oils
- Two-stroke cycle engine oils, air and water cooled
- Transmission fluids
- Gear oils
- Anti-wear hydraulic oils
- Industrial oils
- Marine engine oils
- Natural gas engine oils
- Rail road, stationary power and inland marine oils

These additives can be listed as follows.

1. *Acid neutralizers* that converts acid contents in the lubricants into harmless salts by chemical reaction, for example as in high sulphur fuels.
2. *Emulsifiers* that permit the emulsification or surface foam of the oils to improve the lubricating property of the oil.

3. *Wear Resistant* additives which inhibit the tendency of the wear between mating surfaces.
4. *Anti-sludge* agents that inhibit the tendency of sludge formation in the lubricant of continuous use.
5. *Anti-foam* additives that prevent the tendency of foam formation in the lubricants that are subjected to high revolutions.
6. *Anti-oxidization elements* that react with the oxygen before the latter reacts with lubrication as well as the mating surface, thereby preventing the oxidization of the lubricant and mating parts in the environment of lubrication. This is in a way similar to the galvanizing of steel components, wherein the zinc absorbs the oxygen from the environment before it reacts with the steel components.
7. *Anti-rust additives.*
8. *Corrosion inhibitors.*
9. *Detergents* which accelerate the smooth flow of the lubricant over the mating surfaces, reducing the coefficient of friction.
10. *Pour point additives,* that lower down *pour point,* the lowest temperature at which the oil can flow, so that at all temperatures oil flow can be smooth.
11. *High Pressure stabilizers* that stabilizes the parameters and characters of the lubricant at extremely high pressures.
12. *Oiliness additives* that help the lubricants to retain their oiliness at all working conditions to enable effective lubrication.
13. *Tackiness additives* that inhibit the tendency of oils to stick to the metal surfaces.
14. *Viscosity stabilizers* that ensure that the lubricants retain their viscosity at all working temperatures and pressures, and do not thin out.

An effective way of remembering these several additives, from the examination point of view is to use the acronym developed by the author, ***'NEWS FOR CD POP TV'*** whose initials would represent each additive as follows:

**N** Acid **N**eutralizers
**E** **E**mulsifiers
**W** **W**ear resisters
**S** Anti **S**ludge formants
**F** Anti-**F**oam additives
**O** Anti-**O**xidant elements

| | |
|---|---|
| **R** | Anti-**R**ust additives |
| **C** | **C**orrosion inhibitors |
| **D** | **D**etergents |
| **P** | High **P**ressure stabilizers |
| **O** | **O**iliness additives |
| **P** | **P**our point additives |
| **T** | **T**ackiness additives |
| **V** | **V**iscosity Stabilizers |

## 10.10 Methods of Lubricant Application

There are several methods of applying lubricant to moving parts, which generally depend on:

- The accessibility of the moving parts.
- The loads imposed
- The speed of the moving parts
- Design considerations of the components.

The normal lubrication systems adapted in engineering industry can be illustrated as follows,

*Oil Bath lubrication*: The moving parts are located in open containers filled partly with the oil. This type of lubrication can be adapted for gears of low speeds like the worm gear.

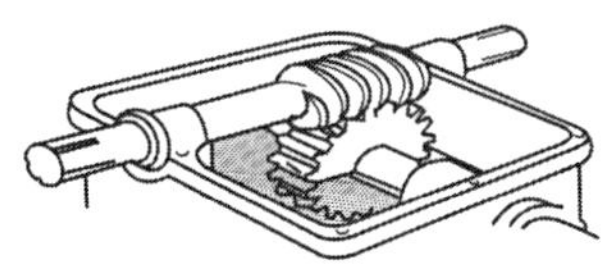

*Splash lubrication*: The moving parts are inside a fully sealed protective housing. During operations the reciprocating components dip into the oil at high speeds and splash oil around to other parts. This is generally used in crank shafts of engines and compressors as illustrated.

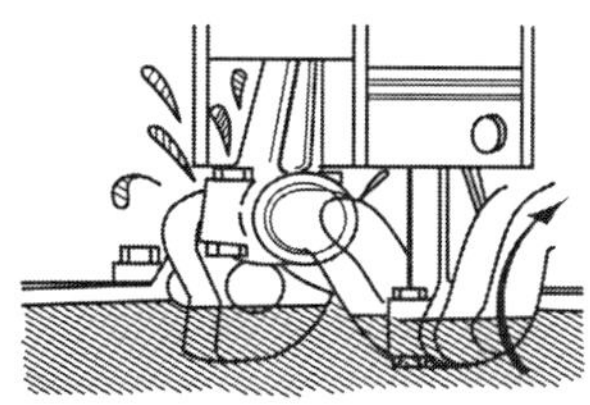

*Circulatory system:* The moving components are totally enclosed in a protective housing partly filled with the

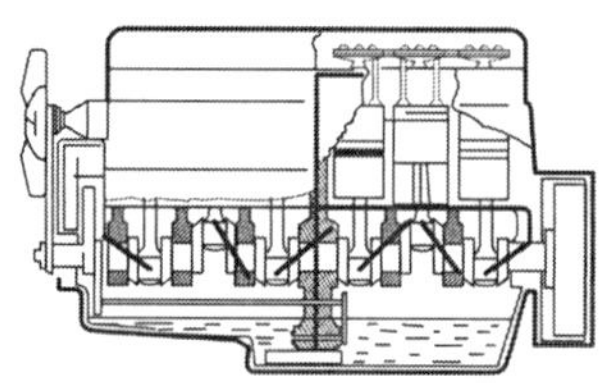

lubricant oil. This container generally in 2 parts, the upper and lower to enable periodic maintenance. The joints of these parts as well as the holes for the gear shafts etc. are sealed with oil seals. These seals should be well maintained to prevent oil leakage.

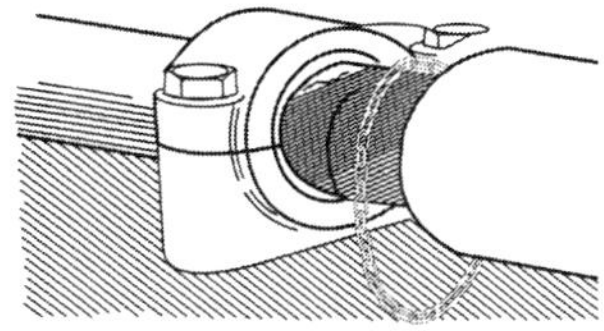

*Ring or Chain lubrication:* The system as illustrated can be used to lubricate plain bearings in low and medium speeds.

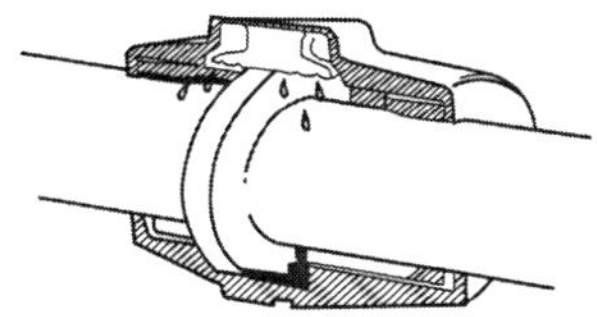

*Collar feed lubrication*: This is similar to the above but the shaft may have a collar as illustrated and a rubber scraper on the container scrapes the oil as it reaches the highest point of travel. This is also called pad lubrication if an oil soaked pad is used inside the collar.

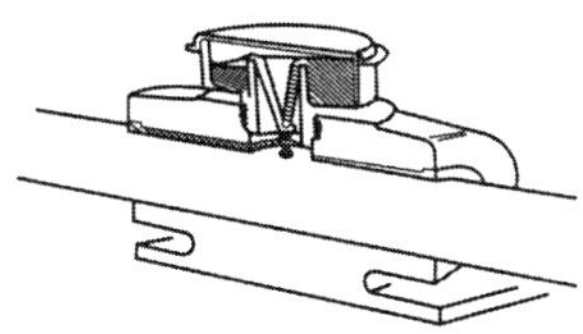

*Wick lubrication:* Components requiring a low rate of oil supply are often lubricated by the wick feed, as shown. One end of the wick is suspended in an oil bath and siphon action feeds the oil to the component to be lubricated. This is generally done for line shaft bearings.

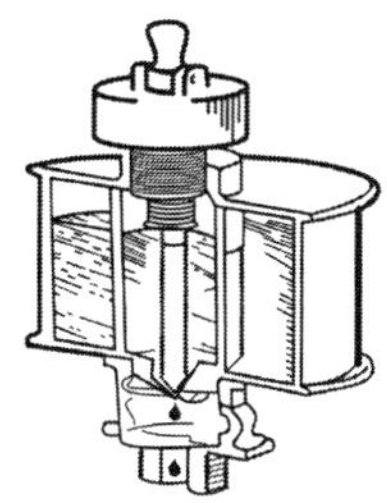

*Drip Feed, also called the cup oilers*: These are used to supply a controlled feed of oil to a bearing. The covered cup containing the oil is connected to a duct leading to the bearing, as illustrated. A needle valve is used to regulate the flow of oil.

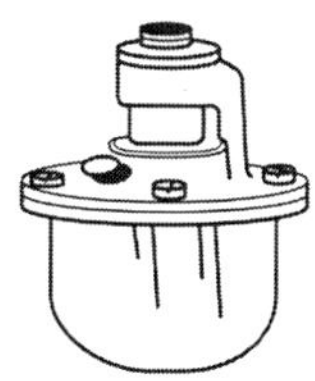

*Oil mist:* Dry compressed air is used to generate a mist and carry the lubricant to the components to be lubricated. This system must have proper vent since a build up of pressure may result in stopping the oil flow. This method is generally used for high speed bearings, enclosed gears and slides.

*Grease cups:* The grease lubrication is generally done by grease cups. These can either be spring loaded or screw operated as shown in the figure.

- *Spring loaded grease cups,* where the grease in the cups is continuously forced under spring pressure. This is suitable for the thinner grades of grease. And used for line shafting
- *Screw operated grease cups* where the grease is forced by screwing the cap of the cup manually wherever required. This is suitable for thicker grades of grease.

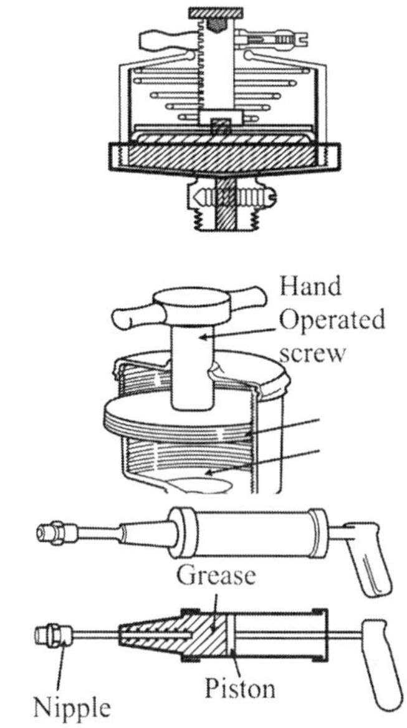


**Figure 10.2** (a to k) lubrication Systems.

*Grease guns*: Grease guns are used to pump grease through a non-return nipple. After greasing the duct is covered with a blanking cap to prevent contamination. They are used for machine tool bearings as well as in automobile servicing.

## 10.11 Lubrication Chart

For every machine, maintaining a lubrication chart (Table. 10.1) and displaying at an apt place near the machine would help in identifying where and when to lubricate and gives information to the operators and/or the patrolling lubrication.

**Table 10.1** A typical lubrication chart.

| Machine: | | LUBRICATION REGISTER | | | | | Location | |
|---|---|---|---|---|---|---|---|---|
| Code | | | | | | | | |
| Sl. No. | Item | No. of points | Lubricant | Application method | Frequ ency | Qty. (kg/ltr) | Inspection | Remarks |
| 1 | | | | | | | | |
| 2 | | | | | | | | |
| 3 | | | | | | | | |

*Contd...*

| Machine: | | LUBRICATION REGISTER | | | | | Location | |
|---|---|---|---|---|---|---|---|---|
| Code | | | | | | | | |
| Sl. No. | Item | No. of points | Lubricant | Application method | Frequ ency | Qty. (kg/ltr) | Inspection | Remarks |
| 4 | | | | | | | | |
| 5 | | | | | | | | |
| 6 | | | | | | | | |
| 7 | | | | | | | | |
| 8 | | | | | | | | |

## 10.12 GM Lubricant Coding

General Motors have coded the various lubricants used by them in 2 letter codes as follows (Table 10.2).

**Table 10.2** General Motors coding for their lubricants.

| | |
|---|---|
| **LA** | Air cylinder/valve, air tool oils |
| **LB** | General purpose/press oils |
| **LC** | Submerged clutch and brake oils |
| **LD** | Dry film, chain/conveyor lubricants |
| **LE** | Environmentally acceptable fluids |
| **LF** | Fire-resistant hydraulic fluids |
| **LG** | Greases |
| **LH** | Antiwear hydraulic fluids |
| **LJ** | Compressor/turbine oils |
| **LM** | Misting oils |
| **LT** | Gear oils |
| **LS** | Spindle oils |
| **LW** | Way oils |
| **LX** | Straight cutting and grinding fluids |
| **LY** | Aqueous cutting and grinding fluids |

## 10.13 Conclusion

Starting from Leonardo da Vinci, friction between two mating surfaces and the solutions for the same has fascinated the maintenance engineers globally resulting in development of specific types of lubricants and several innovations to reduce the friction. This chapter illustrates such developments and enables the maintenance engineers to know what they have in hand to reduce friction of mating surfaces.

## Appendix

**Basic terminology in lubrication:** (adapted from Maintenance Engineering handbook, edited by L. R. Higgins.)

*Ball Bearing*: A rolling type bearing containing rolling elements in the form of balls.

*Barium based grease*: A grease prepared from lubrication oil and barium soap.

*Black oil:* Any dark colored lubricating oil which is used for the lubrication of major parts under exposed condition.

*Bleeding:* The tendency of a liquid component to separate from a liquid-solid or liquid-semisolid mixture, as oil from grease.

*Black grease*: Generally a grease of high dropping point which under normal temperature is firm to the touch and can be handled in block or stick form.

*Boundary lubrication:* A form of lubrication found in the absence of full fluid film or hydrodynamic lubrication. It requires an additive to prevent excessive friction and scouring.

*Bright stock:* A term referring to high viscosity lubricating oils which have been refined to make them clear products of good color.

*Centralized lubrication:* A system of lubrication in which the lubricant or lubricants for the bearing surfaces of a machine or a group of machines are supplied from a central location.

*Circulating lubricant:* A system of lubrication in which the lubricant, after having passed through a bearing or a group of bearings, is recirculated by means of a pump.

*Clearance bearing:* A journal bearing in which the radius of the bearing surface is greater than the radius of the journal.

*Coefficient of friction:* The number obtained by dividing the friction force resisting motion between the two by the normal force pressing the bodies together.

*Cold test:* A test by which the pour point of oil is determined.

*Compounded oil*: A petroleum oil to which animal or vegetable oil or other substances have been added

*Consistency:* The degree to which a semisolid material such as grease resists deformation (per ASTM D217). Sometimes used qualitatively to denote viscosity of liquids.

*Coolant:* A fluid used to remove heat from a process or component.

*Cutting fluid:* Any fluid that is applied to a cutting tool to assist the cutting operation by cooling, lubricating or other means.

*Detergent:* In lubrication, either an additive or a compounded lubricant having the property of keeping the insoluble matter in suspension thus, preventing its deposition where it would be harmful. A detergent may also re-disperse deposits already formed.

*Dispersant:* In lubrication, this term is usually used for detergent.

*Drop-feed lubricant:* A system of lubrication in which the lubricant is supplied to the bearing surfaces in the form of drops at regular intervals.

*Dropping point*: The temperature at which the grease passes from semi-solid to liquid state under specified heat conditions.

*Emulsability:* The ability of a non-water soluble fluid to form an emulsion with water.

*Emulsion:* A mechanical mixture of two insoluble liquids like oil and water.

*EP (Extreme pressure) Lubricants:* Lubricants that impart to rubbing surfaces, the ability to carry appreciably higher loads than would be possible with ordinary lubricants without excessive wear or damage.

*Fibre grease:* Grease having a distinctly fibrous structure which is noticeable when sample of the grease is pulled apart. Greases having this fibrous structure tend to resist being thrown off gears and bearings.

*Filler:* Any substance such as talc, mica or various other powders which are added to grease increase its weight or consistency.

*Filter:* Any device or porous substance used as a strainer for cleaning fluids by removing suspended matter.

*Flash point:* The temperature to which a combustible liquid must be heated to give off sufficient vapour to form momentarily a flammable mixture with air when a small flame is applied under specified conditions.

*Fluid friction:* Friction due to viscosity of liquids

*Force feed lubrication***:** A system of lubrication in which the lubricant is supplied to the bearing under pressure.

*Fretting:* Wear phenomena taking place between two surfaces having oscillatory relative motion of small amplitude.

*Friction:* The resisting force encountered at the common boundary between two bodies, when under the action of external force, one body moves or tends to move relative to the surface of the other.

*Graphite:* A crystalline form of carbon having a laminar structure. It may be of natural or synthetic origin.

*Grease:* Lubricant composed of oil or oils thickened with soap or other thickener to a semi-solid or solid consistency.

*Gum:* A rubber like, sticky deposit, black or dark brown in color, which results from oxidation of lubricant oils or from constituents which deposit during storage or use in gasoline

*Heavy duty oils:* A type of oil having oxidation stability, corrosion preventive properties, and detergent dispersant characteristics, and hence is used in high speed diesel and petrol engines under heavy duty service conditions.

*Hydraulic oil:* Oil specifically used for power transmission medium.

*Hydrodynamic lubrication:* A system of lubrication in which the shape and relative motion of the sliding surfaces cause the formation of a fluid film having sufficient pressure to separate the surfaces.

*Hydrometer:* An instrument for determination of specific gravity as well as API (American Petroleum Institute) gravity of a liquid.

*Hydrostatic lubrication:* A state of lubrication in which the lubricant is supplied under external pressure to separate opposing surfaces by fluid film.

*Imperfect film lubrication*: A condition of lubrication if the lubricant film is not continuous over the bearing area.

*Inhibitor:* A substance which slows down or prevents chemical reactions like corrosion or oxidation.

*Lime based grease*: A grease prepared with lubricating oil blended with a calcium soap.

*Lithium based grease:* Grease prepared with lubricating oil blended with lithium soap.

*Lubricant:* Any substance interposed between two surfaces which are in relative motion for the purpose of reducing the friction and/or wear between them.

*Mechanical lubricator*: A mechanical device for supplying lubricant to various parts of a mechanism.

*Mixed based grease:* Grease in which the thickening agent or base is generally a mixture of sodium and calcium soaps, though other soaps, may be used.

*Neutral oils:* Lubricating oils of low or medium viscosity obtained in petroleum distillation without chemical addition or treatment.

*Oiliness:* The property of a lubricant to produce low friction under conditions of boundary lubrication. The lower the friction, the greater the oiliness.

*Porous bearing:* A bearing made from porous materials like compressed metal powders or polymers, the pores acting either as reservoirs for holding or as passages for the supply of lubricant.

*Pour point:* The lowest temperature at which a lubricant starts flowing under specified conditions.

*Sludge:* Insoluble material formed as a result of deterioration and/or contamination of oil.

*Slushing oil:* An oil or grease like material used on the metals to form a temporary coating against rust, corrosion etc.

*Spindle oil:* Light bodied oil used basically for lubricating spindles and light high speed machinery.

*Sponge grease:* A soda based grease with more fibrous and sponge like structure than the normal greases which is more buttery.

*Strong Acid Number (SAN):* The quantity of base, expressed in milligrams of potassium hydroxide, required to neutralize the strong acid constituents present in one gram of a sample (ASTM No. D664-58).

*Synthetic lubricant:* A lubricant produced by synthesis of organic elements rather than extraction or refinement.

*Viscometer or viscosimeter*: An instrument for measuring the viscosity of oil.

*Viscosity:* The property of a fluid to cause it to resist flow. It is defined as the shear stress on a fluid element divided by the rate of shear. Its unit is poise or dynes/$cm^2$.

*Viscosity index:* This is the measure of a fluid's change of viscosity with temperature. The higher the viscosity index, the smaller the relative change in viscosity with temperature.

*Yarn grease:* Cup, sponge or residual greases containing strands of wool or cotton yarn.

# Chapter 11

# Cutting Fluids

*Chapter synopsis:* *The coolant fluid forms an effective medium to remove the heat generated during the metal cutting process. In view of the significant role played by them in the modern day manufacture, this chapter analyses various types, parameters and requirements of these fluids giving as overview of the heat generation during the metal cutting. A significant part of this chapter is the role played by the tramp oils that leak into these fluids and a case study presented in this connection.*

*Keywords:* *Cutting fluids, coolant oils, lubricant oils, heat generation, heat of deformation, heat of chip distortion, chip welding, heat of burnishing, cooing action, coolant materials, synthetic and semi-synthetic materials, emulsions, gels, mists, dielectric fluids, thin oil film, tramp oils, monday morning stench, biocides, decantation, sponging, slow speed skimming, specially designed coolant sumps, electrostatic separation, belt type skimmers, drum type slow speed skimmers, vortex type oil separator, tramp oil remover*

## 11.1 Introduction

Cutting fluids are used in metal cutting machine tools to cool and lubricate the cutting tool. Hence these are sometimes called coolant oils, which are different from the lubricating oils. Every kind of metal cutting and machining operation uses these fluids, be it turning, boring, drilling, reaming, tapping, milling, broaching, grinding, sawing, shaping or planning. These cutting fluids are used in case of almost all materials except cast iron, plastics etc., which are machined dry, and also sometimes when carbide tools are used. In view of the significant role played by the cutting fluids in the maintenance of industrial machinery, the aspect of maintaining the coolant sump and the replenishment is generally handled by the maintenance department. Hence this aspect is dealt in detail in this chapter.

There are several types and forms of cutting fluids that include oils, oil-water emulsions, pastes, gels and mists. They are made from petroleum distillates, animal fats, plant oils or other raw ingredients. Depending upon context and on which type of cutting fluid is being considered; they may also be referred to as cutting fluids, cutting oils, cutting compounds or coolants.

To a lesser extent they may be referred to as lubricant fluids depending on the type of fluid, but they should be clearly distinguished from the lubricants and lubricating oils, which are basically used to lubricate the other moving parts like the gears and other bearings. The lubricating oils are described more in detail in the previous chapter.

## 11.2 Sources of Heat Generation during Metal Cutting Action

Metal cutting action involves generation of heat, not only due to shear deformation of the cut metal, but also due to the friction between the cutting tool, the work piece and the chip. These are represented in Fig. 11.1. The coolant fluid forms an effective medium to remove this heat.

(a) *Heat of deformation:* Cutting tool exerts high pressure on the work piece grains which start slipping. The friction between the grains generates heat called the heat of deformation, occurring in zone $Z_1$ called the shear zone or the primary deformation zone. Parameters like speed, feed, depth of cut and material contribute to the quantum of this heat. 75 to 80% of this heat is carried away by the chip, 10 to 15% is passed on to the work piece by conduction while 5 to 10% is passed onto the cutting tool.

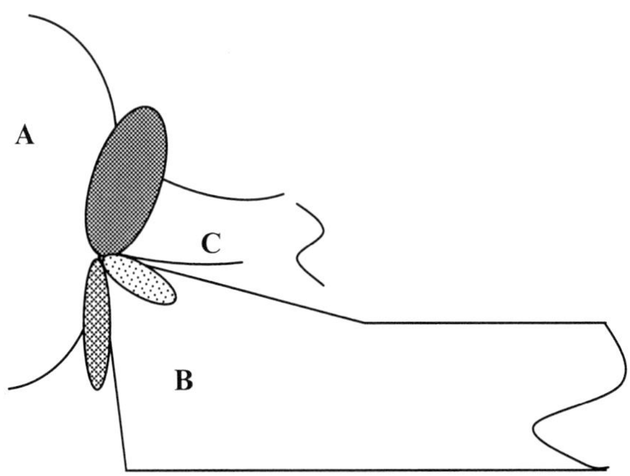


A = Work piece — $Z_1$ – Heat of deformation

B = Tool — $Z_2$ – Heat of chip – tool burnishing

C = Chip — $Z_3$ – Heat of work – tool burnishing

**Figure 11.1** Sources of heat generation during metal cutting action.

(b) *Heat of chip distortion*: As the chip flows out it rubs against the cutting tool and further gets distorted. The additional heat generated due to this distortion as well as the friction with the cutting tool is called the heat of chip distortion and occurs in zone $Z_2$, zone of chip distortion. This heat causes the chip to break into segments which gets welded to the tool face by friction as they pass over the face of the cutting tool. This is called chip welding.

(c) *Heat of burnishing*: This is due to the friction between the work piece and the cutting tool and occurs in $Z_3$, the zone of burnishing. The heat increases on a time scale as the tool wears increasing the contact area.

On an average the total heat generated is distributed to the chip, the working speed and the cutting tool in proportions of 80%, 10% and 10% respectively, while at lower speeds the work piece tends to become hotter than the chip as illustrated in Fig. 11.2.

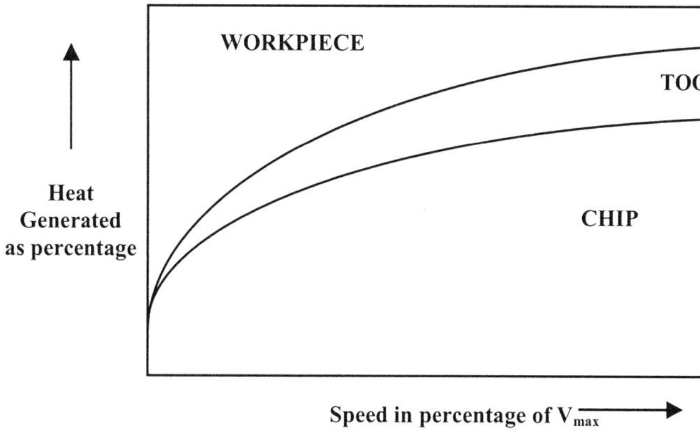


**Figure 11.2** Distribution of the proportion of heat generated.

## 11.3 Cooling Action

Needless to say, that the high temperatures generated at the point of metal cutting cause considerable damage to the cutting tool, work piece and more significantly to the work accuracy and surface finish, necessitating the cooling of the three zones during the cutting action itself. Ambient air cooling is inadequate for normal and heavy cuts and continuous production where more heat produced than removed by ambient air. To minimize the adverse effects of this high temperature generation, cutting fluids are applied to the point of cutting as a steady stream in sufficient quantities. Besides cooling, these fluids aid the cutting action by lubricating the interface between the cutting tool and the chip as well as the cutting tool and the work piece reducing the heat of chip distortion and that of burnishing. This prevents the welding of the chip on the cutting tool or on the work piece.

As indicated above coolant fluids can serve as lubricants and lubricants can serve as coolants, but there are some basic differences between these two types of fluids as indicated in Table 11.1.

**Table 11.1** Coolant fluids vs. lubricants.

| | Cutting fluid | Lubricant |
|---|---|---|
| **Application** | Generally uses for cutting operations | Generally used between mating surfaces without cutting action |
| **Primary function** | To carry away the heat generated. | To reduce friction between mating surfaces |
| **Secondary function** | To reduce friction like the burnishing of the chip against tool or work piece as explained in Fig. 11.1 | To cool the heat generated by mating components |
| **Cooling** | More cooling than reducing friction | More reducing friction than cooling. |
| **Recirculation** | Recirculated through sump and pump | Remains between the mating surfaces, or inside gear boxes, crank cases etc. |
| **Pump** | Pump is required for recirculation | Not, excepting rare cases. Mobility is induced by gravity, wicks etc. See chapter on lubrication. |
| **Material** | Mostly water based or emulsions | See para 10.5 and 10.6 |
| **Appearance** | Mostly white | Mostly yellowish tan |
| **Viscosity** | Low | High |

## 11.4 Materials for Cutting Fluids

1. *Air*
    - Works well with brittle materials like cast iron and glass.
    - Does not tend to stick to the work piece.
2. *Water*
    - It is a good conductor of heat.
    - Its drawbacks are low boiling point, promoting rust formation and not lubricating well.
3. *Mineral coolants*
    - They are petroleum based and vary between light and clear oils used in light industries to thick dark sulphur rich cutting oils used in heavy industries.
    - Were in use since late 1800's.

4. *Semi-synthetic coolants*
    - They form an emulsion with water.
    - They blend the good properties of oils with those of water.
    - They inhibit rusting.
    - By maintaining a stabilized pH value of 9 to 20, they have the ability to work with a variety of metals.
    - They also resist thermal breakdown.
    - Promote environmental safety.
    - Used since around 1930's.
5. *Synthetic coolants*
    - They have properties similar to semi-synthetic coolants.
    - Being developed as late as 1950's.
6. *Pastes or gels*
    - They are generally useful in hand operations like drilling or tapping.
7. *Mists*
    - They enable optimal use of less quantity by weight of coolants.
    - They use aerosol.
    - However, they may cause breathing and other health problems when inhaled.
8. *Other fluids used*
    - Kerosene.
    - Rubbing alcohol.
    - Lard, which was extensively used in the past.
    - Red lead or white lead sometimes mixed with lard, now discontinued.
    - Trichloroethylene used as an additive till 1980s, now discontinued due to its ozone depleting property.
    - WD40 a type of semi synthetic.
    - Dielectric fluids used in electric discharge machines (EDM's). They are usually deionized water or kerosene with high flash point.

## 11.5 Characteristics of Cutting Fluids

1. Should have high specific heat, high heat conductivity and high flash point.

2. Should possess good lubricating properties.
3. Should have low viscosity to permit free flow.
4. Should impart antiweld properties.
5. Should inhibit corrosion and rusting.
6. Should be odourless and nontoxic.
7. Must be non-poisonous and cause no skin irritation.
8. Should not produce fog or mist or smoke during use.
9. Should not discolor the work surface.
10. Should be capable of washing away small chip pieces along with it from the bed.

## 11.6 Other Points to be noted while applying the Cutting Fluids

1. The coolant shall be applied as a continuous stream in sufficient quantities, directed towards the point of cut.
2. The flow should not be at high pressure.
3. The coolant tank must be large enough to keep the coolant at low temperatures.
4. The coolant should start flowing as soon as the operation starts.
5. With the increase of the cutting speeds, the amount of flow should be increased.

## 11.7 Tramp Oil

Among the characteristics of coolant oil, the last point is that it should wash away small chip pieces along with it from the bed, to provide a clean work environment. However, the same property makes the fluid wash away the gear oils and other lubricants leaking from the gaskets and oil seals. While the solid wastes like chips can easily be screened off from entering the coolant sump, these oils pass through most stringent filter pads and enter the sump. Hence they are called tramp oils. These oils being lighter and immiscible form a thin layer of black or brown film floating over the milky white coolant fluids.

## 11.8 Monday Morning Stench

Since the coolant pump is located fairly below the ground level, the tramp oil may enter the circulating fluid stream and may not pose any major problem as long as the machine and the coolant system are working. But during the weekends and rest days, the floating oil film cuts off the oxygen supply to the water based fluid, whereby the microorganisms in the sump decay. When the machine is restarted after the rest day, the decayed microorganisms enter the circulating coolant fluid and give out a foul smell for some time. This is aptly called the Monday Morning Stench. It is a common experience that the first thing done in the morning by the workers is to simply switch on the cooling system for about 5 minutes during which they stay away from the machine.

## 11.9 Methods of Overcoming the Problem of Tramp Oils

The problem of tramp oils has been present ever since water soluble coolants are used for machining and grinding. The following are some of the methods adapted to overcome this problem.

(a) *Use of biocides:* Sometimes chemical additives are added to prevent the presence of the microorganism in the system. But this causes skin irritation and results in constant resentment from the workers against use of these biocides.

(b) *Use of chemicals:* To break up the oils into molecules that can precipitate has been thought of, but so far no right chemicals could be developed

(c) *Periodic cleaning of the sumps*: Quite often the coolant tank of each machine is drained once in a fortnight and replaced with fresh coolant fluid. For an average medium scale factory with about 50 such machines, the coolant drained per month is about 10,000 litres which certainly would not be permitted to be drained into public gutters, unless recycled in effluent treatment plants. This is in addition to the cost of the wasted fluid.

(d) *Decantation:* Though theoretically it is possible to decant off the tramp oils, the film is so thin that more cutting fluid than the oil would come out by this method (Fig. 11.3). Sometimes oil film is manually pushed out by hand while decanting, but this is very ineffective.

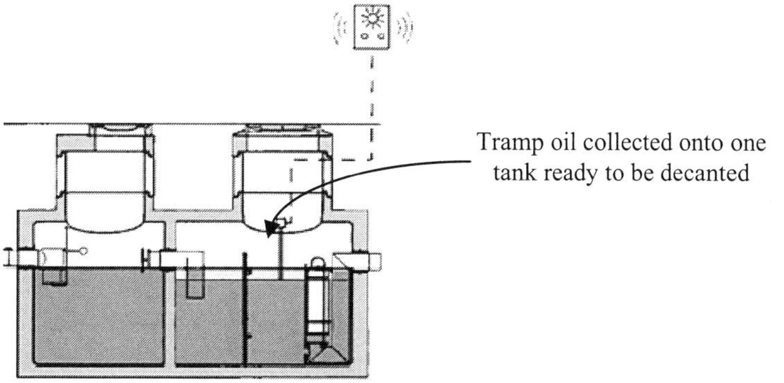


**Figure 11.3** Arrangement for decanting tramp oils from cutting fluid tank.

(e) *Use of sponges or gunny cloth:* Sometimes a sponge or a gunny cloth is spread over the tramp oil layer and the oil sticking to it is squeezed out. This is purely manual and time consuming process, resented by workers as being dirty and irritating to the skin. This cannot be done while the machine is working. Sometimes the coolant pump is centralized for a group of similar machines to aid the process of decantation or use of gunny cloth.

(f) *Use of specially designed coolant sumps:* The upper part of the coolant sump can have narrow neck so that the film can become thicker and its upper layer can be made to flow into separate chambers (Fig. 11.4). But this is very expensive and cannot be done while the machine is working.

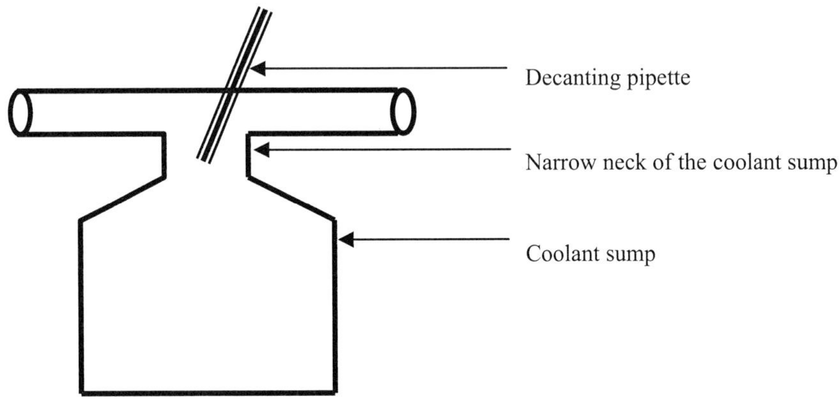


**Figure 11.4** Coolant sump with narrow neck for easy decantation.

(g) *Electrostatic separation:* This has been tried but could not be successful in practice.

(h) *Slow speed skimming:* This works on the principle that oil sticks on to the laminated wheels more than the water based fluids. A specially laminated wheel with microgrooves turns slowly in the coolant tank so that the oil sticking to it gets raised to a higher level and scooped by wiper blades that scrape against the wheel, into a separate container. This is a cost effective gadget that can work even when the machine is working and is illustrated in figure 11.5 (source: Wikipedia) and also by the following case study.

(i) *Belt type skimmers:* Similar arrangement to that of the disc type but using a longer belt so that the total quantity of oil collected per revolution is more (Fig. 11.6).

**Figure 11.5** Disc type skimmer.

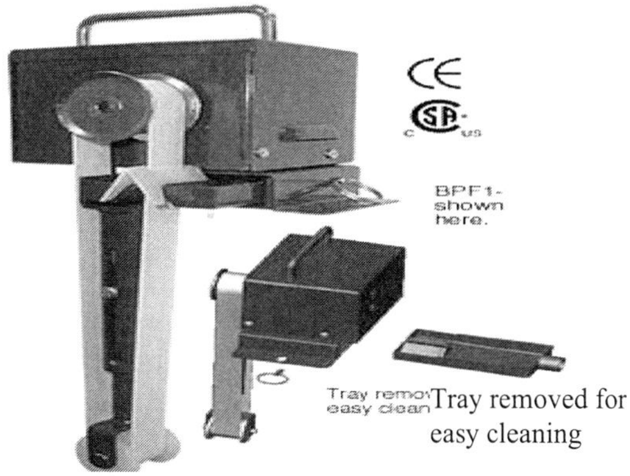


**Figure 11.6** Belt type skimmer.

(j) *Drum type slow speed skimmers:* With a similar principle and construction, the skimming can also be done by using a cylindrical drum in place of the laminated wheel. This is called drum skimming.

(k) *Vortex type oil separator:* The coolant fluid together with the tramp oil is pumped tangentially into an inverted cone shaped separator rotating a high speed. This creates a spinning vortex, by which the strong centrifugal forces separate the heavier water to the cone surface, while the lighter oil moves to the center. This separated oil is removed through an orifice located at the bottom orifice of the cone. The remaining coolant fluid is discharged through a larger outlet located at the top.

## 11.10 Advantages of the Tramp Oil Remover

1. Removes the floating oils continuously leaving clear white coolant.
2. It works without stopping the parent machine. Hence no time is lost.
3. Monday morning stench is eliminated and there is no need for idle circulation of the coolant during Monday mornings.
4. Frequency of coolant change is reduced from fortnightly to once in 5 to 6 months, cutting down the coolant cost by 80%.
5. Life of coolant pumps and other systems are increased.
6. Operators are not subjected to skin irritation.
7. Fewer problems in disposing off the coolant in large quantities.

8. It may be possible to recycle the tramp oil itself and use it as a lubricant where necessary and appropriate.

## 11.11 Case Study

The author, who was a Planning Manager during the earlier years, was quite intrigued at the problem created by the coolants both in terms of cost and the disposal procedure. During his visit to Europe in 1983, he noticed that a majority of machines were having built-in tramp oil skimmers. On his return to India he worked on the development of similar disc type skimmers indigenously.

**Fig. 11.7** The indigenous tramp oil remover

The photograph Fig. 11.7 shows a demonstration model of the tramp oil remover. The cutting fluid with its tramp oil is seen in a tank simulated by the bucket at the center. The skimmer is removing only the black tramp oil that can be seen dropping into a circular tub in a small stream. What can be strikingly seen here is that the fluid in the tub has streaks of white lines, which indicates that the film is being skimmed to show up the white fluid underneath.

The production commenced in 1990 and fitted to machines in some leading automobile industries of Chennai. The respective industries reported significant cost savings by the use of these skimmers.

One specific study of the equipment at one of the heavy duty CNC machines was as under:

1. The coolant tank had a capacity of 200 liters. With a surface area of 1.5 sq. meters, the film thickness was 2 mm, indicating that the tank had about 3 liters of tramp oil and needed replacement of the coolant fluid unless the tramp oil is removed.
2. Within 30 seconds of starting the equipment, the tramp oil was seen flowing out and within 5 minutes white streaks of coolant fluids could be seen through the tramp oil indicating the clearing of the film.
3. Within 90 minutes, 95% of the oil was removed at an average removal rate of 0.5 to 2 liters per hour. This showed that the oil removal rate was far higher than the oil collection rate in the coolant tank.

## 11.12 Conclusion

Among the several methods discussed in paragraph 11.9, the ***slow speed skimming by disc skimmer*** happens to be the most ideal for effectively removing the tramp oils from the coolant fluids and has the benefit of on-line cleaning, that is the machine need not be stopped for the cleaning process. This is evidenced by the appreciation letters received from some of the Indian Automobile giants. Though the production of these equipment is now discontinued, the technology is still available and if any young entrepreneur is interested, this author would be too glad to assist.

# Chapter 12

# Maintenance Costs

*Chapter synopsis:* *Maintenance activities affect the cost of operation and include the direct costs for breakdown maintenance work and indirect costs for preventive maintenance works. The old misconception about maintenance costs was that is impossible to estimate the costs of repair jobs is disproved in this chapter which describes the various costs and cost factors involved in maintenance jobs and how we can estimate and budget them fairly correctly.*

*Keywords:* *Maintenance costs, Collins dictionary, Organisation for Economic Co-operation and Development (OECD), universal measurement of maintenance performance, cost elements, direct maintenance costs, indirect maintenance costs, maintenance labor, maintenance materials, maintenance overhead, total cost of preventive maintenance, total cost of unplanned repairs, net total cost of maintenance, cost categories, non-capital plant improvements, capital improvements and replacements, disposal of obsolete and surplus stock, maintenance budget, six big losses, down time, speed losses, world class maintenance, optimal cost control, maintenance benchmarking, maintenance budgets.*

## 12.1 What are maintenance costs?

The maintenance costs mostly include the direct costs for breakdown maintenance work and indirect costs like preventive maintenance works. For example, the maintenance expenses for automobiles include gas, oil changes, alignment, tire replacement, brake fluid and car washes; while the maintenance expenses for a house include lawn care, plumbing repairs, roof repairs, hazard insurance premiums and replacement of worn out appliances. Let us examine some definitions offered on maintenance costs.

***The maintenance costs*** *are the costs associated with keeping a road, building, vehicle, or machine in good condition and/or good working order by regularly checking it and repairing it when necessary.*

*– Collins Dictionary*

***The maintenance costs*** *are hypothetical costs that would have been incurred in order to meet hypothetical environmental standards using current costs and technologies. They comprise avoidance costs – structural adjustment costs and abatement and restoration costs.*

*- Organization for Economic Co-operation and Development, World Bank.*

***The maintenance costs*** *are the costs incurred to keep an item in good condition and/or good working order.*

*-Investopedia*

***The maintenance costs*** are the costs of up keeping machinery and equipment for the use of b business operations or the upkeep of rental properties.

*– Business Dictionary*

While preventive maintenance improves the machine reliability factor, thereby increasing the productive hours, breakdown maintenance often results in extra for rescheduling and expediting the delayed work as well as the customer dissatisfaction if the deliveries are delayed. When purchasing an item that requires upkeep, consumers should consider not just the initial price tag, but also the item's ongoing maintenance expenses. Maintenance expenses are a major reason why home ownership can be more expensive than renting.

One universal measurement of maintenance performance, and perhaps the measure that matters most in the end, is the cost of maintenance. Unfortunately maintenance costs are often used to compare maintenance performance between companies or between plants within the same company. Equally unfortunately, there is no standard for measuring maintenance costs. Each company, or each plant within a company or more often each department within a plant, develops their own definition of "maintenance costs." For this reason, maintenance cost comparisons should always be accompanied by a clear definition of what is included and what is not, for each plant considered for the comparison.

## 12.2 The Cost Elements

These costs are illustrated in Fig. 12.1 which graphically represent the direct and indirect costs illustrating how they affect the total costs.

**Direct Maintenance Costs**

- Maintenance labor
- Maintenance materials
- Maintenance overhead

**Indirect Maintenance Costs**

- Equipment availability
- Lost capacity
- Equipment life cycle cost
- Production overtime
- Idle production personnel
- Scrap and rework
- Expedited shipments
- Late deliveries
- Lost customers

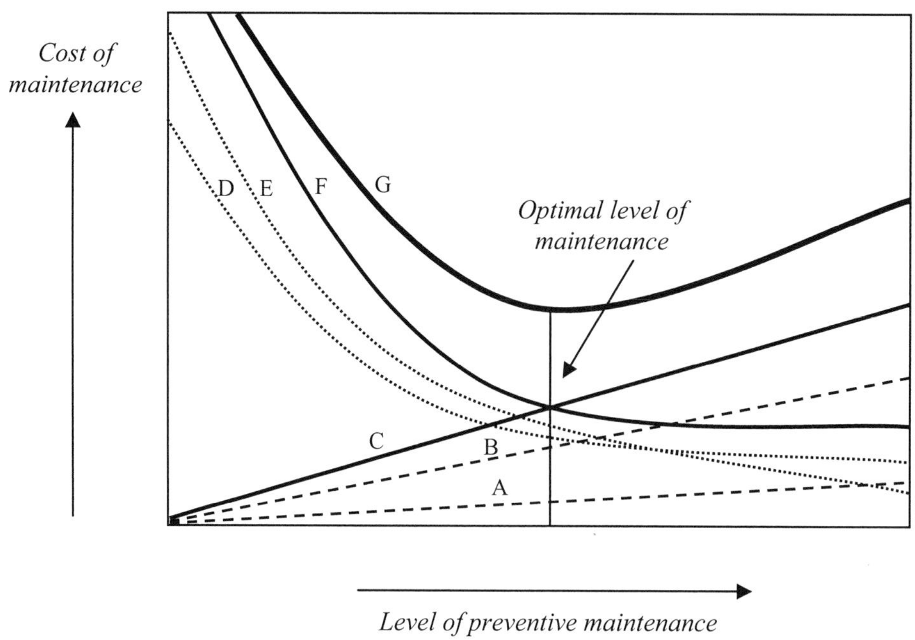


*A = Material cost for PM, B = Labour cost for PM, C = Total cost of preventive Maintenance, D = Stoppage losses, E = Repair costs, F = Total Cost of unplanned repairs, G = Net Total cost of maintenance*

**Figure 12.1** Maintenance costs.

## 12.3 Cost Categories

The website of IDCON, http://www.idcon.com/resource-library/articles classifies various costs involved in maintenance operations in the following cost categories.

| Type of cost (materials and labor) | Recommended cost category |
|---|---|
| Preventive maintenance | Maintenance |
| Corrective maintenance (repair or replacement of failed components) | Maintenance (unless it is a capital replacement) |
| Lubrication (a specific PM task) | Maintenance (In some plants this is a production cost) |
| Contracted preventive and corrective maintenance | Maintenance |
| "Maintenance" work done by production employees. This can include cleaning, inspections, replacement of "production" components (e.g. filter media, etc) and perhaps some lubrication. | Production - but remember to take these costs into account when making comparisons |
| "Non-working" maintenance labor (e.g. maintenance safety meetings, waiting time, etc) | Maintenance |
| Maintenance supervision, planning and administration | Maintenance |
| Non-capital plant improvements. This includes both process and reliability improvements | Probably maintenance, however it is a good idea to include an "improvement" category in work order codes to allow improvement costs to be identified |
| Capital improvements and replacements | Capital (but remember that the definition of "capital" also varies widely) |
| Disposal of obsolete and surplus stock and inventory adjustments (where inventory is working capital) | A special expense account, separate from other costs and not in the maintenance budget |

## 12.4 Effect of Condition Monitoring Integrated Preventive Maintenance on the Maintenance Costs

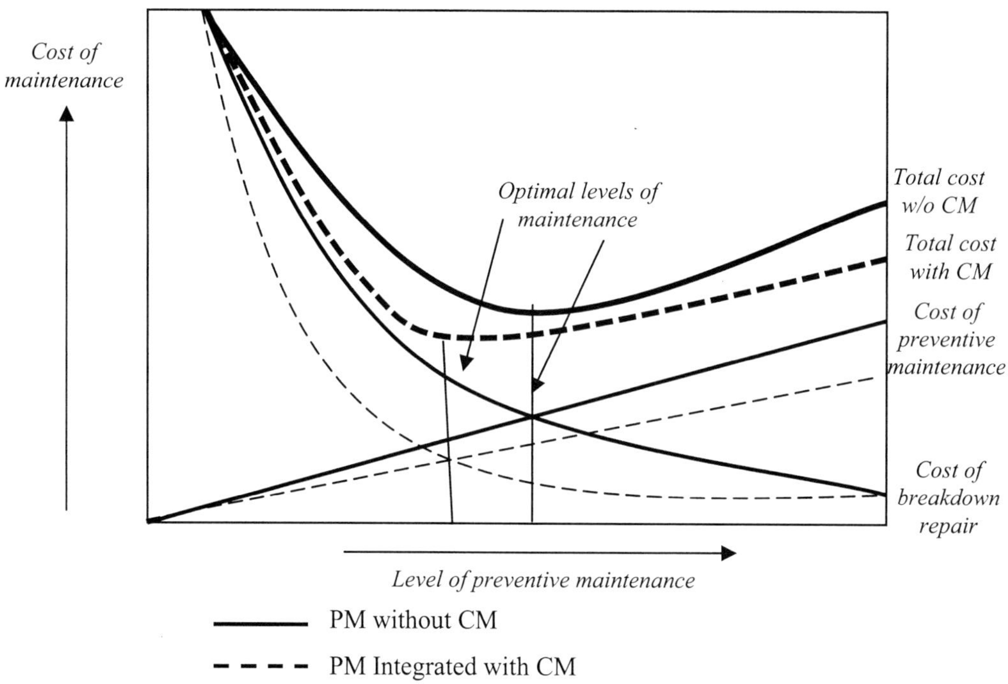


**Figure 12.2** Effect of Condition monitoring integrated PM on the maintenance costs

In Chapter 6 we have stated that condition monitoring when integrated with preventive maintenance to reduce the overall costs. Fig. 12.2 highlights this point illustrating how each cost factor is reduced affecting the total costs for the same preventive maintenance level.

## 12.5 Six Big Losses in Equipment Effectiveness

**A: Down Time**

1. Equipment failure - from breakdowns
2. Setup and adjustment -style change, etc.

**B: Speed Losses**

3. Idling and minor stoppages - due to abnormal operation of line components
4. Reduced speed - due to discrepancies between design and actual speed of equipment

C: Defects

5. Process defects - due to waste/quality defects
6. Reduced yield - from machine start-up to stable production

## 12.6 World Class Maintenance Requirements

1. Clear vision and mission for maintenance
2. Well defined equipment strategies
3. Top management support
4. A proactive and NOT reactive approach
5. Accurately managed costs
6. Total facility participation
7. Effective maintenance planning
8. Strict cost control

## 12.7 Elements Needed for Effective Maintenance with Optimal Cost Control

1. Supervisory control routines
2. Operator and trades training
3. Management training
4. Methodology for sharing information
5. Agreement on maintenance structure,
6. Procedures, and formats (des, pm, pdm, sop, planning, staffing, etc.)
7. Consistent and accurate equipment histories
8. Consistent MRO (maintenance, repair and operations) procedures
9. Implementation of effective maintenance
10. Planning and scheduling
11. Consistent metrics to evaluate progress
12. Consistent application of proven techniques
13. Commitment to make it work

## 12.8 Maintenance Benchmarking

To determine your opportunity for improvement you need to know where you are today. Maintenance benchmarking addresses the following areas:

1. Organization
2. Procedures
3. Planning and scheduling

4. Planned maintenance and lubrication
5. Equipment records and CMMS
6. MRO inventory
7. Training
8. Maintenance costs
9. Maintenance performance measurement
10. Planning & scheduling

## 12.9 Checklist for Benchmarking

1. Are maintenance activities planned?
2. What percentage of maintenance time has been planned atleast 24 hours in advance?
3. Who plans and coordinates for labor, material, and equipment availability?
4. Are estimates made for maintenance actions?
5. What is the current backlog of work waiting to be done? (labor-hours)
6. How are major upgrade/overhaul schedules determined?
7. Is a flow chart of the planning and scheduling process available?
8. Are operations and maintenance schedules coordinated?
9. Does the organization have a long-range (3-5 years) spending plan?
10. Does the department have an annual budget?
11. Are sub-accounts maintained for different types of maintenance activity?
12. How are indirect maintenance costs accounted for?
13. Are cost comparisons made of repetitive jobs?
14. Does the organization have a maintenance cost reduction program?
15. Is maintenance cost performance data shared with hourly employees?
16. Are maintenance costs charged back to the requesting departments?
17. Is there a capital equipment acquisition plan?

## 12.10 Minimizing Total Maintenance Cost

The major emphasis of each department would be to minimize the total costs for their operations at the same time sustaining the optimum performance. In addition to what was specified in paragraph 12.7, this can be achieved by the following three stages which are interlinked.

- Determine the best relationship between preventive and breakdown maintenance
- Determine the level of repair work to be done vis-a-vis the replacement policy and,
- Determine the optimum size of the maintenance staff.

## 12.11 Preparing Maintenance Budgets

Every company prepares individual maintenance budgets peculiar to itself. There cannot be an objective yardsticks or formulae like production cost budgets, in view of the probability of occurrence of major maintenance activities. However for enabling the management to set budget the maintenance expenditure, the department normally submits budgets in two levels.

- One level for the routine maintenance expenditure including the activities that are expected to occur with a high percentage of probability and,
- A second level of budget giving the proposed investment for certain maintenance equipment as well as the probable expenditure in case certain major faults occur. Here the details of the major faults and the likelihood of their occurrence have to be indicated, together with the reasons for expecting them to occur.

## 12.12 Conclusion

The maintenance engineer should not only prevent the equipment failure, but should also be able to perform it at an optimal cost. Hence, the knowledge on the several costs involved in the maintenance function would go a long way in his performing his function efficiently.

# Chapter 13

# Failure Modes and Effects Analysis

***Chapter Synopsis:*** *In general despite high quality control exerted in manufacture and operation any equipment may have inherent weaknesses resulting in their failure. Instead of analyzing why the failures have occurred as a post mortem, we should always anticipate such failures and provide for their corrective action during the design stage itself. This is the principle behind Failure Modes and Effects Analysis and this chapter details the several aspects including the principles and procedures for this essential maintenance management tool.*

***Keywords:*** *Before-the-event action, history of FMEA development, Apollo Space Program, Pinto car project, Advanced Product Quality Planning, Automotive industry Action Group, Design Review Based on Failure Mode, Concept FMEA, Design FMEA, Process FMEA, Service FMEA, Software FMEA, failure mode, failure causes, failure effect, severity factor, probability of occurrence, ease of detection, risk priority number, FMEA procedure, functional block diagrams, fail safe measures, responsibility of action, FMEA software.*

## 13.1 Importance of FMEA to Maintenance Management

Failure mode and effects analysis **(FMEA)** is a procedure for analysis of potential failure modes within a system emphasizing on the probability of occurrence of that failure and the severity of its effect on the system. It is used to identify potential failure modes, determine their effect on the operation of the product, and identify actions to mitigate the failures. It analyses potential reliability problems early in the development cycle where it is easier to take actions to overcome these issues, thereby enhancing reliability at all stages, whether design or operation or maintenance. FMEA should always be done whenever failures would mean potential harm or injury to the user of the end item being designed. According to Besterfield et al, *FMEA is a 'before-the-event' action requiring a team effort to easily and inexpensively alleviate changes in design and production.*

## 13.2 History of Development of FMEA

- FMEA has its origin during the late 1940s for military usage by the US Armed Forces. This is incorporated in the military document MIL-P-1629 dated November 9, 1949, titled *Procedures for Performing a Failure Mode, Effects and Criticality Analysis*, and subsequently ratified as MIL-STD-1629A.
- Its effectiveness in identifying and reducing any unseen problems encouraged its application for space research and design specifically in the Apollo Space program and in developing the means to put a man on the moon and return him safely to earth.
- Its industrial application came during the early 1970s when Ford Motor Company, reeling after the failure of its Pinto Car project introduced FMEA to the automotive industry to improve design and for safety and regulatory considerations. Society of Automobile Engineers (SAE) have documented this as SAE J1739.
- FMEA methodology is now extensively used in a variety of industries including semiconductor processing, food service, plastics, software, and healthcare.
- It is integrated into Advanced Product Quality Planning (APQP) to provide primary risk mitigation tools and timing in the prevention strategy, in both design and process formats.
- The Automotive Industry Action Group (AIAG) requires the use of FMEA in the automotive APQP process and has published a detailed manual on how to apply the method. Each potential cause must be considered for its effect on the product or process and, based on the risk, actions are determined and risks revisited after actions are complete.
- Toyota has taken this one step further with its Design Review Based on Failure Mode (DRBFM) approach. The method is now supported by the American Society for Quality which provides detailed guidelines on applying the method.

## 13.3 Types of FMEAs

There can be several forms of FMEA based on which aspect of the activity is analyzed and can be illustrated as below. Other forms of FMEA like system or service can be interpolated into this illustration.

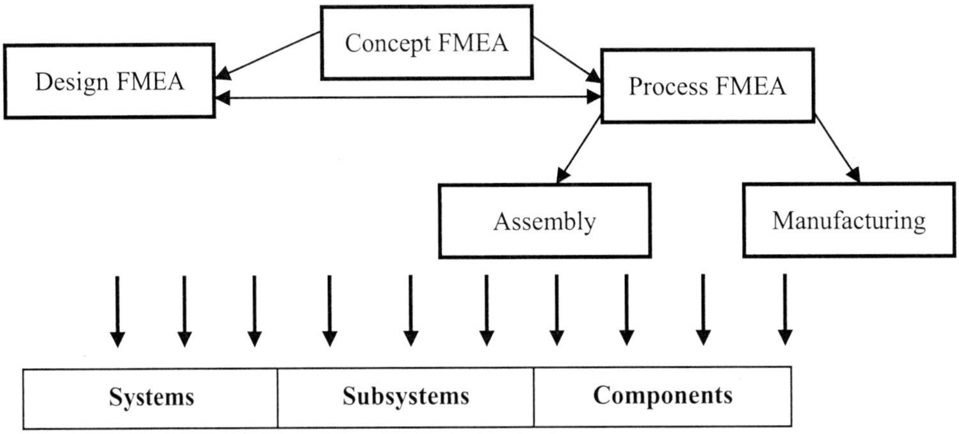


**Figure 13.1** Relationship between the three forms of FMEA.

*Concept FMEA (CFMEA)*

- The Concept FMEA is used to analyze concepts in the early stages before component is defined (most often at system and subsystem)
- It focuses on potential failure modes associated with the proposed functions of a concept proposal
- This type of FMEA includes the interaction of multiple systems and interaction between the elements of a system at the concept stages.

*Design FMEA (DFMEA)*

- The Design FMEA is used to analyze products before they are released to production.
- It focuses on potential failure modes of products caused by design deficiencies.
- Design FMEAs are normally done at three levels – system, subsystem, and component levels.
- This type of FMEA is used to analyze hardware, functions or a combination.

*Process FMEA (PFMEA)*

- The process FMEA is normally used to analyze manufacturing and assembly processes at the system, subsystem or component levels.
- This type of FMEA focuses on potential failure modes of the process that are caused by manufacturing or assembly process deficiencies.

Other forms of FMEA can be

- *System FMEA*, which focuses on global system functions

- *Service FMEA*, which focuses on service functions
- *Software FMEA*, which focuses on software functions.

## 13.4 When to use FMEA

- Whenever a new product or process is being initiated.
- Whenever changes are made to the product design, process or the operating conditions. The product and process are inter-related. When the product design is changed the process is effected and vice-versa.
- Whenever new regulations are being incorporated.
- When the customer feedback indicates problems in the product or process.

## 13.5 Basic Terms of Reference in FMEA

1. *Failure Mode*: It is the manner by which a failure occurs and is observed like occurrence of electrical short-circuiting, corrosion, cracking or deformation. It may be noted that a failure mode in one component can lead to another failure mode in the same or another component. Therefore each failure mode should be listed in technical terms and also giving due consideration for their interrelations.

   IEC 812-1985 enumerates the Generic Failure Modes as below:

   1. Structural failure (rupture)
   2. Physical binding or jamming
   3. Vibration
   4. Failing to remain in position
   5. Eccentric rotation
   6. Failed interlocking system
   7. Failing to open
   8. Failing to close
   9. Internal leakage
   10. External leakage
   11. Fails out of tolerance (high)
   12. Fails out of tolerance (low)
   13. Inadvertent operation
   14. Intermittent operation
   15. Erratic operation
   16. Erroneous indication
   17. Restricted flow
   18. False actuation
   19. Failing to stop
   20. Failing to start
   21. Failing to switch
   22. Premature operation
   23. Delayed operations
   24. Erroneous input (increased)
   25. Erroneous input (decreased)
   26. Erroneous output (increased)
   27. Erroneous output (decreased)
   28. Loss of input
   29. Loss of output
   30. Shorted (electrical)
   31. Open (electrical)
   32. Leakage (electrical)
   33. Other unique failure conditions as applicable to the system characteristics, requirements and operational constraints

2. *Failure Cause:* The product or process defects or any other quality imperfections that would initiate further deterioration leading to a failure. Some failure modes may have more than one cause or mechanism of failure and each of these shall be listed and analyzed separately.
3. *Failure Effect*: The immediate consequences of a failure on operation, function or functionality, or status generally as perceived or experienced by the user. Some of the effects can be cited as, injury to the user, inoperability of the product or process, deterioration in product quality, non-adherence to the specifications, emanation of odors, noise etc. Also the effect of this failure on other systems in immediate contact with the system that failed, has to be considered. If a component fractures, it may cause vibration in the subsystem that is in contact with the fractured part. FMEA is the technique in analyzing the potential failures and their effect on the system.
4. *Severity Factor*: A symbolic measure of the failure effect is the severity factor, which is the assessment of the seriousness of the effect of the potential failure. It is noteworthy that the severity represents the seriousness of the failure and not the mode of the failure. Bester field emphasizes in this connection that no single list of severity criteria is applicable to all designs and the team should agree on evaluation criteria and on a ranking system that are consistent throughout the analysis. As illustrated in Table 13.1, the severity of the effect is given a *severity number(s)* from 1 (no danger) to 10 (critical).

**Table 13.1** Rankings of severity of effect.

| Effect | Severity of effect | Severity Factor |
|---|---|---|
| **Hazardous without warning** | Very high ranking with potential failure mode affects safe operation and regulation non-compliance. Failure occurs without warning. | 10 |
| **Hazardous with warning** | Very high ranking with potential failure mode affects safe operation and regulation non-compliance. Failure occurs with warning. | 9 |
| **Very high** | Fairly high ranking with potential failure mode affects regulation. Generally failure occurs with warning. | 8 |
| **High** | Item is operable, but with loss of performance. Customer is dissatisfied. | 7 |
| **Moderate** | Product is operable but with loss to comfort/convenience. Customer experiences discomfort | 6 |
| **Low** | Product is operable but with loss to comfort/convenience. Customer has some discomfort | 5 |
| **Very Low** | Certain item characteristics do not conform to specifications, but noticed by most customers | 4 |

**Table 13.1** *Contd...*

| Effect | Severity of effect | Severity Factor |
|---|---|---|
| **Minor** | Certain item characteristics do not conform to specifications, but noticed by average customers | 3 |
| **Very Minor** | Certain item characteristics do not conform to specific satins, but noticed by some discriminating buyers (referred to as dissatisfiers in chapter no. 3). | 2 |
| **None** | No effect | 1 |

5. *Probability of occurrence:* Probability of occurrence is the chance that one of the specific failure causes will occur. This recording of probability of occurrence must be done for every cause indicating the probability of occurrence of that cause. This can be done by looking at the occurrence of failures for similar products or processes and the failures that have been documented for them in technical terms. For FMEA, this occurrence rate can be assigned a numerical value from 1 to 10, the least frequent being 1 and the most frequent being 10 (Table 13.2). If this value is more than 4, it implies that the actions needed to identify and analyze them shall be more meticulous. Bester field et. al. suggest the following guideline questions for evaluation.
   - What is the service history or field experience with similar systems or sub-systems?
   - Is the component similar to a previous system or subsystem?
   - How significant are the changes in the component?
   - Is the component completely new?
   - Is the component application any different from the previous?
   - Is the component environment any different than before?

**Table 13.2** Rankings of probability of occurrence.

| Probability of occurrence | Explanation | Possible failure rate | Ranking no. |
|---|---|---|---|
| **Very high** | Failure is almost inevitable | > 1 in 2 | 10 |
| | | 1 in 3 | 9 |
| **High** | Generally associated with processes similar to previous processes that have often failed | 1 in 8 | 8 |
| | | 1 in 20 | 7 |
| **Moderate** | Generally associated with processes similar to previous processes that have experienced occasional failures | 1 in 80 | 6 |
| | | 1 in 400 | 5 |
| | | 1 in 2,000 | 4 |
| **Low** | Isolated failures associated with similar processes | 1 in 15,000 | 3 |
| **Very Low** | Only isolated failures associated with almost identical processes | 1 in 150,000 | 2 |
| **Remote** | Failure is unlikely. No failures ever associated with almost identical processes | < 1 in 1,500,000 | 1 |

6. *Ease of detection:* This is the ability of the inspecting mechanism and/or design control to detect the potential cause or the subsequent failure mode before the component or the subsystem is completed for production. The proper inspection methods need to be chosen. First, an engineer should inspect the current controls of the system which are designed to prevent failure modes from occurring, or atleast detect the failure before it reaches the customer. Hereafter one should identify testing, analysis, monitoring and other techniques that can be or have been used on similar systems to detect failures. From these controls an engineer can learn how likely it is for a failure to be identified or detected. This parameter too is given a numerical value between 1 and 10, called the Detection Rating (Table 13.3). This ranking measures the risk that the failure will escape detection. A high detection number indicates that the chances are high that the failure will escape detection, or in other words, the probability of detecting a failure is high when the number is low.

**Table 13.3** Rankings of ease of detection.

| Ease of detection | Explanation | Ranking no. |
|---|---|---|
| **Absolutely impossible** | No known controls available for detection of the failure mode | 10 |
| **Very Remote** | Very remote likelihood that the current controls will detect failure mode | 9 |
| **Remote** | Remote likelihood that the current controls will detect failure mode | 8 |
| **Very Low** | Very low remote likelihood that the current controls will detect failure mode | 7 |
| **Low** | Low remote likelihood that the current controls will detect failure mode | 6 |
| **Moderate** | Moderate remote likelihood that the current controls will detect failure mode | 5 |
| **Moderately high** | Moderately high remote likelihood that the current controls will detect failure mode | 4 |
| **High** | High remote likelihood that the current controls will detect failure mode | 3 |
| **Very High** | Very high remote likelihood that the current controls will detect failure mode | 2 |
| **Almost certain** | Reliable controls are known with similar processes and current controls almost certain to detect the failure mode | 1 |

## 13.6 Risk Priority Number

Risk Priority number is a function of the three parameters discussed above, viz the severity of the effect of failure, the probability of occurrence and the ease of detection for each failure mode. RPN is calculated by multiplying these three numbers as per the formula below,

$$RPN = S \times O \times D$$

RPN may not play an important role in the choice of an action against failure modes, but will help in indicating the threshold values for determining the areas of greatest concentration. In other words, a failure mode with a high RPN number should be given the highest priority in the analysis and corrective action. The relationship between the above mentioned parameters of FEMA may be represented as per the Figure 13.2.

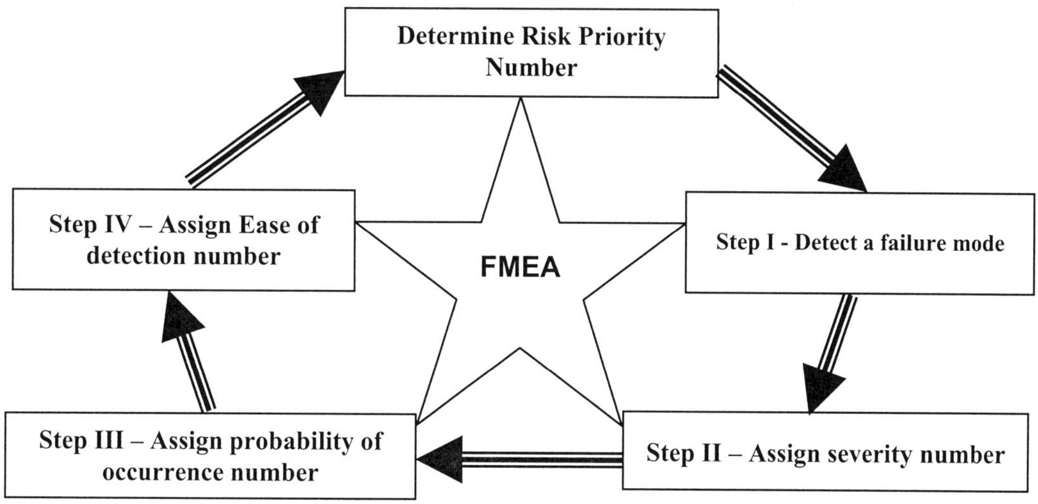


**Figure 13.2** The 5 basic steps of FMEA.

## 13.7 Procedure for FMEA

*Princeton Plasma Physics Laboratory* suggests the following basic steps for FMEA, as recommended by IEC Standard 812 - Procedure for Failure Mode and Effects Analysis (FMEA).

1. Define the system and its functional and operating requirements.
   (a) Include primary and secondary functions, expected performance, system constraints, and explicit conditions that constitute a failure. The system definition should also define each mode of operation and its duration.
   (b) Address any relevant environmental factors such as temperature, humidity, radiation, vibration, and pressure during operating and idle periods.

   (c) Consider failures that could lead to noncompliance with applicable regulatory requirements. For example, a failure that could result in a release that exceeds environmental permit limits.

2. Develop functional block diagrams showing the relationships among the elements and any interdependencies. Separate diagrams may be required for each operational mode. As a minimum, the block diagram should contain
   (a) A breakdown of the system into major subsystems including functional relationships;
   (b) Appropriately and consistently labeled inputs and outputs and subsystem identification;
   (c) Any redundancies, alternative signal paths, and other engineering features that provide "failsafe" measures.

   Existing drawings developed for other purposes may be used for the FMEA if the above elements are adequately described.

3. Identify failure modes, their cause and effects.
   (a) Identify the possible causes associated with each postulated failure mode. For example, a power supply may have a specific failure mode "loss of output" and a failure cause "open contacts (electrical)".
   (b) Identify, evaluate, and record the consequences of each assumed failure mode on system, element operation, function, or status. Consider maintenance, personnel, and system objectives as well as any effect on the next higher system level.

4. Identify failure detection and isolation provisions and methods. Determine if other failure modes would give an identical indication and whether separate detection methods are needed.

5. Identify design and operating provisions that prevent or reduce the effect of the failure mode. These may include:
   (a) Redundant items that allow continued operation if one or more elements fail;
   (b) Alternative means of operation;
   (c) Monitoring or alarm devices;
   (d) Any other means permitting effective operation or limiting damage.

6. Identify specific combinations of multiple failures to be considered. The more multiple failures considered the more complex the FMEA becomes. In many such cases it would be advantageous to perform a FMECA using the guidance of IEC Standard 812 or MIL-STD-1629A. Failure Mode, Effects and Criticality analysis (FMECA) is an extension of FMEA, by including criticality analysis which is useful in charting the probability of failure modes against the severity of their

consequences. Using the FMECA, the severity of failure effects are categorized, the probability is determined, and the number of redundant mitigating features needed to keep probability of failure acceptably low are better determined.

7. Revise or repeat, as appropriate, the FMEA as the design changes. Changes may be direct response results of the previous FMEA or may be due to unrelated factors.

## 13.8 Responsibility for Action

FMEA is a team operation. Everyone should feel fully involved in the process and in moving towards the goal. Nevertheless it is always advisable to delegate certain responsibilities to specified persons, so that the monitoring and reporting can be effective. It is suggested that some of the responsibilities be allocated between the line manager, the analyst and the reviewer as follows:

*Line Manager*

1. Assign individuals to perform FMEA (analyst) and another individual to review it (reviewer). The reviewer shall have as much expertise and technical experience as the analyst.

*Analyst*

2 Describe the system under analysis, prepare system diagrams and use existing documentation to depict all major components and their performance criteria. The level of assembly may vary with the level of the analysis.

3. Perform FMEA as per the procedure described above.
4. Document results using the standard analysis chart.
5. Sign FMEA and provide it to the reviewer.

*Reviewer*

6. Discuss with analyst and review FMEA for technical content and sign if no significant problems are identified.
7. Ensure that the full FMEA documents are filed in the operations center.

## 13.9 Benefits of FMEA

- Effective prevention planning programme.
- Identification of change requirements
- Cost reduction
- Increased throughput

- Decreased waste
- Decreased warranty costs
- Reduction of non-value added operations.

While the general benefits of FMEA can be listed as above, the category wise benefits can be summarized as under

*Concept FMEA:*

- Helps selecting the optimum concept alternatives, or determine changes to design specifications.
- Identifies potential failure modes caused by interactions within the concept.
- Ensures that all potential effects of a proposed concept's failure modes are considered.
- Identifies system level testing requirements.
- Helps to determine if hardware system redundancy may be required within a design proposal.

*Design FMEA:*

- Aids in the objective evaluation of design requirements and design alternatives.
- Aids in the initial design for manufacturing and assembly requirements.
- Increases the probability that potential failure modes and their effects have been considered in the design/development process.
- Provides additional information to help plan, thorough and efficient test programs.
- Develops a list of potential failure modes ranked according to their effect on the customer. Establishes a priority system for design improvements.
- Provides an open issue format for recommending and tracking risk reducing actions.
- Provides future reference to aid in analyzing field concerns.

*Process FMEA:*

- Identifies potential product related process failure modes.
- Assesses the potential customer effects of the failures.
- Identifies the potential manufacturing or assembly process causes and identifies process variables on which to focus controls or monitoring.
- Develops a ranked list of potential failure modes, establishing a priority system for corrective action considerations.

- Documents the results of the manufacturing or assembly process.
- Identifies process deficiencies.
- Identifies confirmed critical characteristics and/or significant characteristics.
- Identifies operator safety concerns.
- Feeds information on design changes required and manufacturing feasibility back to the designers.

## 13.10 FMEA Software

Some of the software developed for industrial and dedicated application of FMEA can be listed as below. It may be noted that the term software FMEA mentioned in Para 13.3 refers to FMEA analysis as applied to any software programs. The term FMEA software mentioned here refers to the type of software developed for solving FMEA problems. (*Source - internet* :)

1. *ASENT FMEA Software* - Raytheon's premiere reliability and maintainability tool suite. Includes a very powerful FMECA tool that combines FMECA, RCM Analysis, and Testability Analysis.
2. *Byteworx* - Powerful, cost-effective software for Failure Mode and Effects Analysis. The global choice of the Ford Motor Company, Byteworx FMEA is fully compliant with SAE J-1739 Third Edition.
3. *FMEA-Pro* - FMEA/FMECA software from Dyadem. An all-in-one software solution that provides corporate consistency and assists with corporate compliance.
4. *Isograph Software* - Their Reliability Workbench contains a FMEA/FMECA tool.
5. *Item Software* - FMEA/FMECA/FMEDA - Failure Mode Effects Analysis tool.
6. *Quality Plus* - FMEA software from Harpco Systems, Inc. Performs both Design and Process FMEAs.
7. *RAM Commander Software* - ALD's integrated FMEA/FMECA modules have been adopted by many civil, military, aerospace, energy and pharmaceutical organizations worldwide.
8. *Relex Software* - Explains FMEA tools & FMEA software with FMEA method to process FMEA and meet all functional FMEA standards for criticality matrix.
9. *XFMEA* - FMEA software from ReliaSoft. Provides expert support for all types of Failure Mode and Effects Analysis (FMEA).

# Chapter 14

# Lean Maintenance

***Chapter Synopsis:*** *Lean Maintenance is the latest concept put forward by IDCON Inc of US. This author remembers to have attended as a university observer, the training sessions of Idhammer Konsult, during the late 1970's in Dar es Salaam mostly on the latest trends in maintenance practices. Having recently seen this new concept by the same consultant Inc., now based in USA, this author feels it relevant to adapt from their website on this latest concept in maintenance.*

***Keywords:*** *Lean manufacturing, lean maintenance, just-in-time, Root cause failure analysis, Toyota Production System, Toyotism, manufacturing reliability, pit shop maintenance of racing cars.*

## 14.1 What is Lean Maintenance?

Lean manufacturing can be defined as the production practice that considers the expenditure of resources for any goal other than the creation of value for the end customer to be wasteful, and thus would be targeted for elimination. In a similar manner, we can explain Lean maintenance as the proactive maintenance practice employing planned and scheduled maintenance activities through several modern and optimal trends in maintenance management that we have been studying in this book. This is further explained in para 14.2. They include

1. Planned and scheduled maintenance,
2. Total productive maintenance (TPM),
3. Predictive maintenance (PdM) analysis,
4. Trending and analysis of condition monitoring results,
5. Reliability centered maintenance (RCM) decision logic,
6. The 5S process,
7. Weekly Kaizen improvement events,

8. Logical fault location,
9. Empowered (self-directed) action teams with multi-skilled, maintenance technician,
10. Well planned and organized spare part stocking policy that provides parts and materials on a just-in-time (JIT) basis,
11. Supply chain management,
12. Root cause failure analysis (RCFA),
13. Failed part analysis,
14. Maintenance procedure effectiveness analysis,
15. Optimization of test and inspection methods,
16. Computer managed maintenance system (CMMS),
17. Enterprise asset management (EAM) system, and
18. Similar new maintenance management techniques being developed.

## 14.2 Original Concept of Lean Manufacturing System

Lean is similar to the reduction of fats from the body in the process of weight reduction and to making one's body attractively Lean. This concept is developed in 1990's mostly from the Toyota Production System (TPS) and so is also called Toyotism or Simple Lean and was successfully applied both in Japan and the Western world.

This concept in the manufacturing activity, which has increased in popularity in the past ten years is now applied to maintenance activity as explained by IDCON Inc. in their website as in the following paragraphs, which also detail the benefits of this concept. When the similar concept is applied to maintenance, preventing excessive maintenance needs and performing the essential maintenance activities more effectively would result in lean maintenance this also include spare part inventory and purchase of maintenance materials.

Simply put, it can be said that the thought process in the lean concept should result in those things that need to be delivered are maintained and repaired on time with less resources. That is to say that it doesn't always have to do with always doing more, that can result in waste, but that things that have been sold should be produced when they need to be delivered.

## 14.3 Increased Reliability with Lean Maintenance

Increased manufacturing reliability will increase product throughput and reduce the time between incoming raw materials to the finished product. Better reliability is the foundation to a faster and safer manufacturing flow. This will result in lowered losses in delayed deliveries, over production, work in progress and energy expenditure. Here lie the biggest gains if you can't sell everything you produce. This cannot materialize if you don't have equipment with very high reliability. This will increase the production reliability and reduce the production costs, including maintenance costs and costs for storage.

To become lean one must prevent maintenance needs and perform remaining maintenance more effectively. It is a well-documented fact. If the previously mentioned are implemented, then the production reliability will increase and thus the production costs, including maintenance costs and costs for storage, will decrease.

To become lean all losses in the manufacturing and delivery stages of raw material and delivery to the customers must be eliminated.

## 14.4 Improvement Opportunities Leading to Lean Maintenance

The biggest losses in maintenance, and thus also the biggest improvement opportunities include

### 14.4.1 Manufacturing Reliability

- Loss in quality.
- Stop times.
- Loss in speed.

### 14.4.2 Partnership between Operations – Maintenance – Engineering

- Reliability and maintenance related design.
- Operator based maintenance.

### 14.4.3 Elimination of Root Cause of the Problem

- Choose problem to eliminate.
- Eliminate problems.
- Educate and teach.

### 14.4.4 Storage

- Reduce the store value at the same time as you preserve service level to maintenance.

### 14.4.5 Integration and Application of Increased Knowledge and Skills

- Education and training of crafts people to enable multi craft or multi skills.
- Implementation of flexible work systems.

### 14.4.6 Over Manufacturing

- Make more than what has been sold.
- Manufacture too early.

### 14.4.7 Over Maintenance

- Perform too much preventive maintenance than optimally required.
- Perform preventive maintenance before it is needed.
- Do corrective maintenance with higher priority than needed.

### 14.4.8 Use of New Technology

- Less need for maintenance.
- Better maintainability.
- Smart tools and methods.

## 14.5 Illustration of a Pit Shop Maintenance Situation

IDCON offers a nice illustration of the significance of lean maintenance when it comes to the pit shop maintenance of racing cars.

> *The drivers of the racing cars are in constant contact with the pit stop crew, they do not show up suddenly telling the pit crew "I think I have a problem with the right front tire" and the crew answering "We will go to the store and check if we have any replacement tire". But this happens daily in most plants. In a car race there is a strong motivation to win the race, in a plant there might be other factors driving motivation.*

In addition to driving, planning and scheduling to precision and excellence, pit crews of race tracks are continuously working on improving the basics. This includes;

- Lug nut tool is turning at 24000 rpm.
- All parts sorted and placed on where they are used.
- Steering wheel straight position is marked.
- Wheel bolts with rounded tops.
- Wheel nuts glued to wheel.
- Measurement of tire temperature.
- Measurement of tire wear.

- Marked wheel position to speed up wheel change.
- Wheel nuts painted yellow for visibility.
- Analysis of problems and successes.

## 14.6 Conclusion

This chapter emphasizes the fact that preventive maintenance, total productive maintenance and reliability based maintenance are the three aspects of lean maintenance and an optimal integrated application if these would result in increased production reliability which in turn will reduce the production costs, including maintenance costs and costs for storage.

# Chapter 15

# Scientific Inventory Management

*Chapter synopsis:* *This chapter discusses the inventory management principles and practices for the production items. The three stages of achieving inventory reduction, viz. the codification, ABC Analysis and inventory parameter fixing are highlighted, emphasizing that the selective control systems like VED are most suitable for maintenance stores.*

***Keywords:*** *MRP, material planning, classification, codification, selective control, ABC analysis, annual consumption value, inventory control parameters, consumption rate, safety stock, maximum stock, lead time, ROQ, ROL, reorder interval, economic order quantity, inventory carrying costs, ordering cost, Q order system, P order system, $I_{max}$ order system, $I_{min}$ order system, Two-bin inventory control, ABC, FSN, MNG, VED, SDE, HML, GOLF and SOS systems of selective control, comprehensive material control project, zero inventory drive and supplier partnership, CPFR.*

## 15.1 What is Inventory?

It may be noted that though several inventory concepts and terminology have been given in the previous chapter, they are repeated as being significant to this chapter.

Inventory is the physical stock of items that a business or production organization keeps on hand for efficient running of affairs or its production. As the name suggests scientific inventory management, unlike the practice of *'buy whatever you need whenever you need'*, involves the following functions:

- Material planning
- Material control
- Material Resources Planning (MRP)
- Stock control
- Store keeping
- Purchasing

- Vendor development
- Make or buy decision and
- Other technological analysis and projects like classification, codification and ABC analysis.

## 15.2 Types of Inventories

1. Raw materials
2. Consumables
3. General supplies
4. Bought out components
5. Maintenance spares
6. Work in progress
7. Finished goods stores
8. Miscellaneous inventory.

## 15.3 Conditions Leading to Increased Inventory

1. Requirements for production or maintenance scheduling,
2. To protect against sudden price increase,
3. To take advantage of price fluctuations,
4. Economics of quantity purchasing to take advantage of quantity discounts etc. especially in case of electric bulbs,
5. Some suppliers insist on minimum supply quantity especially vital items, though the requirement is less,
6. To level out variations in production cycles by producing to inventory,
7. To insure against seasonal demand or availability,
8. Lack of part standardization, a situation diametrically opposite to the previous factors, yet would contribute stocking of large inventories, more in terms of varieties than in terms of quantity per variety,
9. Wrong anticipation of inventories especially for new projects,
10. Multiple storage depots like decentralised maintenance stores,
11. Lack of dependable suppliers,
12. To protect for delays in receipts from the suppliers due to increased lead time,
13. Inadequate attention to inventory control and
14. Size, nature and condition of production facilities of line organizations like:

- Large investment in plant and equipment, continuous production equipment like refineries, steel mills, chemical industries,
- Age of the machinery with low MTBF and
- Lack of planned spare sets supplied by the manufacturer of the equipment.

## 15.4 Disadvantages of Excessive Stocks

1. Blocking of working capital,
2. Occupying space,
3. Increased risk of obsolescence, damage, deterioration etc.,
4. Risk of theft, pilferage etc.,
5. Increased maintenance and preservation costs,
6. Higher insurance premium,
7. Increased inventory carrying costs.

## 15.5 Selective Control in Inventory Management

Selective Control in Inventory Management is very vital for cost efficient but meticulous inventory control systems. As seen in the previous chapter, the following are the features and systems of selective control.

*Why selective control?*

- In any inventory situation some items are used regularly with high annual consumption value and some with very low annual value. On an average, in production situation, only 10% of the number of items constitutes 90% of the total annual value, though it is somewhat different for maintenance items.
- The time and energy spent for meticulous control on any item is almost same, whether a high value or a low value item.
- Hence we should exercise stricter control for some items and looser control for a majority of items. This is called selective control.

Some of the selective control systems are detailed below

1. *ABC control:* Based on the total annual consumption value. The parts are classified into A, B or C items and separate control policies are developed for each, as described more in detail in paragraph 15.8.
2. *FSN control:* Fast moving, Slow moving or Non-moving based on the usage of an item in the past 1 or 2 years, helping us to decide on disposal action.

3. *MNG or MND control:* Moving, Non-moving or Ghost items (Dead items). This is almost a synonym for FSN control.
4. *VED control:* Vital, Essential or Desirable, based on their need for the present and future maintenance works. This is more suitable for maintenance items.
5. *SDE control:* Scarce, Difficult and Easily available, based on their availability. This too is useful for maintenance items.
6. *HML control:* High cost, medium cost and low cost, based on the pricing of each item, useful for identifying items like the gold tipped or diamond tipped items which have to be stocked separately in lockers.
7. *GOLF control:* Government, Local or Foreign, based on the availability, so that we can plan the procurement of heavily rationed Government items like steel rods during the late nineties or imported items.

It may be noted that whether it is a production situation or maintenance situation, more than one of the above controls can be adapted and integrated suitably to suit the individual situation.

## 15.6 Scientific Material Planning

The crux of the scientific inventory management is the meticulous planning of the materials starting from the process of deciding when and how much to purchase. This process involves three stages,

- Classification and codification
- ABC analysis and
- Fixing of inventory control parameters.

## 15.7 Classification and Codification

This is an essential start up point for material planning, especially the first 5 steps without which either ABC analysis or the stock level planning would be futile. The study and analysis of the materials consumed in the company involves 4 steps are cited below.

1. List of all the items forming the inventory of the company.
2. Decide upon the basic classification of materials in accordance to the policy of the company.
3. Classify all the items into these groups and subgroups.
4. Study their specifications and recorded descriptions, to analyses for any duplicated descriptions or specifications by which 2 or more identical items are specified separately causing duplication.

Having completed the 90% of the codification work by step 4, it is as well better to continue with the 5th and 6th step, to complete the codification project as below:

5. Assign code numbers to each item
6. Prepare the company code books for reference.

The advantages of the codification project, in addition to those from the ABC analysis and stock level planning and control, are illustrated in the report append at the end of this chapter.

## 15.8 ABC Analysis

The second stage is ABC Analysis as illustrated in the Fig. 15.1 and the following paragraphs.

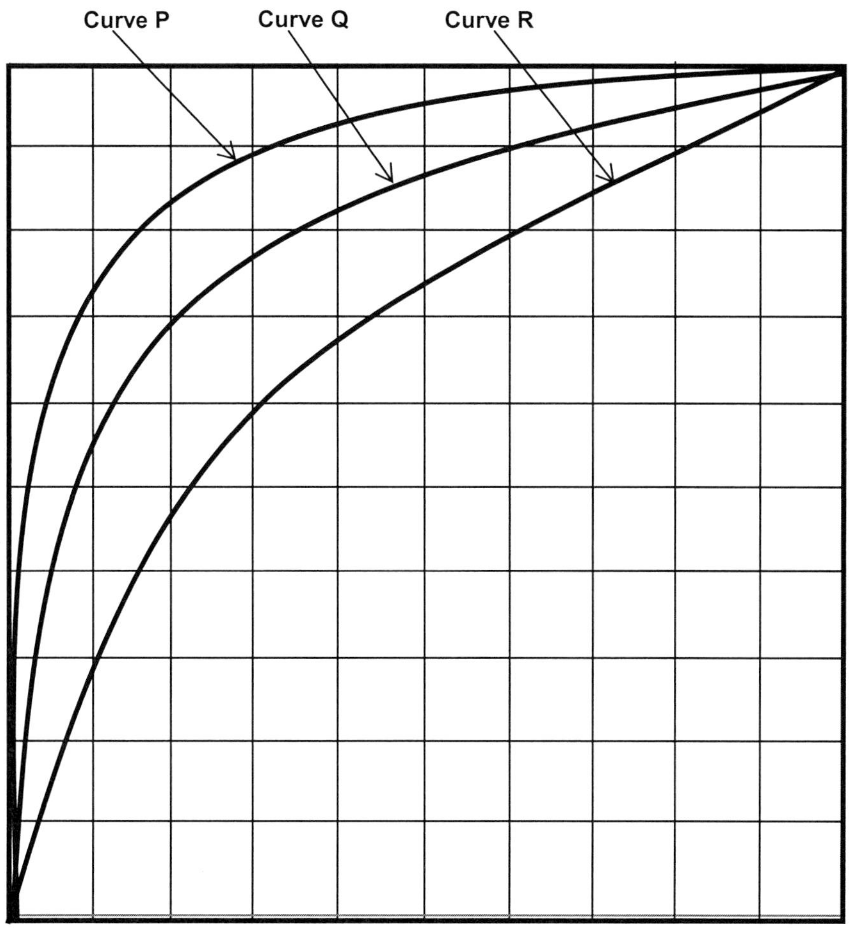


**Figure 15.1** ABC categories of production materials.

ABC analysis is most effective and applicable for all production items. This is based on the principle that in any production situation only 10% of the items forms 70% of the total annual consumption value, as illustrated by curve Q in the figure. These are categorized as 'A' items, for which the stock level planning and control shall be meticulous. The next 20% of the items form 20% of the total annual consumption value and are the 'B' items. The remaining 70% of the items form only 10% of the total annual consumption value and are called the 'C' items. Since the time and energy spent for meticulous control on any item is almost same, whether a high value or a low value item, any effort put in for 'A' items would yield 9 times higher results than putting the same effort on the 'C' items. That means we control 'A' items more meticulously and 'C' items more loosely. This is main principle behind ABC analysis.

While this is true in most cases of production materials in an engineering industry, there can be some variations. In industries like yarn spinning mills, there are only 3 or 4 items like specific qualities of cotton which have too high consumption value compared to all other items. For example only 5% of the items have an annual value of 70% of value, with the next 25% having 25% of the annual value while the balance 70% of items have an annual value of only 5% of the total annual consumption value. In this case, the curve P would be more bent, and ABC system of selective control will yield much more benefits.

On the other hand items of a typical maintenance stores would have 20% of items with an annual total consumption of 50%, the next 20% with 20% consumption value and the remaining 60% with 30% of value. This would be a less bent curve like R of the figure. In any case it may be noted that for maintenance items the ABC system is not as suitable as VED or other selective control systems as cited in chapter 16.

### 15.8.1 Procedure for ABC Analysis

It may be stated here that during the early 1970's when this author conducted ABC analysis at Larsen Toubro Limited that had over 35000 items. It was a 2 week job, especially due to the manual calculation of cumulative value and arranging them value-wise, in absence of a computer. Even the calculator available then was the mechanical one. But today with the computers the whole job may take hardly few seconds.

1. List out all the items as given on the code books.
2. Indicate the annual consumption quantity and the procurement price for each item.

3. Compute the annual consumption value in rupees for each of say 35000 items of an average industry.
4. Now arrange all the items in order of the annual consumption value starting with the highest. Today in excel format, only the first and second steps are to be fed manually, while the third to sixth steps below can be done by the computer in seconds.
5. Write down the cumulative consumption values for each item.
6. Thereby determine the total annual consumption value of 35000 items put together.
7. Draw a line under that item whose cumulative value is 70%.
8. Similarly draw another line under the item whose cumulative value is 90%.
9. All the items above the first line are 'A' items, those between the two lines are 'B' items and those under the second line are 'C' items.

Such computation is illustrated in Fig. 15.2.

| Sl. No | Code number | Item | Annual consumption | unit | Price in Rs./unit | Annual value in rupees | Cumulative annual value in rupees |
|---|---|---|---|---|---|---|---|
| 1 | xxx-xxx-xxxx | Aaaa1 | 25000 | pcs | 5.00 | 1,25,000 | 1,25,000 |
| 2 | xxx-xxx-xxxx | Aaaa2 | 60000 | pcs | 1.50 | 90,000 | 2,15,000 |
| 3 to 9 | ……… | ……. | ……. | ….. | …… | …… | 7,00,000 |
| 10 | xxx-xxx-xxxx | Aaa10 | 2500 | kg | 5.00 | 12,500 | 7,12,500 |
| 11 to 29 | ……… | ……. | ……. | ….. | …… | …… | 9,20,000 |
| 30 | xxx-xxx-xxxx | Aaa30 | 12000 | pcs | 0.20 | 2,400 | 9,22,400 |
| 31 to 99 | ……… | ……. | ……. | ….. | …… | …… | 10,65,000 |
| 100 | xxx-xxx-xxxx | Aa100 | 10 | Ltr | 5.00 | 50 | 10,65,050 |

**Figure 15.2** A typical tabulation for ABC Analysis.

## 15.9 Inventory Control Parameters

The third stage of scientific inventory management is the fixing up of inventory parameters.

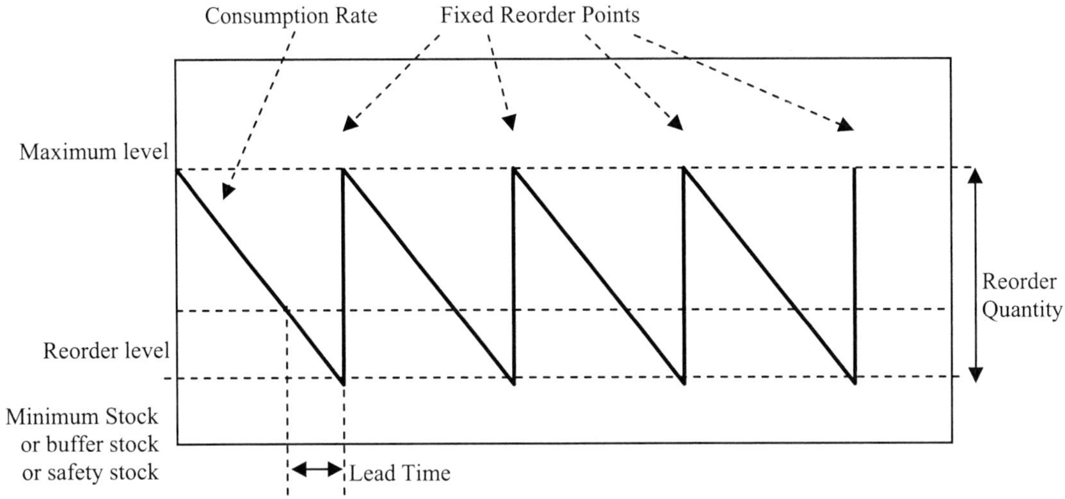


**Figure 15.3** A typical inventory model.

Fig. 15.3 illustrates a typical model for the inventory of maintenance spares, with the following parameters:

1. *The component* itself, each of the item having its own inventory parameters, code number and stores card.
2. *Consumption Rate (r),* the rate of consumption or the mean expected rate of consumption of the item given in its units per year, say kg per year or Lit. per year.
3. *Minimum stock level ($Q_{min}$),* is the minimum stock that must be available at any time in the stores. It is called as buffer stock since it acts as a buffer against mismatch between forecasted consumption and actual consumption or demand, between expected and actual delivery time, and for other unforeseen emergencies.
4. *Maximum Stock level ($Q_{max}$),* the stock immediately after the receipt of goods, which should be kept as low as possible, by effective practice of economic order quantity and other techniques.
5. *Lead Time (L),* the time lapse between indenting and the actual time it is received at the stores.
6. *Re-order Quantity (ROQ),* the quantity for which the order is placed depending upon the purchase and stocking policies.
7. *Re-order Level (ROL),* the stock level at which the reordering must be done, for all production and fast consumed maintenance spares like lubricants and common bearings.

8. *Reorder interval (T):* The usual period between the subsequent orders. This may be constant or vary depending upon the item itself and the system adapted.

The actual inventory ordering systems under uncertainties are explained in paragraph 15.13.

## 15.10 Inventory Carrying Costs or Costs Resulting from Owning the Item

1. Commercial costs like interest, insurance and taxes,
2. Storage costs due to cost of space, record keeping, clerical and overheads use of storage and handling equipment etc.,
3. Risk, costs like obsolescence, deterioration, damage, loss or theft, decrease in the rate of use, decrease in replacement cost (like the computers spares in recent times),
4. Opportunity costs like blocking of the capital from purchasing other essential items,
5. Procurement costs including issuing and closing orders, handling of negotiations etc., and other purchase overheads,
6. Order follow up, collection and transport.

## 15.11 Stock-out or Down Time Costs

On the other hand the following costs may be involved by not stocking an item:

1. Production down time costs which occur when an item is not available when needed and down time costs when a spare is not available when the machine actually breaks down, adding to the MWT and MTTR,
2. Cost of expediting delivery of the inbound spares including communication, traveling, expediting etc.,
3. Cost of panic purchases like surcharge for smaller lot quantities,
4. Expediting of emergency repairs by alternative methods and means which might generally be more expensive,
5. Cost of rescheduling the work,
6. Cost of additional materials and labour.

## 15.12 Economic Order Quantity

Let $D$ = Annual demand in units per year

$C_o$ = Ordering cost per order

$C_i$ = Inventory carrying cost in $ per piece

$C_p$ = Price per unit

$Q$ = Ordering quantity

$Q_o$ = Economic order quantity

$N$ = No. of orders per year

$T_c$ = Total cost per year

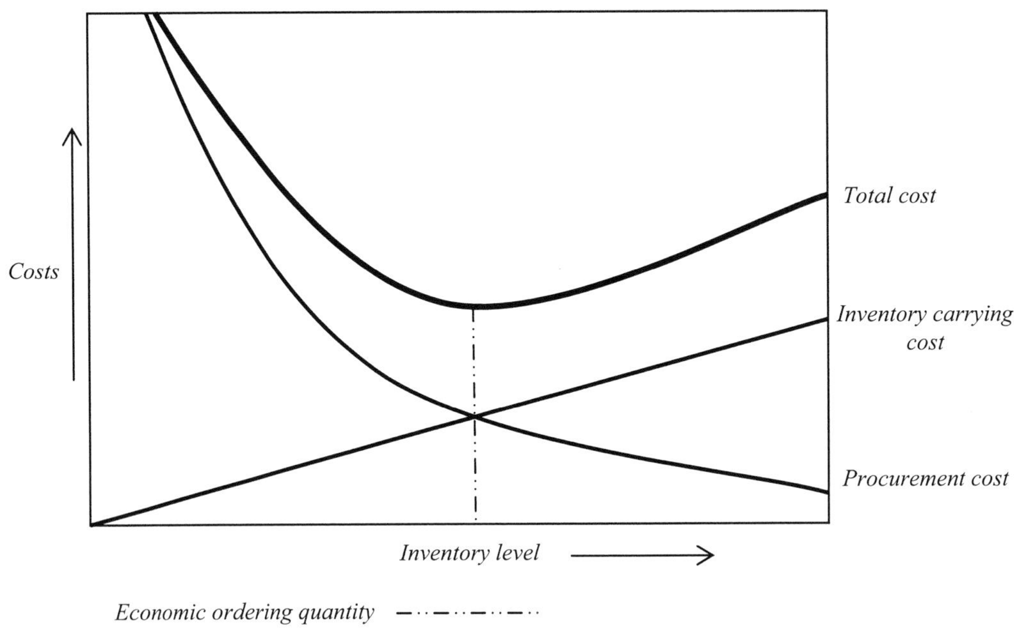


**Figure 15.4** An EOQ model.

Annual ordering cost = no. of orders per year × cost per order

$$\frac{\text{Annual Demand}}{\text{Order Quantity}} \times \text{Cost per order} = \frac{D}{Q} \times C_0$$

Annual inventory carrying cost = Average inventory x Inv. carrying cost per unit

$$\frac{(\text{maximum inventory} - \text{minimum inventory})}{2} \times \text{Inventory carrying cost per unit}$$

$$\text{ICC} = \frac{\text{Ordering Quantity}}{2} \times C_1 = \frac{QC_i}{2}$$

hence Annual total costs $= \dfrac{DC_o}{Q} + \dfrac{QC_i}{2}$

If this cost has to be minimum, then $\frac{dT_c}{dQ}$ must be zero

or $$\frac{d}{dQ}\left[\frac{DC_o}{Q}+\frac{Q_oC_i}{2}\right]=0$$

$$\frac{-DC_o}{Q_o^2}+\frac{C_i}{Q_o^2}=0$$

or $$\frac{Ci}{2}=\frac{DCo}{Q_o^2}$$

or $$Q_o^2=\frac{2DC_o}{C_i} \text{ or } Q_o=\frac{\sqrt{2DCo}}{C_i}$$

## 15.13 Inventory Models under Uncertainties

Due to uncertainties in lead time and rate of consumptions, the ordering schedules and quantities especially for A items have to be varied as per the models illustrated below

### 15.13.1 Fixed Order Quantity (Q system) by Varying the Re-order Level

Here the supplier would be given a supply/quantity schedule, but few days before the supply that is before the lead time the schedule for the next supply is re-computed based on the variations in the consumption rate or lead time and at the reorder date the supplier would be asked to supply the scheduled quantity after due lead time (Fig. 15.5).

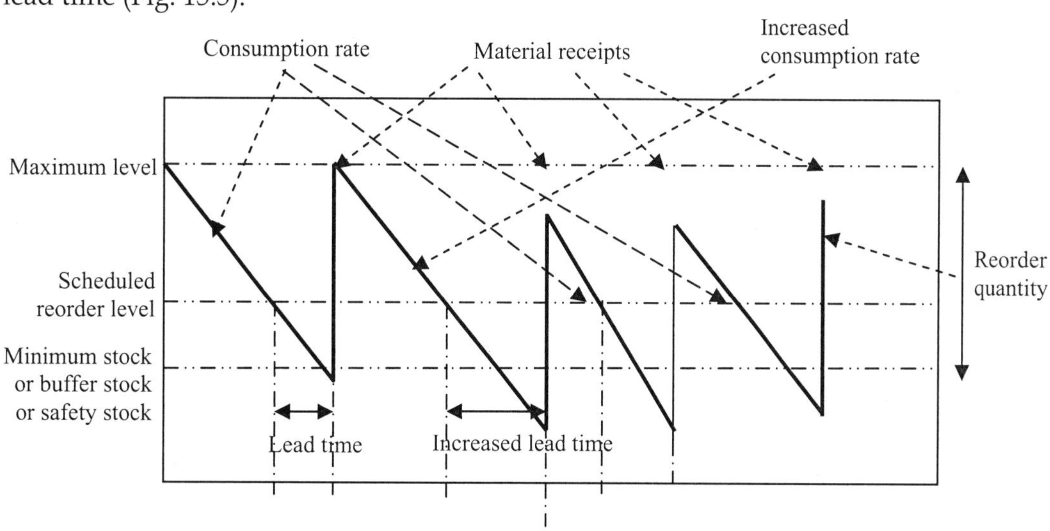


**Figure 15.5** Fixed order quantity system.

### 15.13.2 Fixed Period Ordering (P system) by Varying the Order Quantity

Here the supplier would be given a supply schedule at the time of placing the order, but few days before every supply that is before the lead time the quantity required to be supplied is re-computed so that the maximum level is not exceeded at the time of receipt and the supplier would be asked to supply only the required quantity. (Fig. 15.6).

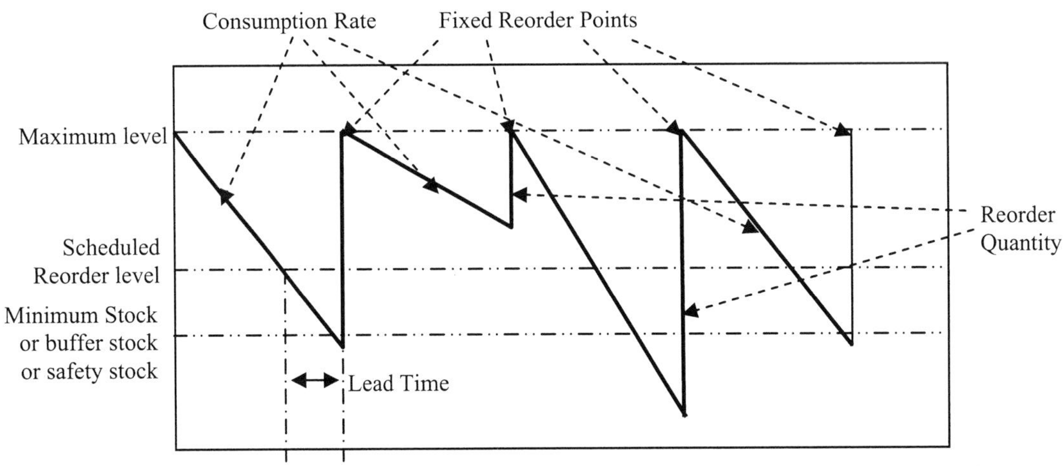


**Figure 15.6** Fixed period ordering system.

### 15.13.3 Fixed Maximum Stock Ordering ($I_{max}$ system) by varying Order Quantity and Re-order Level

This is similar to fixed period ordering, where the re-order quantity and the reorder level are varied depending upon the expected lead time and the rate of consumption, so that at the time of receipt the stocks are not more than the maximum stock level.

### 15.13.4 Fixed Safety Stock Ordering ($I_{min}$ system) by Varying Order Quantity and Re-order Level

This is similar to the above, where the re-order quantity and the reorder level are varied depending upon the expected lead time and the rate of consumption, so that at the time of receipt the stocks are not less than the safety stock level.

## 15.14 Two-bin Inventory Control System

This system is adapted for very low annual value items, even the lower among C category. These are generally some screws etc. which are consumed in fairly large quantities at regular intervals, but being of low individual prices say few paise each, the annual value is too low to justify elaborate system if maintaining individual stock register entering each and every issue. Even when the stock card is required to be maintained as a company policy only the monthly, quarterly or annual issues as computed from the warrants are entered in the card.

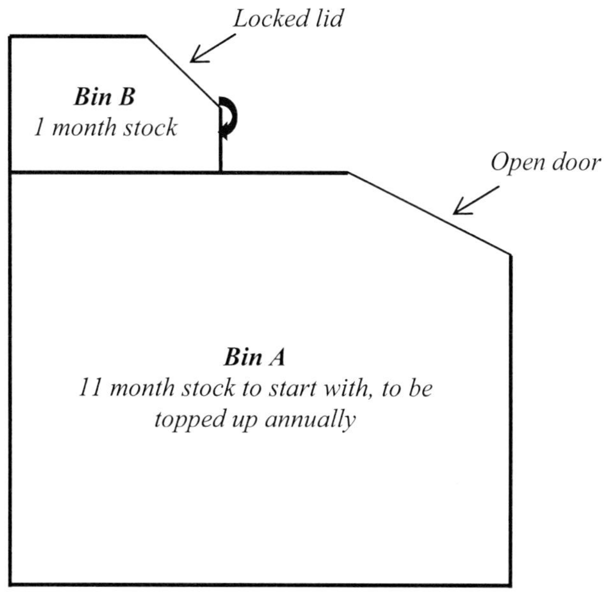


**Figure 15.7** Two bin system.

Two bins are maintained for each such item as shown in the Fig. 15.7. The smaller one containing one month stock is normally locked and is kept on the top of the larger one that contains the balance of the items is kept open. Items in required quantities are issued as and when required on material warrant, and entered into the bin card, without maintaining a separate stock register for that item.

When this larger bin becomes empty, the smaller bin is unlocked and a fresh order is placed. The new supplies arrive before the smaller bin becomes empty and the one month stock is topped up in the smaller one and locked up. The remaining items are kept in the larger to continue the above process. This is like the reserve petrol in the petrol tanks of motorcycles.

Advantages of two bin control:

1. The costly inventory control procedure of recording each and every issue and be on lookout for reorder level for each item is avoided.
2. The lock for the smaller bin will not be opened until the new indent is raised. So the chance of omission on the part of the stores-in-charge in watching for the reorder level is less.

3. The discrepancies between the physical stocks and card stocks do not generally create stock-out since the control is based on physical check.
4. Any unskilled helper can easily identify the empty bins and alert the officer. It is simpler to distinguish between *empty* and *few* more easily than *few* and *many*.

## 15.15 Recent Trends in Inventory Management

- During 1960's the norm for average production inventory (excluding finished products) in engineering industry was 25% of the total annual turnover.
- During the early 1970's this author undertook a comprehensive material control project that includes Codification, ABC Analysis and Inventory parameter fixing for each of the 35000 items as explained in the previous paragraphs. By this project, the total inventory level fell down to 18% of the turnover, which was said to be a big achievement.
- By the 1990's due to additional and more stringent controls this norm was lowered to 5%, especially in the TVS group companies of Chennai.
- The zero inventory drive started around the turn of the century and companies like Brakes India devised systems so that the receiving inspection is conducted at the vendor's place under the supervision of the buyer's staff, and on the specified date, the goods equal to one or two days requirement go directly to the shop floor for assembly, eliminating the inwards store inventories.

## 15.16 Suppler Partnership

The zero inventory drive principle of vendor place inspection is extended to the latest concept of suppler partnership which is defined as an extended relationship between buyers and sellers based on confidence, credibility, and mutual benefit. The buyer provides long-term contracts and assurance of only a small number of competing suppliers. In reciprocation, the seller implements customer's suggestions and commits to continuous improvement in quality of product and delivery. By this the seller assumes full responsibility for the quality of the product he supplies, that is, any defects found out at the assembly stage would be his liability and he has to finally pay a price. This motivates him to be absolutely quality conscious. This also provides confidence to the buyer in doing away with separate inward inspection. This concept has reduced the inventory norms to less than 1% of the turnover, tending to zero inventory.

## 15.17 Collaborative Planning, Forecasting and Replenishment (CPFR)

Collaborative Planning, Forecasting and Replenishment aims to enhance supply chain integration by supporting and assisting joint practices. Originated by Walmart in 1995. It is a form of cooperative inventory management by sharing data on the material required and material available to aid replenishment of products throughout the supply chain. In case of supermarkets like Walmart, information is shared between suppliers and retailers aids in planning and satisfying customer demand. This allows for continuous updating of inventory and upcoming requirements, making the end-to-end supply chain process more efficient.

## 15.18 Conclusion

For effective spare part inventory control, a basic knowledge of the scientific inventory control as applied to the production materials is essential as illustrated in this chapter.

# Appendix

## Codification Program for Stores Materials

1.0 A codification system identifies each of the unprocessed, semi-processed or fully processed materials that are used before, during or after the manufacture, whether the material goes into the product or used for the manufacture of the product. Broadly speaking the codification systems are

### *1.1 Sequence coding*

As the name indicates the system assigns the number serially to each item introduced irrespective of the category or classification. While this is practical for a few or non-recurring items like invoice numbering, this is inapplicable for stores materials.

### *1.2 Block coding*

This is a small modification of the above, assigning a block of numbers to each group of the items, so that all items of the same group can be numbered closer. This is good enough for well-defined basic categories where the necessity for the addition of new categories is limited. Nevertheless it is inefficient for the codification of production stores material due to the large number of varieties involved.

### *1.3 Mnemonic coding*

Both letters and numbers are used in this system with or without significance. For example, a spare part dealer may have an item say, B type Horn, size 55 mm of R make for Fiat Cars, which can be coded as F-02-B-55-R (Fiat-Horn-type-size-make). The registration number of Indian vehicles is a classic example for Mnemonic coding like TN 10 R 8841 indicating the state (TN), the RTO office code (10) and the serial number (R 8841), the prefixed letter meaning A 0001 will start after 9999 and B 0001 will start after A 9999.

### *1.4 Decimal coding*

In this system the signifying digits are assigned progressively and are used generally for library coding. In view of the possibility of unlimited number of digits, this method becomes complicated in a computerized inventory system.

### *1.5 All numeric or group numeric coding*

The classification of the materials is given high importance in this system. The digits are grouped as major or minor groups. Each digit or a group of digits is made to signify a parameter.

For example in a 10 digit code of 716 - 12- 2 – 0265

- 716 is the major group with 7 signifying Raw material and 16 signifying stainless steel
- 12 is the sub-group, signifying flat
- 2 is the subgroup signifying AISI 316, and
- 0265 is the minor group signifying 10 mm thick and 30 mm wide.

This system is perhaps the most suitable for the industrial material especially in view of the rapidly progressing data processing systems. This is hence discussed in more detail in the following paragraphs.

## 2. Basic categories of industrial materials

The all numeric system will be successful only if all material additions, existing or future, are properly classified. A well planned definition for the basic categories as well as the detailed groups is hence essential before taking up a codification program. In general, the basic categories are:

- Capital goods
- Machinery spares
- Raw materials
- General supplies,
- Components and products etc.,

## 3. Definitions of the basic categories

### 3.1 *Capital goods*

Generally same for all types of industries and include land, buildings, machinery etc. However in certain cases especially in engineering industry, small machines like portable grinders may be listed as general supplies, since they are purchased and distributed as consumables.

### 3.2 *Machinery Spares*

These include machinery spares bearings etc. It may be noted that bearings are 100% spares in a process industry, while in engineering industries like fan manufacturers they may go to the final product. This may necessitate the former classifying it as a spare, while the latter as general supplies since they are purchased and distributed like consumables. Again small companies that buy components and assemble into fans may call it as a component itself, as they may not have any other bearings to classify otherwise. However it is normally better to classify bearings as consumables or general supplies irrespective of the type of industry.

3.3 *Raw materials*

These include all items that go into the final product as primary items after due processing. Here again colors, paints etc., that go into the final product may be classified as general supplies in engineering industry where as some processing industries like the textile mills may classify them as raw materials for the sole reason that the total number of all other items under this category is small.

3.4 *General supplies*

They consist of the following

- Consumables that donot go into the final product but are consumed during the manufacturing process like the tools,
- Items that go into the product as secondary items for binding or beautifying the primary components like fasteners, colors and welding rods,
- Items that are used as spares for regular maintenance works, other than machine repairs like small motors, bearings, pipe fitting.

While these are self-explanatory, they also explain the variations discussed in previous paragraphs. In any case the pre-determination of the categories of items is necessary to adapt such variations that are clearly defined in the code books.

3.5 *Components and products*

Assemblies and sub-assemblies which are either finished products or go into the assemblies as final products. They may either be purchased or factory made.

## 4. Objectives of a codification program

4.1 To classify, list and code in an orderly and logical manner to

- Uniquely identify each item,
- Eliminate unnecessary varieties,
- Restrict addition of new varieties, and
- Facilitate greater utilization of existing materials.

4.2 To classify all materials by their property and composition, rather than the brand names, so as to achieve interchangeability, and reduce the confusion among the designers and purchasers because of superfluous detailed descriptions.

4.3 To develop a rationalized description and specification for each item so as to eliminate incomplete and ambiguous descriptions used at various stages of manufacture, and the consequent confusion in ordering and issuing, aggravated by improper communications and interpretations.

4.4 To assist the designers and other users in selecting the preferred sizes and varieties of items so as to enforce standardization effectively.

4.5 To facilitate data processing and inventory control systems and to achieve mechanized accounting and computer programming.

4.6 To ensure that when the item is sorted in their numerical sequence all the like items are grouped together. The primary consideration for such a grouping would be

- Basic classification of the item
- Purchasing and stores requirements
- Accounting and inventory control requirements.

4.7 In view of the continuous introduction of new items, the system shall be able to absorb them without upsetting the established uniformity and sequence of the numbers. In other words suitable openings shall be provided in all cases where future inclusions are possible. At the same time the number of the digits shall be limited to facilitate clerical work.

4.8 All the codes of a particular variety shall be in ascending order of sizes to facilitate easy identification.

4.9 The code number, once assigned for a particular item shall never be changed in future.

4.10 The system shall include all the items used in the specific concern before, during and after manufacture.

**5. Road map for developing the coding system**

5.1 A codification committee shall be formed consisting of the material control engineer and the concerned personnel from stores, designs etc.

5.2 A detailed study of all the materials to be included in the program as stated in Paras 2 & 3 shall be made.

5.3 Grouping and classification of all materials according to the Indian Standards (foreign standards where Indian standards are not available) shall be done. Equivalents of Indian/foreign standards as well as the different brand names shall be developed.

5.4 Basic structure shall be defined. The first digit shall indicate the basic category as per Para 3. The structures for the subsequent digits may be different for each category, but shall be clearly defined.

5.5 The codification committee shall meet periodically to discuss the development of codes at every stage. Suggestions may be pooled regarding the specific cases of improvements needed in the existing system or in the proposed system.

5.6 After finalizing, the code books may be prepared and issued to all concerned departments.

5.7 One master copy shall be retained at the material control department which should be the sole authority to assign the code numbers to the newly introduced items and get them recorded in the master books.

5.8 For all such additions made in the master book, revised sheets may be prepared periodically and distributed to the departments.

**6. Features of the system**

6.1 A well planned and executed codification project would achieve all the objectives stated in Para 4 which is exhaustive enough to enumerate all its advantages. A similar program conducted by the author at a large scale industry in Bombay can be illustrated highlighting the benefits.

6.2 The main feature was the rationalization of the varieties and consequent reduction in the number of material specifications. For example, the total number of specifications among raw materials could be reduced from 301 to 195, that is by 36%, made possible by proper integration of identical materials with different specifications. More specifically it was possible to reduce the number of specifications in

(a) Steels from 199 to 145, that is by 28%

(b) Steel plates from 25 to 12 i.e. by 52%

(c) Tool and alloy steels from 59 to 26 that is by 56%

(d) Copper cables from 11 to 6 that is by 45%

It may be noted that this variety reduction is in addition to further variety reduction achieved by a standardization program involving standardization of the sizes of specific varieties.

6.3 The existing system of classification of steels was based on the major alloying element, like mild steels, Mn alloy steels, Cr alloy steels etc. Due to overlapping of the alloying elements the system was creating difficulties in judging the chief alloying element. The IS system of classifying steels was hence followed, viz

I. Steels specified by tensile properties like mild and structural steels being designated as mild steels

II. Steels specified by chemical composition designated as carbon steels

III. Stainless steels

IV. Tool steels further classified as

  (a) Non-distorting steels

    (i) Cold work quality

      (i) High Carbon High Chromium ($HCC_r$)

      (ii) Low alloy

    (ii) Hot work quality

  (b) High speed steels

  (c) Carbon tool steels

V. Alloy steels.

  (a) Constructional alloy steels,

  (b) Case hardening steels

  (c) Shock resisting steels

  (d) Other qualities.

This system not only conforms to IS 1570 and ASP, but also to many international standards like AISI.

6.4 Many items were not moving for over 3 years since only brand names were known and not their properties. By this program it was possible to identify their characteristics and properties and utilize them by suitably classifying them as above. All obsolete or dead items and most non-moving items could be disposed off.

6.5 Chemical compositions, mechanical properties and the corresponding Indian and International standards could be indicated in the code books itself. Also the units like kg and meter for each item and size could be indicated where applicable. These had facilitated further planning and design.

6.6 All the non-standard varieties and sizes were bracketed out resulting in progressive elimination of non-standard items

6.7 All raw materials were assigned a single unit (kg) except in cases like tubes that were supplied and issued in meter lengths. By this the number of units prevailing and the errors due to this mix-up could be minimized.

6.8 It may be added here that a similar program conducted in another medium scale industry revealed that a single identical bearing was purchased under 8 different brand names, as illustrated in paragraph 5 of the next chapter. This was eliminated by introducing a New Item Indent Requisition (NIIR) form, wherein for every new item that is to be procured, the maintenance engineer for maintenance items and the design engineer for production items had to certify that this is new one and there is no such item in the existing stock.

# Chapter 16

# Spare Part Inventory Control

***Chapter synopsis:*** *After explaining the different categories and forms of spare parts required by the maintenance department, the several concepts of planning and controlling their stocks are discussed. A feature of this chapter is to make a distinction between the production materials and maintenance materials with respect to inventory control concepts. Special attention is given for the maintenance stores layout and the reclamation of spare parts.*

***Keywords:*** *Spare Part Inventory, simple replaceable items, complex replaceable items, permanent items, regular vs irregular consumption, S-Pilate lathe, Padmini lathe, Spare Part Analysis, Code Numbers, NIIR (New Item Indent Requisition) form, Selective Control, ABC or FSN or MNG or MND or, VED or SDE or HML or GOLF or SOS systems of selective control, layout of the maintenance stores, service spares, centralised vs decentralized stores, Reclamation of spare parts and seiri.*

## 16.1 Importance of Spare Part Inventory

One of the most vital and the more difficult problem in maintenance management is the stocking of spare parts. In spite of well established procedures, random failures do occur and any absence of the required spare item leads to production delays and raised blood pressures. At the same time stocking of each and every spare item without the real need would only increase the inventory costs against all the JIT and other management principles.

The situation is aggravated in developing countries due to the following factors:

- Shortage of foreign currency.
- High delivery time from suppliers.
- Delay in transport.
- Loss or damage at the stores.

A majority of the types of maintenance spares are the specialized machine components that need to be stocked in certain minimum quantities, say 1 or 2 or 10. Nevertheless, there are many maintenance items like bearings, gaskets and bulbs, which are needed to be stocked in larger quantities, not for one particular machine but are to be used at several locations and equipments and hence consumed at regular intervals. Inventory control of maintenance items is hence a complex process, needing differential approaches for the occasional spares and the regularly used maintenance items.

In many companies the cost of such maintenance materials is much more than the cost of maintenance labor itself and so there is a good potential for achieving savings by controlling the overall costs with a well established spare part inventory control system. Fig. 16.1 is an illustration of a check list wherein the answers to all the questions lie in the effective inventory control.

1. Are frequently needed items available or is there a panic buying?
2. Are all materials unambiguously identified and located in the stores so as to eliminate duplication and interpretation in stocking?
3. Are all the catalogues and lists up-to date?
4. Does the store contain obsolete materials?
5. Are all the materials booked out to an account numbers?
6. Are all materials inspected before being accepted into the stores?
7. Are the stores adjacent to the maintenance department?
8. Is storekeeping for spares items a full time job?
9. Does the maintenance manager decide stock, levels, the disposal action etc. of the items held in stock?

**Figure 16.1** Checklist for spares inventories.

## 16.2 Categories of Maintenance Materials

The major categories of maintenance materials basically fall into 3 categories as described below and as illustrated in Fig. 16.2.

- *Simple replaceable items* that are generally short life items. A detailed schedule of maintenance action, frequencies, and resources required are also generally recommended by the manufacturer. The actions are usually of adjustment or replacement on a fixed time or inspection basis.

- *Complex replaceable items* that are the items that cause concentration of the maintenance efforts and planning, because of the uncertainty involved in their failure characteristics and the high cost of replacement or repair. The procedure usually adapted for these items is condition monitoring, though the other forms of maintenance being discussed in this book also shall be adapted in the required proportion. The sophistication and periodicity of the condition monitoring would depend upon the failure characteristics and the consequences of failure.

- *Permanent items* that are not expected to fail during the lifetime of the equipment or machine, and hence no preventive maintenance action is planned. Nevertheless, in the rare event of such a failure, it should be recorded, analysed, and where necessary, appropriate maintenance procedure should be identified. The principle of Design Out Maintenance shall here be applied to eliminate such failures in future designs.

A spare part may be described as a part manufactured to be identical to one or more specific parts of a machine or a plant. Maintenance materials covers all kinds of materials used in maintaining machines or plant but are not specific to a particular machine and hence not classified, under spares. For example the spindle for a S-Pilate lathe is a spare part while a ball bearing is a maintenance material. A special headed bolt or a gear shift lever for Padmini lathe may be a spare item but the standard tool post bolts are maintenance materials. A reference may be made to para 3 of the appendix of chapter 15 which distinguishes between maintenance spares and production items.

*Characteristics of Spares:*

- Relatively expensive compared to normal stock.
- Specialized for use on one or limited number of equipment.
- May be difficult to get promptly from the suppliers.
- Not required frequently like production materials, yet a minimum stock has to be maintained.

*Characteristics of Maintenance Materials:*

There can be overlap between the production materials and maintenance materials like pipe fittings for a pressure vessel manufacturer.

*Regular vs irregular consumption:*

Spares can either be of regular consumption, like gaskets and piston rings. They can also be of irregular consumption, like ball bearings, brackets, pumps, valves and electrical controls.

Similarly maintenance materials can either be of regular consumption like screws, nuts, grease, oil, and gloves. They can be of irregular consumption like bearings, bulbs, heavy pipes and foundation bolts.

## 16.3 Maintenance Procedure for Complex Replaceable Items

In chapter 15 we have stated that the maintenance items need a different approach for the stock control as compared to the production materials. Table 16.2 summarizes these approaches and the guidelines for the selection of maintenance procedures for complex replaceable items while Fig. 16.3 illustrates the guidelines for the selection of maintenance procedures for critical units.

| ITEM CATEGORY | SIMPLE REPLACEABLE ITEMS | COMPLEX REPLACEABLE ITEMS | PERMANENT ITEMS | |
|---|---|---|---|---|
| | | | MAIN | SECONDARY |
| MAINTENANCE PROCEDURE | Adjust or replace by FTM/CBM | Repair or replace by CBM/FTM/OTF/DOM | Repair by CBM | No action |
| PREVENTIVE PROGRAM | Inspection, routine lubrication & routine periodic service | Condition based replacement/overhaul | Routine inspection | None |
| CORRECTIVE POLICY | Refer decision guidelines for corrective action | | | |
| INVENTORY POLICY | Stock mostly those items, where demand for some components is predictable or vital | Stock one or two of most need items (Use past experience since demand is difficult to predict) | No stocking except for insurance spares | |

**Figure 16.2** Basic categories of maintenance materials.

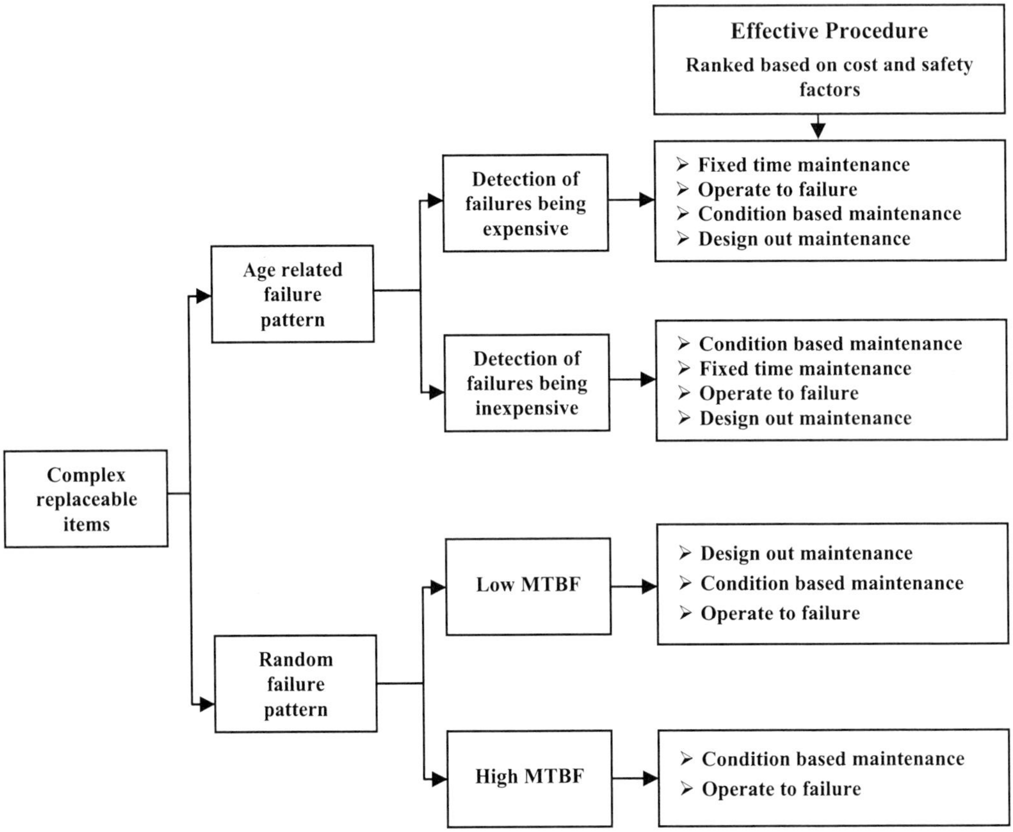


**Figure 16.3** Maintenance procedures for critical units.

## 16.4 Conditions Leading to Increased Maintenance Stores Inventory

1. Requirements for maintenance scheduling.
2. Economics of quantity purchasing to take advantage of quantity discounts etc. especially in case of electric bulbs etc.
3. Lack of part standardization, a situation diametrically opposite to the previous factor, yet would contribute stocking of large inventories, more in terms of varieties than in terms of quantity per variety.
4. Multiple storage depots like decentralized maintenance stores.
5. Lack of dependable suppliers.
6. Increased lead time.

7. Inadequate attention to inventory control.
8. Size, nature and condition of production facilities, like:
    - Large investment in plant and equipment, continuous production equipment like refineries, steel mills, chemical industries.
    - Age of the machinery with low MTBF.
    - Lack of planned spare sets supplied by the manufacturer of the equipment.

## 16.5 Spare Part Analysis

In view of the factors stated above, unwanted spares keep on getting stocked and it is essential to study and analyse them periodically to develop systems both to reduce the inventories as well as to provide guidelines of the future. First, a list of all the components for each machine shall be made. Next, for each complex component and for each separate part the following information shall be noted

1. Identification number for the machine or part
2. Part name and/or description, type and manufacturer
3. Delivery time from supplier including transporting and receiving time
4. Price inclusive of duty etc. i.e., the landed cost of the factory
5. The cost of the loss when the item is not stocked
6. The expected lifetime of the equipment
7. The total production per annum
8. The plant utilisation time in hours per annum
9. The internal rate of interest on investments.

## 16.6 Code Numbers for Spare Parts

Just like the production materials, it is essential that every item that enters the maintenance stores, especially the spare parts, is given a code number, specified by the codification program of the company as illustrated in the previous chapter. It might already have an identification number given by the manufacturer, yet most likely it may be a common item used in several other machines like gears or the bearings.

> *In a certain medium sized machine shop, a particular bearing was used in 6 different machines and each machine manufacturer had his own part number for the same bearing. Occasionally the same bearing also was purchased as a production item from the local market, either of NBC or SKF make whichever*

*was available at that time. As a result, the same bearing was purchased under 8 different part numbers, and stocked separately though all were identical, This was revealed during a codification program conducted in this company and the first remedy was to introduce a New Item Indent Requisition (NIIR), form where for every new item that is to be procured, the maintenance engineer for maintenance items and the design engineer for production items had to certify that this is a new one and there is no such item in the existing stock.*

Though the labour and cost involved in this initial allocation of a code number is high, it is more than outweighed by the following advantages, as further detailed in the Annexure:

1. Elimination of duplication, and reduction of inventory costs
2. Simplification of parts ordering, and identification
3. Easy location of the stocked item at the stores and the consequent reduction in the mean time to repair (MTTR)
4. Increase in accuracy of stores documentation
5. Simplification of machine accounting and data processing.

## 16.7 Inventory Control for Maintenance Items

Though the detailed explanation of the inventory control principles and practices with specific reference to production materials are discussed in chapter 15, what a maintenance engineer must know about basic terminology used in inventory of maintenance materials and the basic costs involved are once again listed and discussed below.

### 16.7.1 What is Inventory?

Inventory is the physical stock of items that a business or production organization keeps on hand for efficient running of affairs or its production.

### 16.7.2 Related Terms and Functions

1. Material control,
2. Material planning,
3. Material Resources Planning (MRP),
4. Stock control,
5. Store keeping,

6. Purchasing.,
7. Vendor development,
8. Make or buy decision etc.

### 16.7.3 Types of Inventories

1. Raw materials,
2. Consumables,
3. General supplies,
4. Bought out components,
5. Maintenance spares,
6. Work in progress,
7. Finished goods stores,
8. Miscellaneous inventory.

### 16.7.4 Need for Inventory

1. To gain economies in purchasing beyond current requirements, like when the prices are expedited to go up,
2. To protect against sudden price increase,
3. To take advantage of price fluctuations,
4. To insure against seasonal demand or availability,
5. To ensure against non-availability of an item in the market,
6. To protect from delays in receipts from the suppliers,
7. To conform certain suppliers' minimum quality requirements.

### 16.7.5 Disadvantages of Excessive Stocks

1. Blocking of working capital,
2. Occupying storage space,
3. Increased risk of obsolescence, damage, deterioration etc.,
4. Risk of theft, pilferage etc.
5. Increased maintenance and preservation costs,
6. Higher insurance premia,
7. Increased inventory carrying costs.

### 16.7.6 Inventory Carrying Costs or Costs Resulting from Owning the Item

1. Commercial costs like interest, insurance and taxes,

2. Storage costs due to cost of space, record keeping, clerical and overheads use of storage and handling equipment etc.,
3. Risk, costs like obsolescence, deterioration, damage, loss or theft, decrease in the rate of use, decrease in replacement cost (like the computers spares in recent times),
4. Opportunity costs like blocking of the capital from purchasing other essential items,
5. Procurement costs including issuing and closing orders, handling of negotiations etc., and other purchase overheads,
6. Order follow up, collection and transport.

### 16.7.8 Stock-out or Down Time Costs

On the other hand the following costs may be involved by not stocking a spare item:

1. Downtime costs which, occur when the spare is not available at a time when the machine actually breaks down, adding to the MWT and MTTR,
2. Cost of expediting delivery of the inbound spares including communication, traveling, expediting etc.,
3. Cost of panic purchases like surcharge for smaller lot quantities,
4. Expediting of emergency repairs by alternative methods and means which might generally be more expensive,
5. Cost of rescheduling the work,
6. Cost of additional materials and labour.

## 16.8 Inventory Control Systems for Production Materials vs Maintenance Components

The inventory parameters cited in paragraph 16.6.2 are suitable more for the production stores materials, than the maintenance stores materials for the following reasons.

- The total annual consumption value of production stores materials forms the bulk of the total cost of production, and comparatively the total annual consumption value of maintenance stores materials is far less.
- But the total number of varieties of items in the maintenance stores is several times more than that of production stores.
- As can be seen in paragraph 16.8, the selective control for the production materials is purely based on the total annual consumption value whether A, B or C item. But in the maintenance stores the importance to an item is not based on the annual consumption value, but due to its criticality as a spare, its availability etc.

- In production stores all non-moving items can be taken off the stores, but a maintenance item cannot not be disposed off, even though it might not have been required for use in more than a year. If a machine breaks down, the non-availability of any spare part, even a small one like a screw will cause a major hold up of the repair and you do not know when such a need for the small screw occurs. Thus the maintenance stores constitute much more varieties of low value items.

For this reason the ABC system of inventory control cannot be applied for maintenance stores. Nevertheless the chapter on scientific inventory control discusses all aspects of inventory control applicable to the production situation including a case study on ABC analysis. Some of the information relevant to maintenance is repeated here for illustration.

## 16.9 Selective Control in Inventory Management

*Why selective Control?*

- In any inventory situation some items are used regularly with high annual consumption value and some with very low annual value. On an average, in production situation, only 10% of the number of items constitutes 90% of the total annual value, though it is much different for maintenance items.
- The time and energy spent for meticulous control on any item is almost same, whether a high value or a low value item. This is due to the fact that the safety or buffer stock can be as low as 1 in case of maintenance spares, unlike the production items.
- Hence we should exercise stricter control for some items and looser control for a majority of items. This is called selective control.

Some of the selective control systems are detailed below

1. *ABC control*: Based on the total annual consumption value, the parts are classified into A, B or C items and separate control policies are developed for each, as described in the previous chapter.
2. *FSN control*: Fast moving, Slow moving or Non-moving based on the usage of an item in the past 1 or 2 years, helping us to decide on disposal action.
3. *MNG or MND control*: Moving, Non-moving or Ghost items (Dead items). This is almost a synonym for FSN control.
4. *VED control*: Vital, Essential or Desirable, based on their need for the present and future maintenance works. This is more suitable for maintenance items.
5. *SDE control:* Scarce, Difficult and Easily Available, based on their availability. This too is useful for maintenance items.

6. *HML control*: High cost, medium cost and low cost, based on the pricing of each item, useful for identifying items like the gold tipped or diamond tipped items which have to be stocked separately in lockers.
7. *GOLF control*: Government, Local or Foreign, based on the availability, so that we can plan the heavily rationed Government items like steel rods during the late nineties or the foreign items.
8. *SOS control*: Seasonal or Off Seasonal like mango pulp for food industries, to help procurement planning 6 months in advance.

It may be noted that these systems are cited already in chapter 15 as they form part of the scientific inventory control system. Nevertheless they are repeated here since barring ABC system, the rest are significant for spare part inventory control system. Whether it is a production situation or a maintenance situation, more than one of the above controls can be adapted and integrated suitably to suit the individual situation.

## 16.10 Layout of the Maintenance Stores

While the classification given in paragraph 16.6.3 is, in general, depending upon the use of the item, the maintenance items can also be grouped into three distinct groups for locating them in the maintenance stores so that can be located and retrieved easily.

### 16.10.1 Plant Spares, Sub-grouped into

- Mechanical items like gears, bolts, shafts, bearings and seals,
- Electrical items like electric cables, starters, relays, small motors and winding coils,
- Electronic items like PCB's and mother boards and
- Service spares (like hydraulic and pneumatic pipe fittings and valves). Reference may be made to paragraph 16.7 where a machine spare is distinguished from a general maintenance item. The former can be located in exclusive racks as distinct from general maintenance items even though they look similar, so that wrong item may not be picked up for maintenance jobs.

### 16.10.2 Tools, Sub-grouped into

- Hand tools like hammers, screw drivers etc. normally in the general tool kit, portable electric and pneumatic tools.
- Machine tools like twist drills, lathe tools, milling cutters and grinding wheels that are exclusive for the maintenance department, though similar duplicated tools used by the production department in their own stores.

However in a small industry it is common to find a single stores used by both maintenance and production operatives.

- Test instruments and gages, both electrical and, mechanical like the multimeters.
- Lifting tackle items, like slings, chains, jacks, ladders and gantry scaffoldings that can be located in convenient spots.

### 16.10.3 Consumable Items, Sub-grouped into

- Sheet metals used for general maintenance works. Since an average industry uses several types and quantities of raw materials for its production jobs, sheet metals required for maintenance in smaller quantities should be stored separately under the exclusive control of the maintenance department so that they can draw these as and when they need without the meticulous production material issues.
- General hardware like abrasive papers, screws, nuts and bolts.
- Pipe fittings and sanitary items needed mostly for building maintenance, including some needed for machine repairs.
- Lubricants and oils stored in accordance to the local fire regulations.
- Building materials like cement and sand, purchased and stored bulk, as distinct from the smaller sanitary items and pipe fittings.

## 16.11 Centralised vs Decentralized Stores for Maintenance Department

In chapter 19, we shall see that the organisation for maintenance can either be centralized or decentralized, each having its own merits and demerits. Similarly the location for maintenance stores too can be either of the below:

- Centrally controlled by central stores along with the production materials.
- Centralised, maintenance under the maintenance department having all spares and maintenance items as described under paragraph 16.9.1 at one location catering to the needs of maintenance supervisors of all individual shops.
- Decentralised, where all maintenance materials and spares required for a particular shop being stocked and consumed by the shop superintendent itself for his day to day maintenance works.
- A combination wherein routinely required **C** items like lubricants, grease, cotton waste, simple tools etc. along with, specific machine spares are being stocked by the shop superintendent-himself, while the major items like

bearings, costly spares, special tools, instruments etc. being stocked at central maintenance stores.

*Advantages of centralised maintenance stores:*

- Duplication of items avoided
- Inventory control is easier and more practical
- More efficient space utilization
- Fewer stores attendants and clerks
- Less chances for losses and pilferage
- Easier for cost accounting
- Simplified and mechanised, handling equipment.

*Advantages of decentralised maintenance stores:*

- Less walking and waiting time for maintenance personnel to collect tools etc.
- Close control by the craftsman/foremen
- Easier work, planning due to better knowledge of the available materials
- Chances of getting wrong materials is less
- Tendency for passing the buck is less.

## 16.12 Reclamation of Spare Parts

Even though one of the 5S's of Japanese principles is Seiri, that is to scrap unwanted items, it is always found to be profitable to reclaim dissembled spare parts if the maintenance department follows a systematic routine in reclaiming the spares before scrapping them. For example, a worn out axle may be reduced, in size and used in place of another smaller axle, when needed. While this is in vogue as an unofficial practice in several small and medium scale industries visited by the author, a systematic and established procedure is suggested to make this reclamation really worthwhile.

- Collect all the worn out parts.
- Select the parts that can be restored/reduced in size etc. and those that are beyond scope of reclamation and have to be scrapped.
- Sort them out in groups depending upon the reclamation operations.
- Establish the standards for these reclaimed parts.
- Perform the reclamation operation.
- Subject them to vigorous quality tests with respect to dimensional accuracy and mechanical properties.

- Return the parts to normal circulation.
- Keep a record of these parts.

In some companies the reclaimed parts, subject to quality satisfaction, are 'sold' to the stores at the original value, the profit being credited to the maintenances department. This may act as an incentive for the department to pay more attention to the possibilities of organising reclamation in a more systematic way.

## 16.13 Spare Part Planning at the Tendering Stage

On an average, the future requirement of the spares for any equipment requires budgeting of as high as 50% of the purchase price of the equipment, spread over its life span of say 10 to 15 years. A common problem faced now by many companies is that the equipment they purchase especially the electronic items like computer, becomes obsolete before the guarantee period itself and the spares no more becomes available. Hence it is essential that the spare part problem must be considered, fully before deciding on the equipment purchase with reference to the supply of necessary spares along with the equipment itself or to have confirmation from the manufacturer about the availability of the spares for the next 10 years.

## 16.14 Conclusion

As explained in these two chapters, though maintenance spare part inventories cannot be managed by standard inventory control methods and the conditions for applying the underlying models of chapter 15 cannot be applied, the basic aspects of inventory control like the inventory carrying costs, procurements difficulties are to applied in a similar, if not identical, manner for the spare parts. The concepts discussed in this chapter are proved to result in considerable savings to organizations, as indicated in the case studies.

# Chapter 17

# Total Productive Maintenance

*__Chapter synopsis:__ The revival of the 16th century practice of the machine operator himself doing the routine machine maintenance is one of the most significant steps in modern maintenance practices. The operator feels responsible not only for the output and quality of his output, but also for proper working of his machine. This also motivates him to take part wholeheartedly in the quality circle meetings and put forward his suggestions for the improvement of machine performance. This modern practice is hence aptly called Total Productive Maintenance. This chapter discusses the history of the development of this concept and highlights its various aspects in relation to the total production system.*

*__Keywords:__ Routine machine maintenance, TPM philosophy, autonomous maintenance, Nippondenso, pre-war industrial scenario, PEEP, Cleanliness in the office environment Seiichi Nakajima, Business Dictionary, equipment effectiveness, Fault-location, levels of TPM – repair, prevention and improvement, equipment effectiveness, maintenance prevention program, focused improvement, initial control, quality maintenance, office TPM, five zeros of TPM, five goals of TPM, Kobetsu Kaizen, Jishu Hozen, Hinshitsu Hozen, Safety, Hygiene & Environment (SHE, Japan Institute of Plant Engineers (JIPE), tero-technology and logistics, quality circles and maintenance work sampling.*

## 17.1 Introduction

In the earlier chapters we have seen that during the early days of industrial working scenario, the routine maintenance of machines was being carried out by the operator himself, while major repairs were carried out by the supervisor or an external specialist. The development of Taylors' principle of specialization resulted in the creation of a separate specially trained maintenance team to do all maintenance works including lubrication by patrol maintenance gang, together with all minor repairs like screw tightening. The industry did benefit a lot by such specialization in light of the centralized maintenance planning and control the benefits of which we had seen in the previous chapters. However it turned out that whenever small machine hold-ups need minor works like lubrication or screw tightening, the operator simply waits for the maintenance man to come and do that job, thereby losing production time. Hence the logical outcome is the development of the thought that why the operator himself can't be trained to do such minor adjustments. This thinking lead to the development of the TPM philosophy.

In other words TPM philosophy emphasizes auto-maintenance meaning every worker is responsible to the proper working of his machine apart from being responsible for the quality. The production operators share the preventive maintenance efforts, assist the mechanics with repairs when the equipment is down, and together they work on equipment and process improvements on a team activity basis. TPM aims to use all equipments at its maximum effectiveness by eliminating waste and loss incurred by failure of the equipment, increased set up time, reduced speeds, and process defects etc., which finally lead to reduced output.

The objective of TPM is to maintain the plant or equipment in good condition without interfering the daily process. To achieve this objective, preventive and predictive maintenance is required. By following the philosophy of TPM we can minimize the unexpected failures of the equipment.

## 17.2 The Meaning of TPM

**T** represents *Total* employee involvement, indicating the team spirit with well-coordinated work between the production and maintenance operatives. This term 'Total' also means total equipment effectiveness.

**P** represents *Productive,* indicating the production of goods and services that meet the customers' expectation by maintaining the reliability of the outputs, which would be possible if the reliability of the respective machines is kept at a high level.

**M** represents *Maintenance,* keeping the equipment and plant in as good as the original condition at all times

## 17.3 History of TPM

TPM is an evolving process, starting from a Japanese idea that can be traced back to 1951, when preventive maintenance was introduced by Deming in Japan. Nippondenso, an ancillary unit of Toyota, was the first company in Japan to introduce plant wide preventive maintenance in 1960. As a result of high level of automation, the equipment maintenance became complex needing more specialized maintenance personnel.

The main principle behind TPM, that is, the production operator taking care of routine maintenance of his machine can be traced back to the 1950's, and the subsequent development of present TPM philosophy can be traced to 1960's, when Toyota group of companies in Japan made the headway of introducing this practice.

The Nippondenso management led by Seiichi Nakajima, aptly called as the father of TPM, decided to pass on the routine maintenance jobs to the production operator, focusing the maintenance department's attention more towards the major maintenance jobs. This practice gave birth to the concept of TPM, further encouraged by the Japanese Institute of Preventive Maintenance which later became Japanese Institute of Plant Engineers (JIPE), whose member companies refined the TPM concept of operator doing routine maintenance, against the Western countries practice of patrol inspection, lubrication, nut tightening etc. by maintenance gangs.

## 17.4 Definitions of TPM

- TPM is a proactive approach that essentially aims to identify issues as soon as possible and plan to prevent any issues before occurrence. One motto is "zero error, zero work-related accident, and zero loss".
- *Seiichi Nakajima,* the originator of TPM concept had defined it as a process to continuously improve all operational conditions, within a production system, by stimulating the daily awareness.
- Business Dictionary defines TPM as a methodology designed to ensure that every machine in a production process always performs its required task and its output rate is never disrupted.
- TPM can be defined, as per QCFI, as a manufacturing philosophy that pursues production efficiency to its ultimate limits of comprehensive efficiency by
    - Putting together a practical shop floor system to prevent losses before they occur throughout the entire production system's life cycle,
    - Involving all functions like production, development, sales and management,
    - Having employee participation from top executives to front line workers,
    - Achieving zero losses through overlapping small groups

There are several other definitions of TPM and most of them imply the following:

1. TPM aims to maximize the equipment effectiveness
2. TPM establishes a thorough system of preventive maintenance for the equipment's entire life span.
3. TPM is implemented by various departments like operations, maintenance, engineering etc.
4. TPM involves every single employee from the top management to the shop floor workers.
5. TPM is based on the promotion of preventive maintenance through motivation.

## 17.5 TPM is an Extension of TQM

In manufacturing and service industries, improved quality of products and services depend very much on the features and conditions of the company's equipment and facilities. TPM is a maintenance process developed for improving productivity by making processes more reliable and less wasteful. This is where TPM plays a major supplementary role for TQM.

TPM focuses primarily on manufacturing and is the first methodology Toyota used to improve its global position during the 1950s. After integrating TPM with it, the focus was stretched, to include various other aspects of TQM like Supply Chain involving suppliers and customers. The next methodology was called lean manufacturing where TPM plays a critical role. If machine uptime is not predictable and if process capability is not sustained, the process must keep extra stocks to buffer against this uncertainty and flow through the process will be interrupted. This would deteriorate the product quality and also delay the delivery to the customer. In this context TPM contributes a lot to TQM and hence we can say TPM is an extension of TQM (Total Quality Management). Many of the aspects like employee empowerment, benchmarking and documentation are used in TQM to implement and optimize TPM.

While we have stated above that TPM is an extension of TQM, the following distinguishing features of TQM and TPM would enable us to understand the statement better.

1. TPM focuses on the reliability of the equipment whereas TQM focuses on the reliability and quality of the products and services, which however cannot be achieved without achieving TPM's focus on equipment.
2. TPM focuses on improvement of the performance of the equipment, that is the hardware, while TQM focuses on the improvement of systems and standards that is the software.
3. In TPM, personnel training centers on maintenance technology and skill specific to the maintenance, whereas in TQM centers on the personnel training in management technologies like SQC, kaizen, quality circles etc.
4. Like quality circles that are popular with TQM, TPM also advises weekly meetings among small groups of production and maintenance operators, which some companies prefer to call as 'productivity circles'.

## 17.6 Other Concepts of TPM

The other concepts in Japanese terminology for TPM are:

1. *Focused improvement* (Kobetsu Kaizen): Continuous improvement, even though in small steps.

2. *Planned Maintenance*: It focuses on increasing availability of equipments & reducing breakdown of machines.
3. *Initial Control:* To establish the system to launch the production of new product & new equipment in a minimum run up time.
4. *Education & Training:* Formation of autonomous workers who have skill & technique for autonomous maintenance.
5. *Autonomous Maintenance* (Jishu Hozen) – meaning "Maintaining one's equipment by oneself".
6. *Quality Maintenance* (Hinshitsu Hozen)-Quality Maintenance is establishment of machine conditions that will not allow the occurrence of defects & control of such conditions is required to sustain Zero Defect.
7. *5S:* The Japanese concept of 5S which has become a major factory in the industrial scenario for maintaining high quality production, has come to stay in maintenance operations to reduce the wasted elements and eliminate losses. The contemporary textbooks on Total Quality Management may be referred to where this aspect would be explained in more detail.
8. *Safety, Hygiene* and *Environment:* The main role of SHE (Safety, Hygiene and Environment) is to create safe and healthy work place where accidents do not occur, uncover and improve hazardous areas, and do activities that preserve environment.

Fig. 17.1 illustrates the pillars that support the structure of TPM, each pillar representing each of the above concepts.

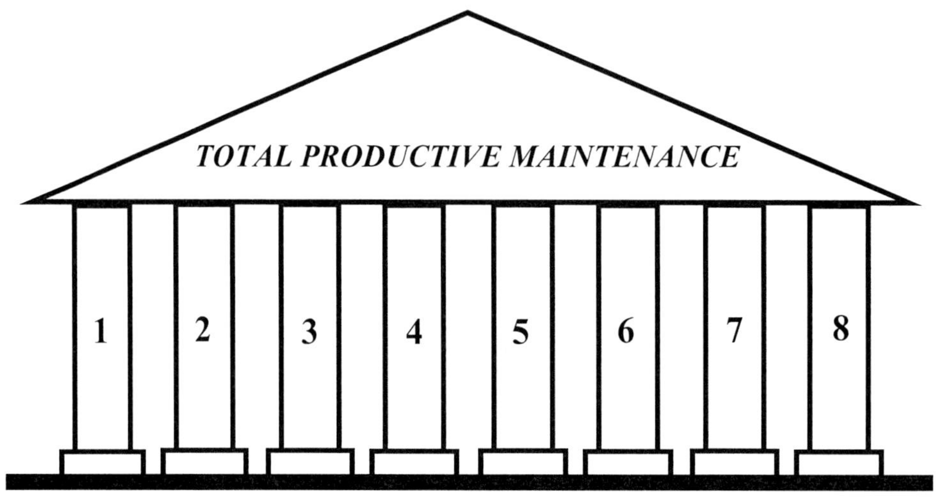


**Figure 17.1** TPM structure.

## 17.7 The Five Zeros of TPM

In any industrial situation the major effort of the management is to minimize the factors that increase production hold ups and cause losses to the profitability of the company. To create the awareness and commitment among all employees, companies successful in TQM practices, publicize the following through posters etc. so as to create a mindset among employees.

- Zero breakdowns
- Zero defects
- Zero accidents
- Zero pollution
- Zero inventory

Actually the last one, viz. zero inventory is applied in general to the production situation wherein the inventory plays a major cost-adding factor, more than for the maintenance function, whose inventory requirements are different and normally TPM books cite four zeroes. Nevertheless, since some of the syllabi specify 5 zero's for Maintenance Management, this concept is explained as above together with the last one. The significance of inventory control is dealt in detail in chapter 15 for production items and in chapter 16 for maintenance items.

## 17.8 What can TPM Achieve?

Wikipedia emphasizes that an accurate and practical implementation of TPM, will increase productivity within the total organization, where:

1. A clear business culture is developed to continuously improve the efficiency of the organization
2. A standardized approach for preventing known or unknown losses.
3. All employees and departments could be involved to form a transparent multidisciplinary organization to reach zero losses
4. Steps are taken on the whole road map and not a quick menu.
5. Finally TPM will provide practical and transparent ingredients to reach operational excellence.

Whereas in most production settings the operator is not viewed as a member of the maintenance team, in TPM the machine operator is trained to perform many of the day-to-day tasks of simple maintenance and fault-location. Teams are created that include a technical expert (often an engineer or maintenance technician) as well as operators. In this setting the operator is enabled to understand the machinery and identify potential problems, correcting them before they can impact production and by so doing, decrease downtime and reduce costs of production. He attends to the daily cleaning of the machine bed and slide ways, and lubricates the lubrication points including the work surfaces. Where an oil well is provided, he checks the oil level, start the machine and then checks the oil flow etc. before commencing the production work. Any defect is reported to the supervisor and ensure his immediate attention. To assist the operator, all lubrication points should be painted in distinctive colors. These maintenance jobs should be incorporated as suitable allowances in the standard output rates.

## 17.9 The Three Levels of Autonomous Maintenance in TPM

1. Routine lubrication and minor repairs like screw tightening or belt tightening, if necessary under instructions from the supervisor. This is called the *repair level of TPM.*
2. In case of any abnormal sound like vibration or bearing noise identify the root cause or notify the supervisor, as a part of the condition monitoring. This is called the *prevention level of TPM.*
3. Not only taking corrective action as above, but also take it up during quality circle meetings, departmental meetings etc. and discuss on improvements. This is called the *improvement level of TPM.*

## 17.10 The Five Goals of TPM

1. TPM's chief goal is to *improve system effectiveness.* It identifies and examines all losses that occur, whether it is down time losses or speed losses or defect losses.
2. TPM achieves *autonomous maintenance* by motivating the operators to take responsibility for routine maintenance tasks as explained in the paragraph 17.8.
3. TPM adopts *systematic approach* to all the maintenance activities. The level and nature of preventive maintenance for machine and equipment is identified and standards developed for condition monitoring. While operators are considered as owners of the machines taking their general care, the maintenance staff is considered as specialists providing supportive role for preventive and corrective maintenance activities.
4. TPM defines the *responsibilities* of the operators and maintenance staff and that each has all the needed skills to carry out their roles. TPM emphasizes on appropriate and continuous training and the maintenance department is given the responsibility of training the operators in routine and minor maintenance.
5. TPM strives to *attain the early design out maintenance* aspects for equipment. Its aim is to move towards zero maintenance through maintenance prevention

program (MPP). This involves considering and analyzing failure causes and maintainability of the equipment during every stage, whether during design, manufacture, installation or commissioning of the equipment.

## 17.11 Procedure for the Implementation of TPM

1. Study the existing equipment history, maintenance records etc. for all equipment and make a preliminary report of the need for TPM.
2. Obtain the consent of top management for the introduction of TPM in the organization.
3. Discuss with the concerned department heads, supervisors and the unions.
4. Set goals and norms for TPM parameters like equipment's effectiveness and availability standards.
5. Segregate the maintenance jobs into three classes :
   (a) Routine maintenance jobs that could be done by the operator:
      (i) Machine lubrication and oil levels monitoring.
      (ii) Cleaning and upkeep of the equipment.
      (iii) Adjustments, set up and minor repairs like screw tightening.
      (iv) On-line condition monitoring with the help of production supervisor.

   (b) Jobs essentially to be done by the maintenance staff:
      (i) Major breakdown jobs.
      (ii) Related works like work order, material indents.
      (iii) Plant shut downs for preventive maintenance.
      (iv) Major checking and off line condition monitoring.
      (v) Major shut downs and capital repair jobs.

   (c) Jobs that are to be done by the planning and technical service groups:
      (i) Overall planning of the preventive maintenance schedules.
      (ii) Simple design alteration in the equipment to suit maintenance.
      (iii) Preparation and issue of periodic checklists.
      (iv) Specialized inspection and condition monitoring.
      (v) Spare part inventory control, salvaging, etc.
      (vi) Subcontracting of maintenance jobs.
      (vii) Design of forms and documentation for the maintenance department.

6. Make necessary reorganization of operation, maintenance and planning groups to suit above segregation and earmark specific persons where necessary and possible.
7. Introduce autonomous maintenance concept for production group.
8. Seek the assistance and cooperation of the production personnel for major shutdown and breakdown jobs along with the maintenance staff. This would not only enthuse a sense of belonging to the production staff but would train them on various aspects of the machine structure and its maintenance.
9. Initiate maintenance prevention measures.
10. Evaluate the effect of TPM with respect to the set goals.
11. Ensure that the implemented system is maintained atleast for one year. Be present to answer and solve any queries from the concerned staff about the system.

## 17.12 TPM, Tero-technology and Logistics – A Comparison

Though all these three have common approach in terms of achieving lower life cycle costs, they differ in their fields of application and the location of responsibility. TPM is applied on the shop floor as a coordinated effort between the production and maintenance departments. Tero-techonlogy is holistically applied in the factory as a coordinated effort of all the departments, whereas logistics includes all other infrastructural activities like manufacture, transportation to markets, warehousing, systems and information flow.

## 17.13 5S vs TPM

One of the significant reasons attributed for the success of Japan in becoming a benchmark for the global TQM movement is 5 S, which originated in Japan. 5 S basically is a method for organizing the workplace in a shop floor or even in an office. It advocates *what* to keep, *where* to keep and *how* to keep (maintaining, cleaning etc.) the needed items at the workplace. A bad and dirty environment or work place distracts the attention of the worker making him produce more defects. This concept is today used not only to improve the work environment but also to improve the awareness, thinking and philosophy of working. Table 17.1 below explains each of the 's'.

While most books profess this 5S environment for the production departments, the author believes that this philosophy is more applicable for the maintenance activities, because an untidy workplace is more prevalent around the machines or equipment that are under maintenance than the machines that are working.

**Table 17.1** Meaning of the 5S's.

| Japanese word | In Japanese script | Equivalent English words generally used | Alternative English words that can be used | Meaning |
|---|---|---|---|---|
| ***Seiri*** | 整理 | Sort | Structurise | Distinguish between wanted and unwanted items. Remove unwanted items. Adapt red-tagging |
| ***Seiton*** | 整頓 | Straighten | Set in order or Systemize | Organize the workplace (***PEEP***- A place for everything and everything in its place). Keep items in their correct place for easy and immediate retrieval. |
| ***Seiso*** | 清掃 | Sweep or sanitize | Shine | Be committed to keep the work place clean and look for ways to keep the workplace clean |
| ***Seiketsu*** | 清潔 | standardize | Standardize | Maintain work place or storage as per the established standards. Exhibit all the procedures posters for everyone to see, understand and follow |
| ***Shitsuke*** | 躾 | Sustain the discipline | Self-discipline | Follow the rules of the company and honour established procedures. Do things in a way they are supposed to be done |

Of the 5S's explained above, the 2$^{nd}$ and 3$^{rd}$ are the most significant philosophies to be understood by the maintenance engineer and practiced to the core.

## 17.13.1 A Place for Everything and Everything in its Place (PEEP)

While spreading out the disassembled machine components around the equipment under repair, ensure that each component is kept at an identifiable location, so that they can be easily located even after one day.

## 17.13.2 Seiso (Shine, sweep or sanitize)

This is the very principle that led to the origin of Total Productive Maintenance. Infact it is a revival of the 16$^{th}$ century (before the industrial revolution) practice of the machine operator himself doing the routine machine maintenance, as discussed in paragraph 17.1. At the end of each shift, the work area is cleaned up and everything is restored to its place.

This should be strictly adhered to, even when the maintenance gang is responsible for the routine cleaning of the machine, which is still in practice in several medium and small industries. This makes it easy to know what goes where and have confidence that everything is where it should be, as professed by PEEP principle.

Few examples of dirty work places that need cleaning are:

1. Dirty machines
2. Dust on products, parts and raw materials
3. Dirty jigs, fixtures moulds and bobbins
4. Dusty walls, roofs, littered floor etc.
5. Untidiness outside the factory

### 17.13.3 Cleanliness in the Office Environment

The cartoon along side which appeared in the New York Times Magazine of 1964 illustrates how a clean work environment is importance even in the administrative offices. This also shows that even in those days before the 5S philosophy evolved, neatness in office or factory was being emphasized.

**Figure 17.2** Cleanliness in office environment.

## 17.14 Quality Circles vs TPM

Quality circles are formed by the workmen of a particular activity location. TPM teams are formed by senior managers, supervisors and workmen and carry out similar exercises on a company-wide basis.

## 17.15 Maintenance Work Sampling

Work sampling is a proven measurement technique of industrial engineering, based on a random sampling of the workforce to determine what types of activities they are performing over the course of the day. It identifies and quantifies maintenance workforce efficiency opportunities. World class maintenance operations can potentially perform the same amount of work with half as many workers. By identifying and eliminating barriers to productivity, the value-added contribution of existing maintenance resources can be significantly increased.

## 17.16 Conclusion

As indicated in paragraph 17.1, the concept of operator taking care of general maintenance of his machine has become more effective and popular than FW Taylor's system of job specialization. The Japanese concept of 5S as explained in paragraph 17.13 has given a fillip to the practice of total productive maintenance.

## Further Reference


1. Nakajima Seiichi (1989) – *Introduction to TPM*, Cambridge, Mass, Productivity Press.
2. Kiran D.R (2015) - *Total Quality Management, An Integrated Approach* = BS Publications
3. Hartman, Edward George (1992) –*Successfully Installing TPM in a Non-Japanese Plant: Total Productive Maintenance*, Cambridge, Mass, Productivity Press.
4. Nakajima Seiichi (1989) – *TPM Development Program & Implementing TPM*, Cambridge, Mass, Productivity Press.
5. Japanese Institute of Plant Maintenance – *TPM for Every Operator (Shop Floor Series)*-Cambridge, Mass, Productivity Press.
6. Leflar, James A (2001)- Practical TPM: *Successful Equipment Management at Agilent Technologies, Portland* - Productivity Press
7. Campbell, John Dixon, James V. Reyes-Picknell– *Uptime: Strategies for Excellence in Maintenance Management*, Cambridge, Mass, Productivity Press
8. Borris, Steven. *Total Productive Maintenance* (1st ed.). New York, New York: McGraw Hill
9. http://reliabilityweb.com/index.php/maintenance_tips

# Appendix

(Source: Total Quality Management by Samuel K Ho)

## Checklist for JIPE's Productive Maintenance Excellence Award

Japan Institute of Plant Engineers (JIPE) sponsored the *Productive Maintenance Excellence Award* recognizing outstanding achievements in TPM field for Japanese industries as indicated earlier. JIPE developed a checklist as indicated below, to decide on the award. This check list also gives the full insight into the salient aspects of TPM and makes us understand better the several factors that contribute for the success of TPM.

**1. Policy and objectives**

1.1 How is equipment management integrated into company policy?

1.2 Are equipment management policy and objectives set and prioritized correctly?

1.3 Are the management guidelines and evaluation criteria good?

1.4 Are long-term and annual plans integrated?

1.5 Are policy and objectives thoroughly understood and implemented?

1.6 Are accurate checks done to make sure objectives are being met?

1.7 Are the year's results considered in formulating goals, objectives, and plans for the next year?

**2. Organization and operation**

2.1 Are the organization and personnel assignments right for managing the equipment?

2.2 Is a good organization in place for promoting TPM?

2.3 Is the TPM promotion organization in close contact with production lines?

2.4 Are all the necessary departments participating fully in TPM?

2.5 Is effective co-ordination being done between the head office and the factories?

2.6 Are there any impediments to the exchange and use of information?

2.7 Are relations good with any subcontractors responsible for equipment, dies, tools, and maintenance work?

**3. Small-group activities and autonomous maintenance**

3.1 Is the organization of the small groups optimal?

3.2 How are small-group activity objectives set?

3.3 Do the groups meet frequently and are their meetings lively?

3.4 Are there lots of good suggestions, and are they handled properly?

3.5 How is goal attainment ascertained?

3.6 Do operators take the initiative in maintaining their equipment?

**4. Training**

4.1 Do the different departments understand TPM?

4.2 Are the training programmes broad enough and their curricula right?

4.3 Do the training programmes follow the curricula?

4.4 Do people take part in outside training programmes?

4.5 How many people have technical or expert certifications?

4.6 How knowledgeable and skilled are people in maintenance techniques?

4.7 How is skill assessment done?

4.8 Is there any way to make sure the training is having an effect?

**5. Equipment maintenance**

5.1 Are the 5-S being implemented?

- Is the machinery clean and free of dust, filings, oil, and other waste material?
- Have policies been instituted to deal with dirt, places that cause foiling, and places that are difficult to clean, inspect, and lubricate?
- Are lubrication labels, gauge limitations, bolt tighten-to-her marks, and other visible indicators used?
- Are all of the tools, materials, gauges, and other things stored neatly and kept clean?

5.2 Are equipment diagnosis techniques used in these cases?

- Cracking, corrosion, and loosening.
- Abnormal vibration, heating, and noise.
- Water, oil gas, and air leakage.

5.3 Are power lines, water lines, hydraulic lines, and other lines neatly and properly handled?

5.4 Are oils properly selected and properly replaced or filtered at the appropriate intervals?

**6. Planning and management**

6.1 Are the appropriate efforts being made to improve maintenance techniques and efficiently?

6.2 Are equipment standards properly set and enforced in a planned manner?

6.3 Are monthly and annual maintenance plans drawn up and implemented properly?

6.4 Are the purchasing plans for spare parts and other maintenance equipment properly drawn up (e.g., how much of what to buy from where) and are such things cared for?

6.5 Are equipment blueprints well cared for?

6.6 Are the dies, tools, and gauges properly cared for?

6.7 Are good records being kept on equipment wear and equipment failures that mandate stoppages or other maintenance efforts?

6.8 Are maintenance records used to improve processes?

6.9 Are the right maintenance techniques properly applied?

**7. Equipment investment plans and maintenance planning**

7.1 Is equipment investment co-ordinated with the development of new products and new processes?

7.2 Is equipment investment cost-effective?

7.3 How is the plant investment budget drawn up and controlled?

7.4 Are maintenance planning suggestions duly reflected in equipment investment standards?

7.5 Are reliability and maintenance duly considered in selecting, designing, and placing equipment?

7.6 Is equipment closely monitored in the start-up stages?

7.7 Is the company good at developing its own dies, tools, and equipment?

7.8 Are policies promptly instituted to keep major problems from recurring?

7.9 Are plant assets properly managed?

**8. Production volume, scheduling, quality, and cost**

8.1 Is equipment control closely integrated with production volume and scheduling?

8.2 Is equipment control closely integrated with quality control?

8.3 Are maintenance budgets drawn up and managed properly?

8.4 Is energy and other resource conservation practiced?

**9. Safety, sanitation, and environmental conservation**

9.1 Are sound policies in place for safety, sanitation, and environmental conservation?

9.2 Are the right organizations in place for safety, sanitation, and environmental conservation?

9.3 Are safety, sanitation, and environmental conservation methods known and practised?

9.4 Is equipment investment integrated with safety, sanitation, and environmental conservation considerations?

9.5 Are safety and sanitation polices paying off?

9.6 Do environmental policies meet all of the legal requirements?

**10. Results and assessments**

10.1 Are results properly measured?

10.2 Are policies being implemented and objectives met?

10.3 Is maintenance paying off in terms of enhanced productivity and other management aims?

10.4 Does the company make an effort to publicize its activities and its successes?

10.5 Does the company know where the problems are?

10.6 Have plans been drawn up for future progress?

# Chapter 18

# System Approach to Management Theory

*Chapter Synopsis: For any manager, irrespective of the function he heads, a comprehensive knowledge of the management principles and managerial skills are essential. This is very true in case of the Maintenance Managers. This chapter hence highlights the two aspects the manager has to deal with, in his day to day affairs, viz the systems approach and the decision theory.*

*Keywords: Management theory, systems approach, input-process-output, 5 M's of input, sustema, systemic, systematic, manufacturing system, business organization system, agricultural system, automobile workshop system, educational system, elements of control, environment, open and closed systems, subsystems, generations of subsystems, information flow, operating data, control data, planning data, series, parallel and integrated combination of subsystems, control systems with automatic and non-automatic correction, cube of management, planning pyramid, decision theory, problem analysis vs. decision making, decision tree, logical fault location, decision making under certainty, risk, uncertainty and under conflicts, systematic decision making,*

## 18.1 Development of System Approach to Management Theory

In the past, the departmental mangers tried to solve their problems by considering them as isolated situations, independent of other activities of the organization. For example, if a certain product manufacturer noticed a sales decline and traced it to the lack of aggressive sales effort, it was treated purely as a sales management problem and action was taken against that particular salesman or the department. The contribution of other departments to this effect like the quality control, design, credit policy, or the advertising, were never given much thought. As early as the fifties, a new integrated approach which considers the management in its totality has been developed and this is called the system approach.

## 18.2 What is a System?

We are surrounded by and live in several systems. In fact, man is a system by himself. Some systems are natural like the planetary systems, animal systems, and environmental systems. Others are man-made, like the business systems, production systems, material handling systems, social systems and total quality management systems. Whichever is the system it is basically characterized by three components which form the input-process-output system (Fig. 18.1), with,

- A set of two or more elements forming the *input,*
- Seeking a common goal operating on certain data or matter or energy, forming the *process,* and
- Yield matter or data in a time frame, forming the *output.*

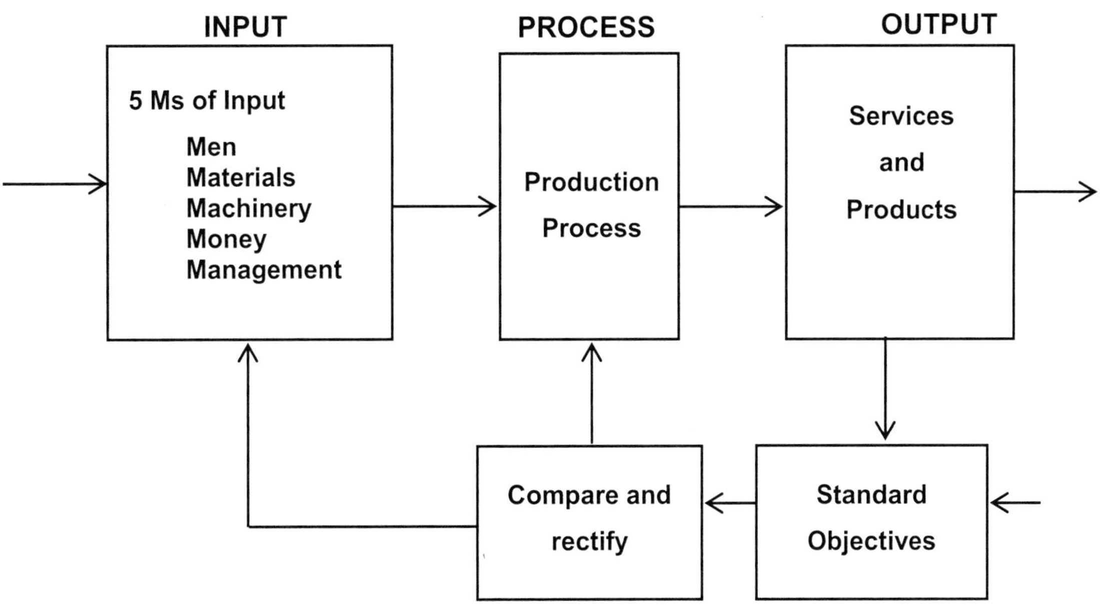


**Figure 18.1** Manufacturing system.

## 18.3 Definition of a System

Originating from the Greek word *sustema,* the word system has three definitions in Oxford dictionary as below,

1. Group of things or parts working together, like the digestive system, railway system or the control system or in computer field, a group of related hardware units or programs or both. Its adjective is *systemic,* meaning related to a system as a whole.

2. A set of ideas, theories or principles or the social order like a political system or a government system.
3. Orderliness or methodical or an organized manner of working. Its adjective is *systematic* and the word *systematize* means arranging things according to a system.

American Management Association in its AMA Management Handbook explains these three groups of definitions as follows:

1. The first group relates systems to living creatures, including the human beings, emphasizing on the impact of personality and behavioral sciences.
2. The second group emphasizes on the general meaning of the system as a set of two or more elements like people, things or concepts, which are related or joined together to achieve a common goal, operating on data, matter and energy, to yield a result with a time reference. For example, a system of Government, a system of measurement or a system of classification.
3. The third group of definitions connects systems with specific science techniques like philosophy, mathematics or biography.

Whatever may be the definition, a system approach is a purposeful and powerful means for accomplishing an objective. It provides the management with analytical framework with which it can identify, describe and inter relate the process and the components that make up a particular system as explained and illustrated in the following paragraphs. The study of any object or process must rely on the methods of analysis involving the simultaneous variations of mutually dependent variables.

## 18.4 Types of Systems

Some specific cases of systems can be explained below, illustrated by the Figures 18.2 to 18.6.

1. *Manufacturing system*: A group of *men* working on *machinery* and *materials* with the aid of *money* and a good *management*, (*5 M's of input*), working per work-methods, schedules and specifications (*process*) to yield the specified products and information by the date the customer wants them (*output*) (Fig. 18.1 as above).
2. *Management information system*: A group of people with a set of manuals and data processing and recording equipment (*input*), working for selecting, storing, processing and retrieving data (*process*) to yield the information needed by the managers (*output*) (Fig. 18.2).

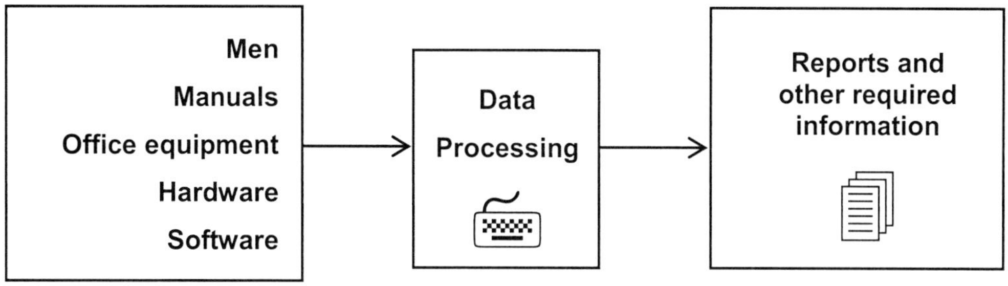


**Figure 18.2** Management information system.

3. *Business organization system*: A group of people with a vision and mission *(input)*, by performing production, design, finance, marketing etc. functions (*process*) towards achieving the goal of optimal profit for the business (*output*) (Fig. 18.3).

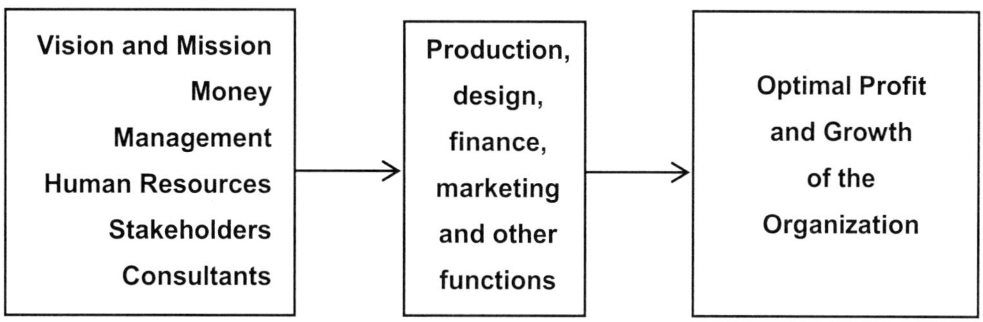


**Figure 18.3** Business organization system.

4. *Agricultural system*: A group of people with a set of implements, fertilizers and land *(input*), performing the farming activity (*process*) to yield the agricultural produce (*output*) (Fig. 18.4).

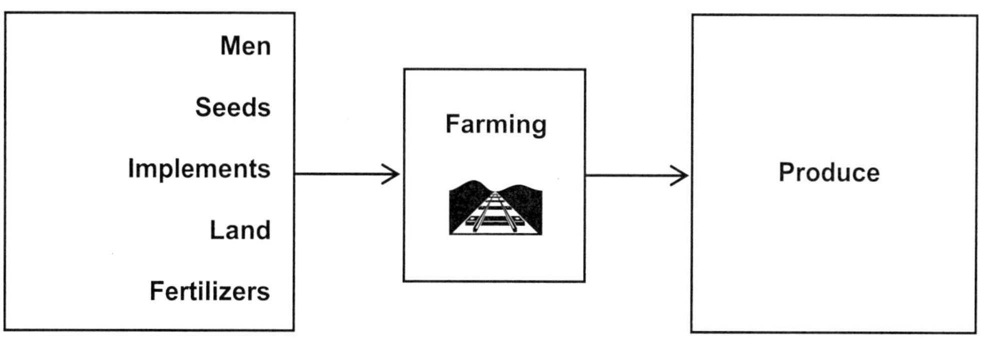


**Figure 18.4** Agriculture system.

5. *Automobile workshop system*: A group of people with spares, materials, tools, equipment and management *(input)*, maintaining and repairing the automobiles (*process*) to provide the required services to the customer *(output)* (Fig. 18.5).

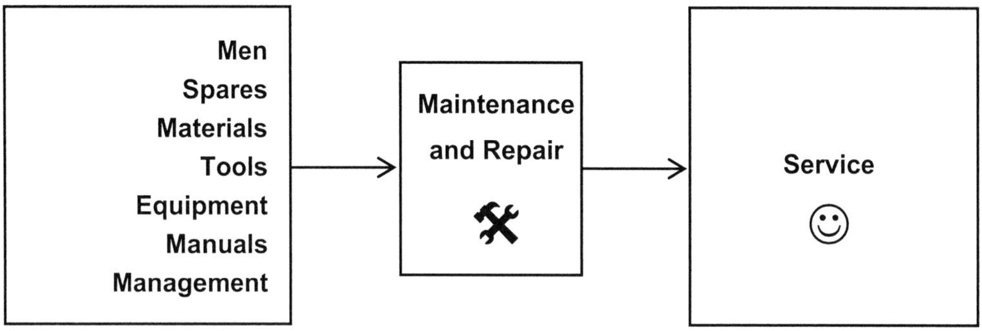


**Figure 18.5** Automobile garage system.

6. *Educational system:* Faculty aided by college infrastructure, textbooks, visual aids etc. (*input*), teaching and training to the students (*process*) to provide enlightenment (*output*) (Fig. 18.6).

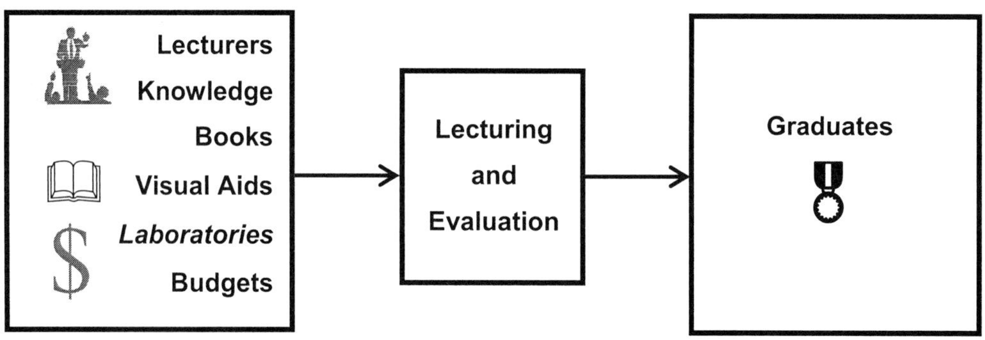


**Figure 18.6** An educational system.

## 18.5 Components of a System

Systems are generally represented in flow charts and block diagrams. The major components of a system are input, process and output. While Figure 18.1 indicates the process in its simple form, Figure 18.7 below illustrates all the other controlling factors.

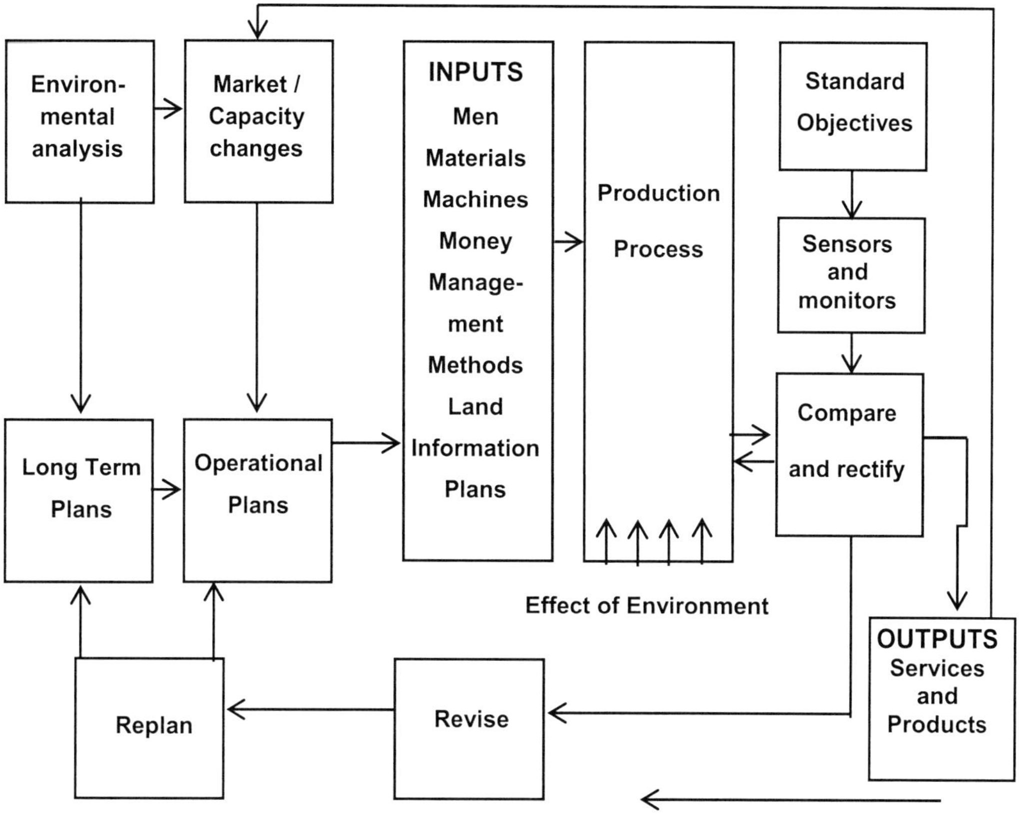


**Figure 18.7** The complete production system.

## 18.5.1 Input

Input to a system can be classified basically into 3 categories. They first category comprise of the physical inputs including the materials that flow through the conversion process where work is performed on them. The second is the non-physical materials but required for the performance of the process, like the management, money, capital, energy, labor and land. The third category is the environment that effects the system's operation. The design of the amount, placement, timing and types of these inputs will have an impact on the conversion facility whether it is a factory or a hospital or an office.

## 18.5.2 Conversion Process

The second basic part of the system is the conversion process through which the inputs flow to produce the desired outputs. To be effective and efficient, systems must be designed so that the correct process acts on the inputs at the proper time.

### 18.5.3 Output

The third major component of the system after conversion process is the output, which comprises of the desired accomplishment of the system. In an automobile industry, the output is the number of completed cars of a desired quality produced within a specified time frame. This input - process - output elements would exist whatever may be the situation, whether agriculture or educational etc., as explained in paragraph 18.4.

## 18.6 Elements of Control in System Approach

Just as goals or objectives are needed in the design of a system, control is required in its operation. The purpose of the control is to maintain the quality and quantity of the output so that it meets the goals of the system. In the ideal case, the control would be self-regulating and holds the output at desired levels without outside intervention. The control has the following elements.

1. A well-defined objective
2. An output standard
3. A measuring device
4. A feedback signal and
5. A corrective course of action.

Unfortunately in most management situations not all of the elements of control are present or not precisely defined. For example, consider a production process in which the standard of the units of output per hour has been established for a product. However when the output from the process is monitored, it is found that only 80% of the units are produced per hour. In this situation, there is an objective, an output standard, a measuring device and a feedback signal, but no corrective action has been specified.

## 18.7 Environment

A system's environment is composed of all activities outside the system which, if changed, will affect the system or which will be changed by any changes in the system itself. In this context the general level of economic activity can be considered to have environmental impact for most production activities since any change in it will affect the production levels.

## 18.8 Open and Closed Systems

Every system has a boundary which functions to maintain proper relationships between the system and the environment. A closed system has no environment around the system. No outside system have any effect on this system and the inputs and outputs have definite mathematical relationships like a chemical process inside a hermitically sealed container (Fig. 18.8a).

The open system has an environment with which it relates, exchanges and communicates, like a human system or a production system (Fig. 18.8b).

**(a)** A closed system

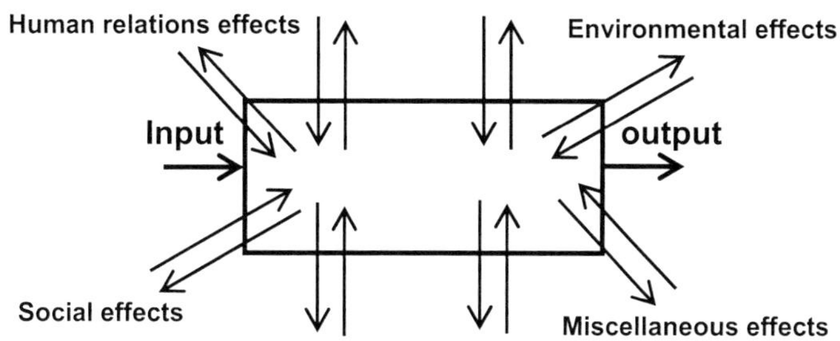


**(b)** An open system

**Figure 18.8** Closed and open systems.

## 18.9 Systems and Subsystems

Because of the complexity of most real systems, it is extremely difficult to work with or to understand the entire system. In order to overcome this problem, systems are broken into smaller subsystems and still smaller sub-subsystems to comprehend them better. For example, the whole universe is a system, while the solar system is its subsystem. We may still break it down to the earth system, world system, a group of nations, a country system, a social system, man as a system, blood circulation system etc., as illustrated in Figure 18.9.

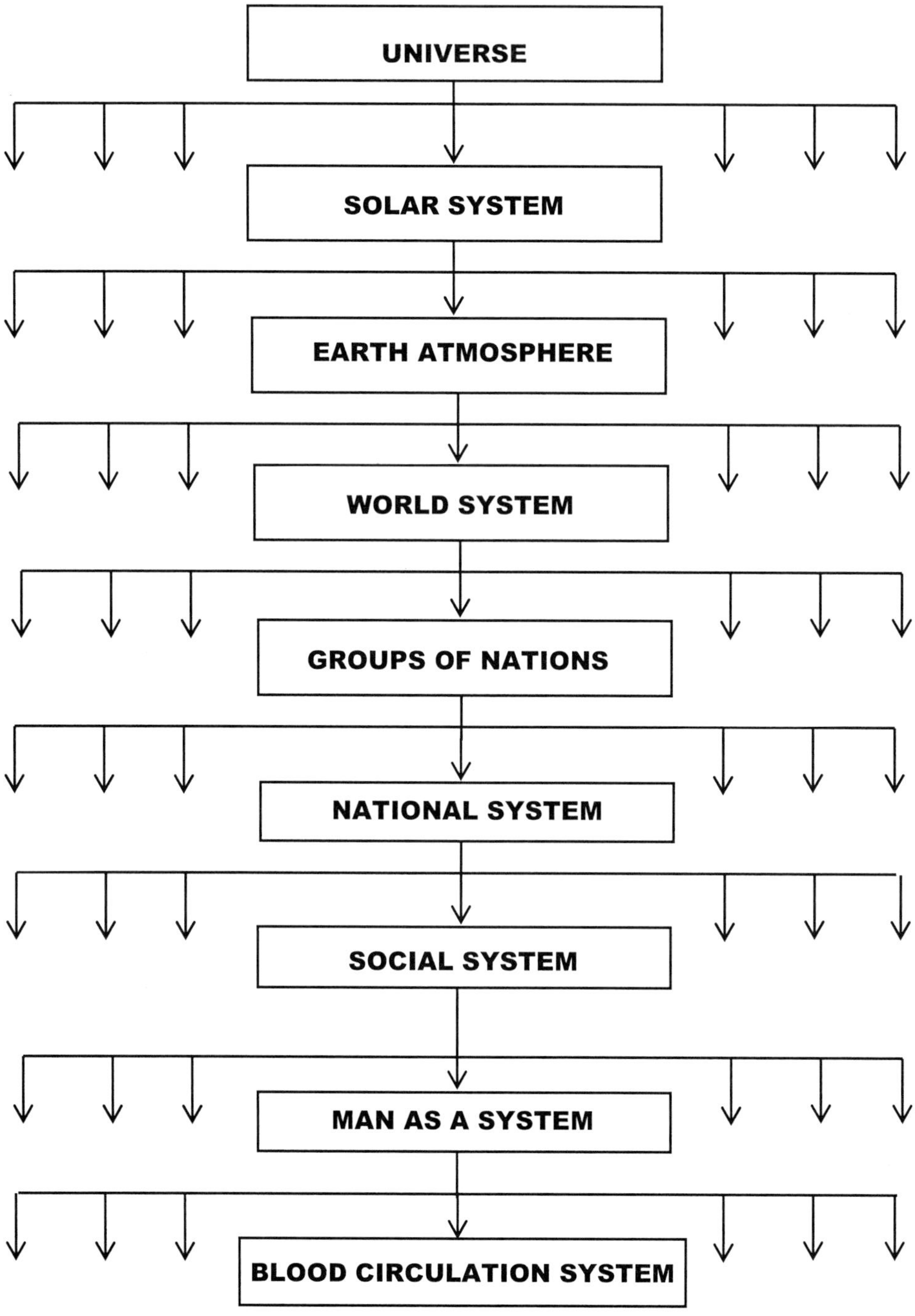


**Figure 18.9** Division of systems into subsystems.

## 18.10 Relationship between the Systems and Subsystems

Each system and its subsystems are mutually related, some more and some less, some directly and some indirectly. This relationship is in the context of the whole and is complex. Any change in one part would affect the other to a varying but a predictable degree. Figure 18.10 illustrates the relationship between the systems and Figure 18.11, the different generations of subsystems.

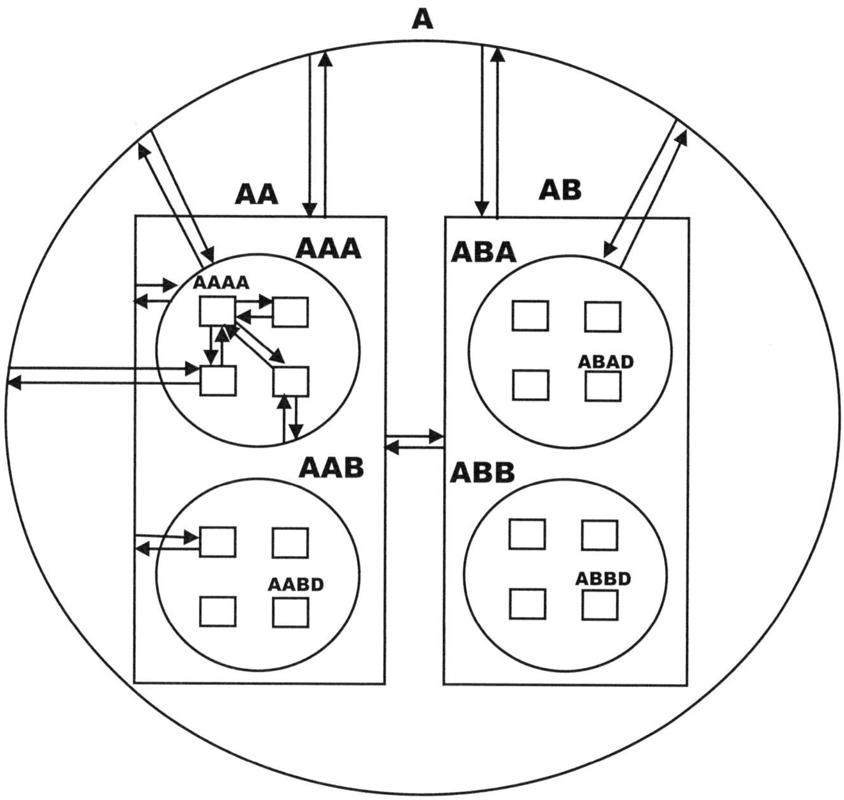


**Figure 18.10** Inter relationships between systems and subsystems.

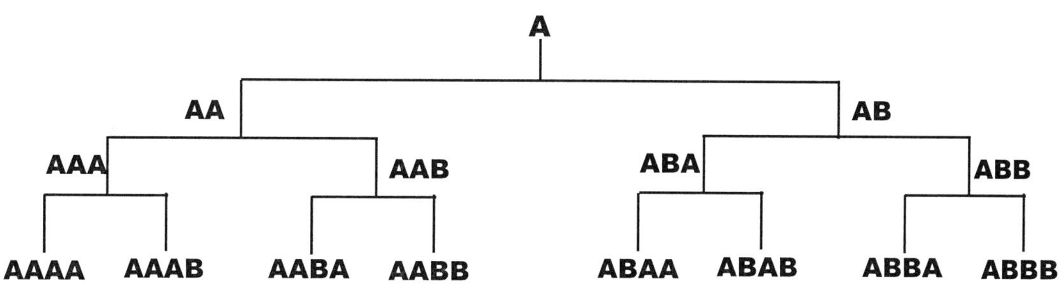


**Figure 18.11** Generations of subsystems.

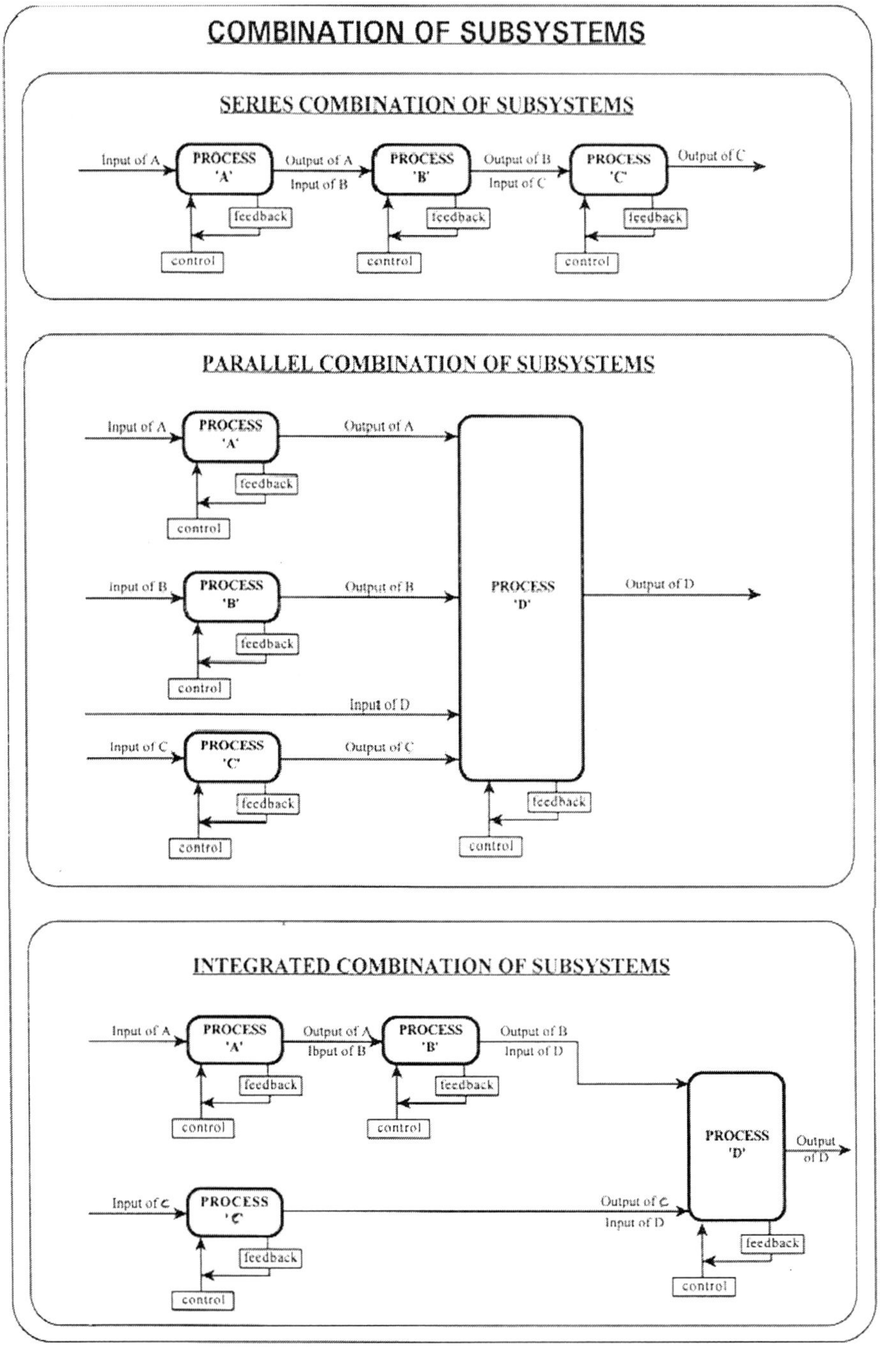


**Figure 18.12** Combination of subsystems.

## 18.11 Combination of Subsystems

Two or more subsystems can be integrated in either series, or in parallel or in combination as illustrated in Figure 18.12.

## 18.12 The Management Cube

The management activity is influenced by three aspects.

1. The management processes including the planning, controlling and execution,
2. The management functions like design, planning, maintenance, marketing, purchasing, personnel and finance,
3. The management level, like operatives, supervisors, engineers, managers, directors, president and chairman.

These three activities can be illustrated by Figure 18.13, which is called the Management Cube. It shows how the managers of different functions perform the processes at different levels.

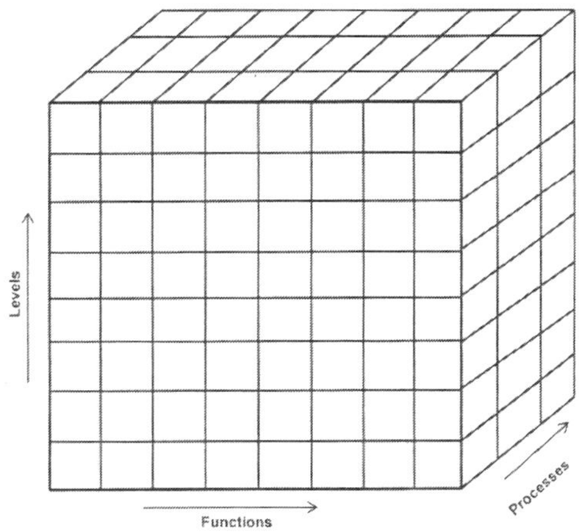


**Figure 18.13** Management cube.

## 18.13 Planning Pyramid

The management cube considers same number of managers at each level. But infact the number of persons in each level becomes less at higher levels because of the span of control. For example, the planning of the vision and mission or the policy making is done at the Chairman level, while the day-to-day operational planning is done at the Manager level. As one ascends higher in the hierarchy, the level of planning is escalated and simultaneously the number of individuals responsible narrows down.

Hence these activities are better represented by a pyramid signifying this aspect and showing what type of planning is done at each level in the organization, as illustrated in Fig. 18.14.

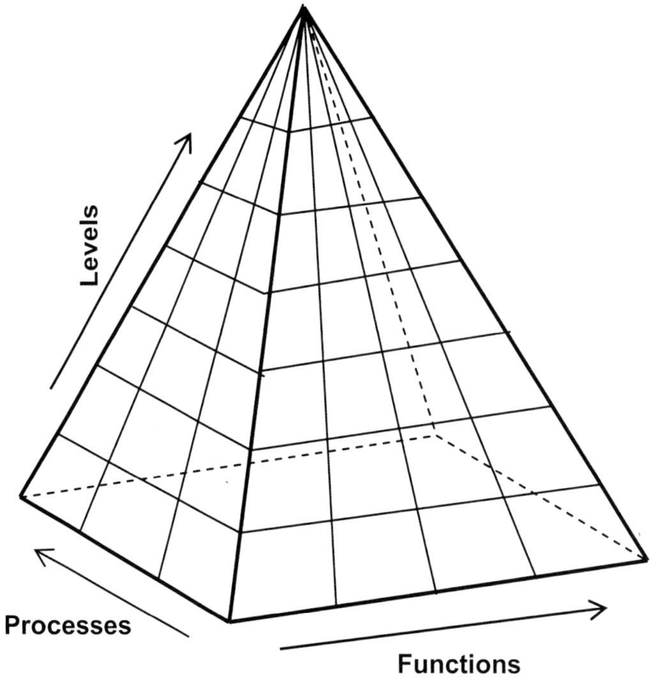


**Figure 18.14** Planning pyramid.

## 18.14 Decision Theory

The day-to-day administration of the affairs of any business enterprise requires an endless sequence of decisions. They may either be routine or unusual, demanding the manager's closest attention and intelligence.

In the past, when it was mostly a seller's market, profit maximization was perhaps the sole motto of an enterprise, and accordingly all decision makings were simple. But today the manager is influenced by a hoard of factors.

1. Increased regulation of business activity by the local or National governments.
2. Growing separation of ownership and management in industry, that is the management becoming more and more professional.
3. Steadily increasing importance given to high level education for managers.
4. Development of new processes and equipments.

5. Expansion of market areas in view of improved communication and transport logistics.
6. Increased competition from other suppliers.
7. Growth of labor unions and more stringent labor laws.
8. Growing social pressure.
9. Importance given to customer satisfaction in all production related functions.

The following paragraphs detail the decision theory with regards to identifying its rationality, uncertainties and other issues relevant in a given decision, as well as the resulting optimal decision.

## 18.15 Problem Analysis vs Decision Making

*Wikipedia distinguishes problem analysis from decision making as follows. Traditionally problem analysis must be done first, so that the information gathered in that process may be used towards decision making.*

*Problem analysis*

- Analyze the performance with reference to the expected results against actual results.
- Problems are merely deviations from performance standards.
- Problem must be precisely identified and described.
- Problems are caused by a change from a distinctive feature.
- Problem analysis can be used to distinguish between what has and hasn't been affected by a cause.
- Causes to problems can be deducted from relevant changes found in analyzing the problem.
- Most likely the cause to a problem is the one that exactly explains all the facts.

*Decision making*

- Objectives must first be established.
- Objectives must be classified and placed in order of importance.
- Alternative actions must be developed.
- The alternative must be evaluated against all the objectives.
- The alternative that can achieve all the objectives is the tentative decision.
- The tentative decision is evaluated for more possible consequences.
- The decisive actions are taken, and additional actions are taken to prevent any adverse consequences from becoming problems and starting both systems (problem analysis and decision making) all over again.

- There are steps that are generally followed that result in a decision model that can be used to determine an optimal production plan
- In a situation featuring conflict, role-playing is helpful for predicting decisions to be made by involved parties.

## 18.16 Decision Tree

Decision Tree is a pictorial representation of a decision situation, normally found in discussions of decision-making under uncertainty or risk. It shows decision alternatives, states of nature, probabilities attached to the state of nature, and conditional benefits and losses.

Investopedia, Financial Dictionary defines decision tree as a management tool to clarify and find an answer to a complex problem. The structure allows users to take a problem with multiple possible solutions and display it in a simple, easy-to-understand format that shows the relationship between different events or decisions. The furthest branches on the tree represent possible end results. The tree approach is most useful in a sequential decision situation and can be illustrated by the logical testing procedure cited in chapter 7.

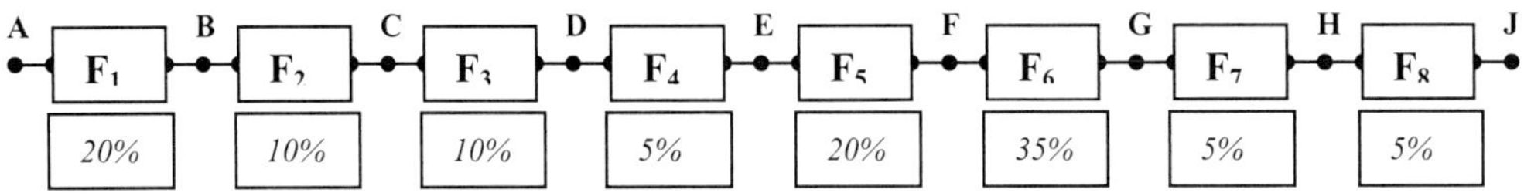


**Figure 18.15** A simple series circuit with known failure probabilities.

If a process consisting of a group of units connected in series as per Figure 18.15 fails, then individual units have to be checked for failure since the exact unit which has failed is not known, the decision tree illustrated in Figure 18.16 would help.

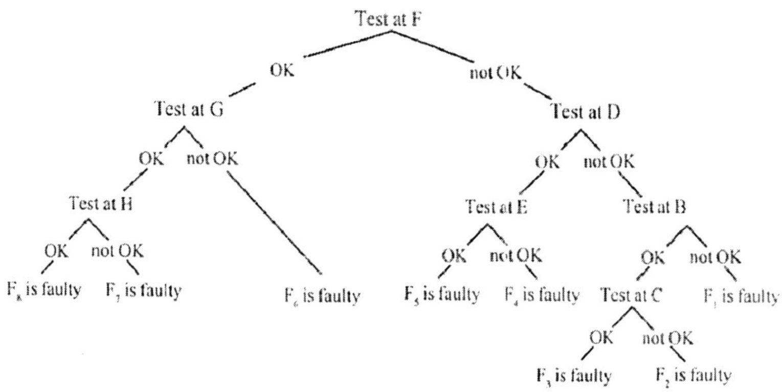


**Figure 18.16** Decision tree for the midway check method for the above situation.

## 18.17 Situations under which Decisions are taken

1. *Decision making under certainty,* when we know with confidence what will occur, in which case we have to consider only one possibility of occurrence for every alternative.
2. *Decision making under risk,* when there are a number of possible alternatives and the decision maker knows the probability of occurrence in each, which is estimated from past data and past experience. This is the most common case in the normal production situations.
3. *Decision making under uncertainty,* similar to the above, but when no past data is available to compute or estimate the probability of occurrence. The market demand for a new and untried product is an illustration for this situation.
4. *Decision making under conflicts,* when the different individuals involved directly in the decision making have contradictory opinion of data.

Illustration on the lighter side for the fourth situation:

> *A sea captain on a rough sea may take a decision depending upon his attitude. An optimist expects it to change and takes no action. A pessimist complains about the wind, prays for the heavenly abode (moksham) for all and keeps the crew on tender hooks. A realist adjusts the sails, takes control of navigational aids and takes emergency precautions.*

## 18.18 Systematic Decision Making

1. Outline your goal and outcome.
2. Gather data.
3. Develop alternatives (i.e., brainstorming).
4. List pros and cons of each alternative.
5. Make the decision.
6. Immediately take action to implement it.
7. Learn from and reflect on the decision.

## 18.19 Information Flow

As just noted to effectively control a system, data relevant to the activity must be transmitted to the decision maker. This is true whether the system is self regulating or requires management for decision making.

For analysis the system data flow can be divided into three classes

- Operating data.
- Control data, and
- Planning data.

***Operating data:*** Operating data include all the routine information that must be generated within the system, in order to keep it in motion. Some of the kinds of information that must be kept to operate a business are the reports, purchase orders, payroll records etc.

***Control data:*** The second type of data is control (or feedback) information. Since systems are designed to achieve specific objectives, they must be monitored continuously to make sure that they are meeting those objectives satisfactorily. This data reflecting the status of operation must be constantly reviewed. For example, in a production of an item with a specific standard time, often the management needs to know if this standard is being met or if there is a significant deviation.

***Planning data:*** The last type of data that should be made available is planning data, This includes all the information that will be required by management to solve anticipated future problems.

## 18.20 Features of Management as a System

All the points described in the previous paragraphs can be summarized as under:

1. *Management is a system*: Management can be defined as a system and unlike the biological and mechanical systems, it has the characteristics of a social system. It has subsystems which are integrated as a whole.
2. *Management is an open system*: It has an environment and with interaction, management takes its resources or inputs, allocates and combines the resources to produce the desired outputs to the environment.
3. *Management is dynamic*: The equilibrium of the organization is ever changing, moving towards the growth and expansion by preserving some of its energy. Thus the organizational effectiveness is determined by the ratio of energy exchanges.
4. *Management is probabilistic*: A deterministic model always specifies the use of a situation in condition with predetermined results. Management being probabilistic, points out only the probability and never the certainty of performance and the consequent results. Nevertheless for aiding decision making, the management assumes certain amount of deterministic models to arrive at some conclusions and then do the analysis in combination with the probabilistic approach.

5. *Management is multi-dimensional*: System approach points out complex multilevel and multi directional characters. At macro level, management can be applied to the business systems as a whole. At micro level, it can be applied to an organized unit or any of its elements.
6. *Management has multiple variables:* There is no simple cause-effect phenomenon like in physics. An event may be a result of so many variables which themselves are interrelated and interdependent.
7. *Management is adaptive*: The survival and growth of an organization in a dynamic environment demands an adaptive system that continuously adjusts to changing environment through feedback mechanism. The feedback mechanism provides information to take corrective action for achieving the desired results.
8. *Management is multi-disciplinary*: Management being a system draws integrated knowledge from several disciplines and schools of thought. Infact integration of the relevant aspects of various disciplines is the real contribution to the management as a system.

## 18.21 Conclusion

The chief purpose of the systems approach is to provide the management with an analytical framework by which it can identify, describe and interrelate the process and components that make up a particular system. In other words, the systems approach enables a manager to maintain a perspective of the whole process while he analyses the parts.

# Chapter 19

# Organisation for Maintenance

*Chapter synopsis: The organization structure of a company forms the foundation for effective control in enabling people to work towards the accomplishment of the goals of all activities of the organization. It is hence essential to incorporate the verifiable objectives and a clear understanding of the role of each and every individual and their relationships with others. This chapter discusses the essential principles and types of organization structures prevalent in industry and deals with the specific organization structures suitable for the maintenance departments.*

*Keywords: Organization, principles and objectives, specialization, coordination, authority, definition, correspondence, span of control and balance, level of reporting, line organization, line, functional and staff organization, committee organization, centralized, decentralized organization, matrix type organization, degree of centralization, F.W. Taylor, Kevan Hall , Context, Cooperation, Control, Community, Visio software OrgPlus software, ERP, HRMS, SAP software, Oracle eBusiness Suite, Rallis India group of engineering industries, Rallifan. Ralliwolf, Rockwell India, JLO, manpower planning and external maintenance services*

## 19.1 Why Organization

Organization, as defined by the American Society of Mechanical Engineers, is the process of determining the necessary activities and positions within an enterprise, department or a group, arranging them into the best functional relationships clearly defining the authorities, responsibilities and duties of each position and assigning them to individuals so that the available efforts can be effectively and systematically applied and coordinated.

The concept of organization can be equated to the structure of the human body which is divided into parts or organs that perform different functions in close cooperation and coordination with each other. The integrating force of whole system is the brain which plans and controls the functioning of each and every organ. So is the Industrial organization.

As per the above definitions and explanations, there are two aspects of the organization. The static aspect of the organization defines the relationship between the various subtasks or subsystems. This view point places emphasis on job content, definition, analysis and job relationships. The dynamic aspect of the organization is the pattern in which a number of people relate themselves to each other in a planned systematic accomplishment of the objective. This view point places emphasis on the human element and personal relationship.

## 19.2 What are to be Organized in the Maintenance Function?

Basically those resources needed to be established and maintaining a facility capable of achieving, the following are to be organized.

(a) Economy in the use of financial resources needed to provide and support the maintenance and repair function.

(b) Economy in the use of the physical resources necessary to house, provide tools and equipment and methods, for the efficient performance of engineering standards required as determined by the management.

(c) Supply and economical use of resources, in the correct numbers with desired levels of technical competence and effectively utilize and develop these resources, both human and equipment, for the purpose of providing and producing the required level of fully maintained equipment.

## 19.3 Principles of Organization

Urwick presents the following ten principles of organization, which are better remembered by their first letters as OSCAR - DCSBC.

1. *Principle of Objective:* Every organization and part of the organization must reflect the very purpose of the undertaking or the department. Otherwise inclusion of that part is meaningless and therefore redundant.
2. *Principle of Specialization:* The activities of every member of the organization should be specialized, for the performance of a single function. This principle put forth by Winslow Taylor, was the most preferred principle during the mid $20^{th}$ century. However the latest trend is the principle of delegation where the shop floor operatives are given the overall responsibility of their operations including machine maintenance, 5S etc. This is further explained in the chapter of Total Productive Maintenance.
3. *Principle of Coordination:* The purpose of organizing the persons, as distinct from the purpose of the undertaking, is to facilitate coordination and unity of effort.
4. *Principle of Authority :* In any organized group, the supreme authority must rest somewhere. There should be a clear line of authority from the supreme

to every individual, of the group. This principle has also been called, as the principle of hierarchy, or the principle of chain of command.

5. *Principle of Responsibility:* The responsibility for the functioning of each and every activity should lie somewhere. The superior also should accept the responsibility of the acts of the subordinates. A factor that influences whether the ultimate responsibility for a certain function to be placed at a higher level reporting directly to the Managing Director or a lower level is generally based on its financial importance to the company, as illustrated in Fig. 19.1.

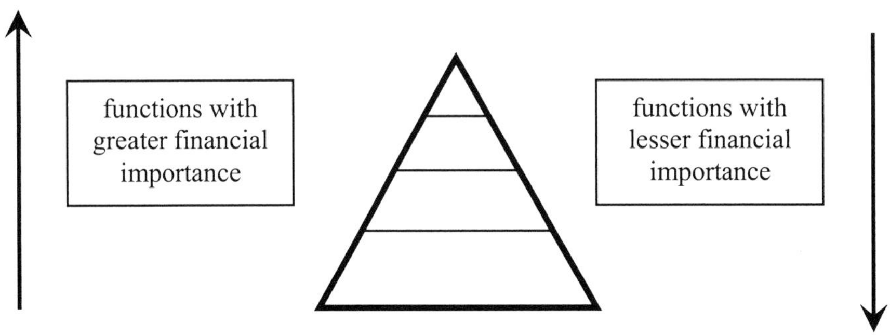


**Figure 19.1** Level of reporting is judged by the functional responsibility.

6. *Principle of Definition:* The extent of each person or position, with reference to the duty involved, the authority, the responsibilities, and the relationships with other positions should, be clearly defined in writing and made known to all concerned.
7. *Principle of Correspondence:* In every position the responsibility should correspond to the authority vested, and vice versa, for the performance of any function.
8. *Principle of Span of Control:* No person should supervise five or more direct subordinates, whose work Interlocks.
9. *Principle of Balance:* It is essential that the various units of the organization should be kept in balance with respect to their importance to the organization and the level of reporting.
10. *Principle of Continuity*: Reorganization is a continuous process. Specific provision must hence be made for future modifications or alterations at every stage.

The Production handbook, edited by Carson, adds an eleventh principle, viz., the principle of staff function, which is an extension of the principle of specialization. The staff managers do the functions not forming the routine of the day-to-day production job, but could have done if they had the time and the specialized knowledge. Maintenance function is a typical example of the staff function and its organization specialties are best understood from the following paragraphs.

## 19.4 Classes of Organizational Structures

For the overall organization for any industrial undertaking, there can be basically 4 types of structures

### 19.4.1 The Line Organization

Each department is a self-sufficient unit and as full authority and responsibility for the performance of its particular product, process or function. This is equivalent to the decentralized organization for the maintenance function as detailed in Figure 19.2. This is also called the military organization.

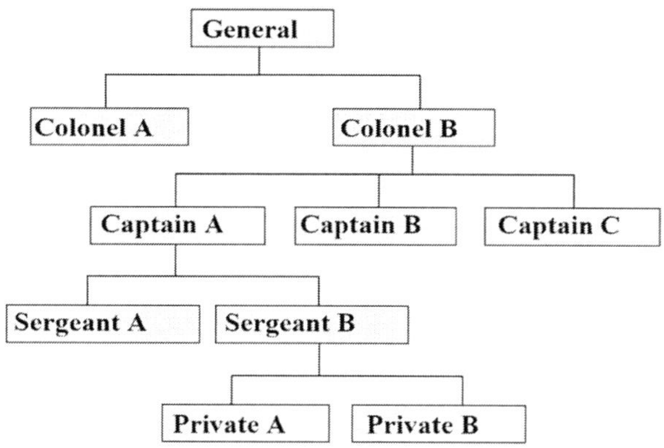


**Figure 19.2** A typical line organization – the military organization.

### 19.4.2 The Functional Organization

Conceived by F.W. Taylor, this is the most popular structure in today's industry. It may be noted that the concept of TPM has in recent days favours the decentralized control prevalent in the line organization. The overall responsibilities of the company is divided based in the functions like production, maintenance, finance, material control and, marketing. Each of these functional heads controls all aspects of his department and reports to the GM. The GM and in his absence the functional head has a centralized control over the functions of all activities of his department. This is hence equivalent to the centralized organization as detailed in Fig. 19.3 for the maintenance function. This type of organization, based on the principle of specialization takes advantage of the individual's proficiency and specialized knowledge.

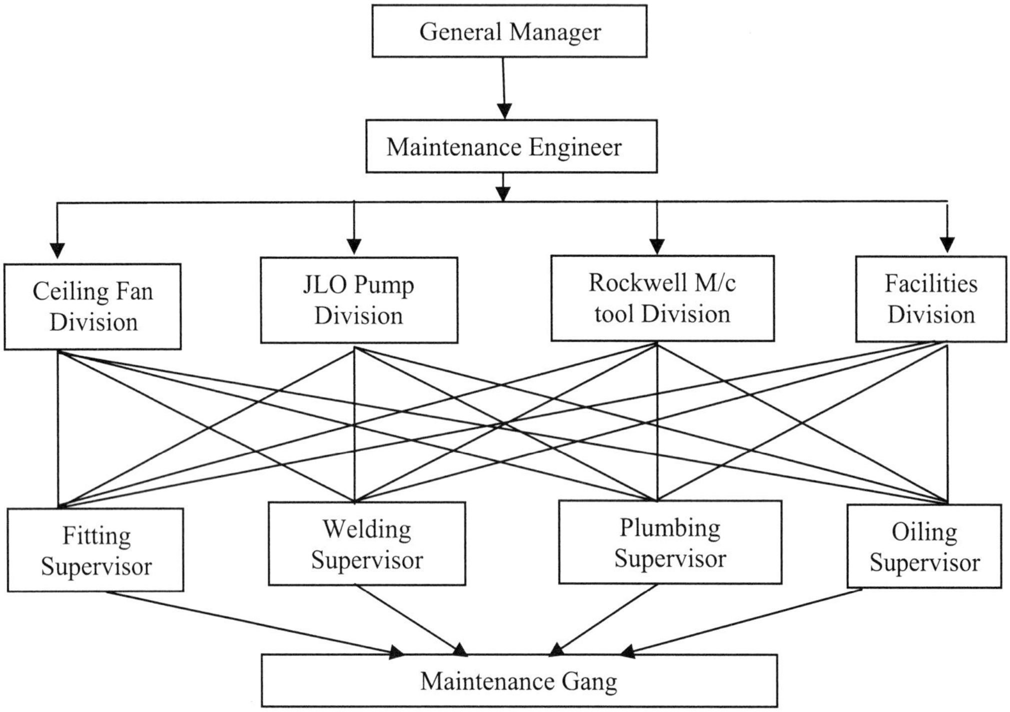


**Figure 19.3** A typical functional type of organization.

### 19.4.3 Line and Functional Staff Organization

Instead of having a special department for each staff function, some of the major departments like the production are provided with experts in certain functions like the maintenance and PP&C given to the production head, inventory control and purchasing given to the material control head. A typical line and functional staff organization as adapted in Ralli Machine Tools of Surat, India is illustrated in Fig. 19.4.

### 19.4.4 Committee Organization

While the organization is headed by the GM or Chief Operating Officer, each of the function like material control and maintenance would be handled by the respective committee. This is prevalent in professional associations and may be in small organizations. This is similar to the functional type of organization as discussed above or the matrix type of organization for maintenance discussed later.

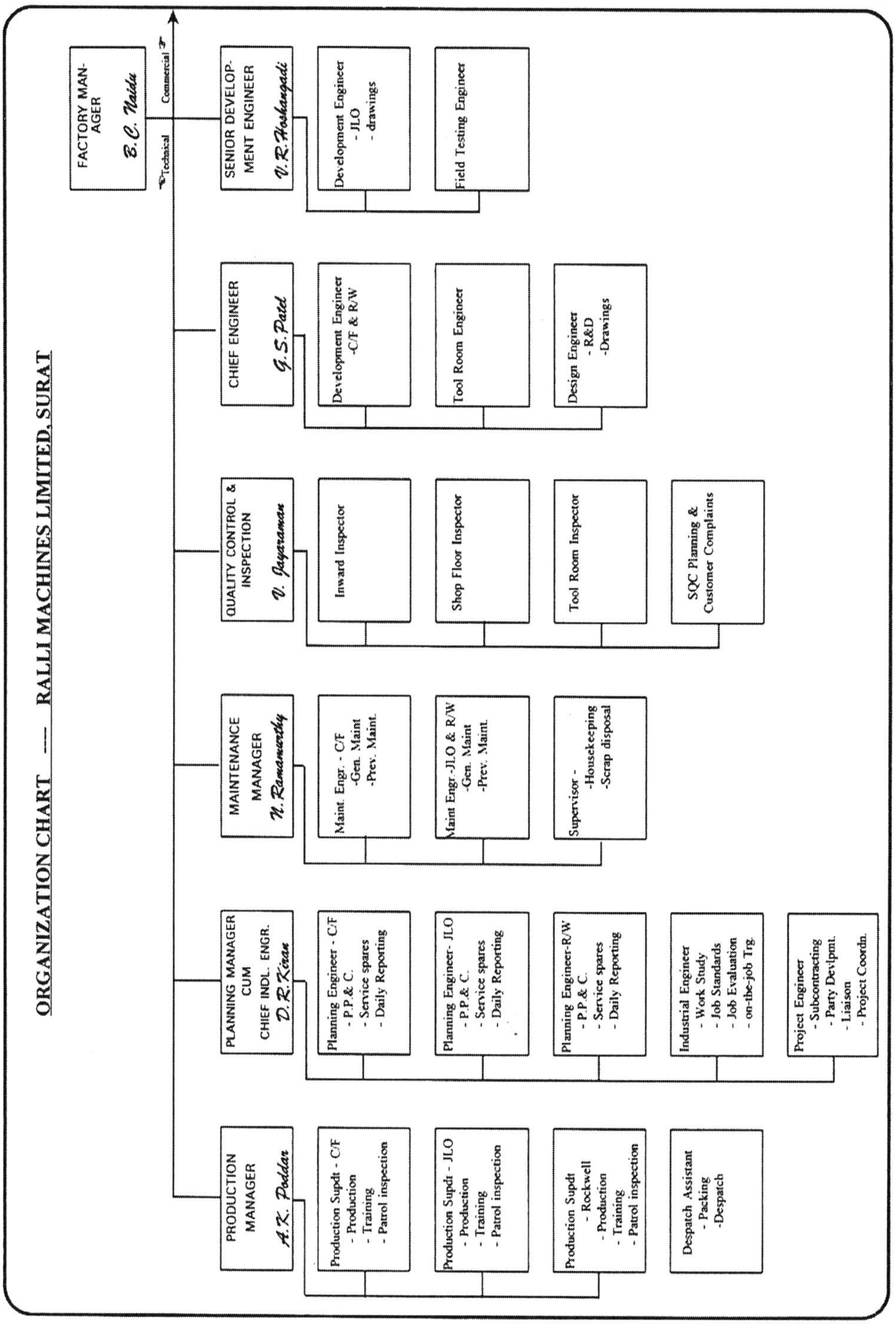


*Fig. 19.4* Example of a centralized organisation

## 19.5 Organization for the Maintenance Function

The organization for the maintenance, just like any other industrial function, is based on the above principles and in view of its high importance, not only technically but also financially, the ultimate responsibility is normally placed, at the highest level, except in the case of decentralized type, as seen in the following paragraphs.

There can be three types of organizations suitable for the maintenance function.

(a) Centralized Organization,

(b) Decentralized Organization, and

(c) Matrix type Organization.

## 19.6 Centralised Maintenance Organization

This is an organization unit having similar organization principles and having the same status as the other departments like the production department. This is equivalent to staff organization as detailed above. This unit is normally headed by a Manager, who is responsible for all the maintenance activities of the company, irrespective of the geographical location or separation of the production units of the company. All maintenance staff and facilities, working anywhere in the factory ultimately report to the Maintenance Head. We can compare this to the line and staff and functional type of the traditional classification of organizations.

*Factors that influence the degree of centralization are:*

1. Size of the company, number of the employees, turnover etc. Actual plant size sometimes dictates the strength of the maintenance staff, and the amount of supervision needed for this staff. Many more subdivisions in both line and staff may be justified since the overhead can be distributed over more departments.
2. Geographical, distances of the individual units, the maintenance that works in a compact area differs from that dispersed in several buildings over a large area.
3. Technology used, and type of production.
4. Number of the working shifts, for example, during the day shifts, the maintenance function may be more centralized, while for the same company's maintenance activities during the night shifts, may be more decentralized, that is done by the production personnel themselves.
5. Possible use of subcontractors.
6. Attitude between the maintenance and production as well as between maintenance and management.

7. Level of training & reliability of the work force. In industries where sophisticated equipment predominates with high wear or failure rate, more skilled mechanics and supervisors would be required, and they generally are under centralized control.
8. Management policy.

*Advantages of the Centralized Organization:*

1. Efficient use of maintenance technology and opportunity for further development of skills of personnel, with good scope for sending employees for higher training. Skill preservation and skill development is possible for different trade personnel.
2. Less total number of maintenance personnel required due to possibility of emergency mobilization of staff from one section to another of the centralized department; this helps in better redistribution of resource.
3. Major repairs and shutdowns can be completed faster by pooled up resources from different shops and departments when needed.
4. Better feedback to the management.
5. Better standardization and variety reduction achieved as a result of good inventory control that can be planned and achieved in a centralized maintenance and spare part situation.
6. Less chance of preventive and shutdown maintenance postponed or bypassed by production managers, who are reluctant to stop production in their eagerness to have more production time.
7. Hiring of external services would be cheaper and broader based.
8. Better utilization of the specialized maintenance installations and tools.
9. Maintenance costs can be better isolated and analyzed for the efficient running of company.
10. Ready availability of data and past history of line equipment.

*Disadvantages of the Centralized Organization:*

1. Might develop into a constant friction between the production and maintenance personnel.
2. Tendency for long downtimes.

## 19.7 Decentralized Maintenance Organization

This is more or less similar to the line organization, each shop or a production section, will have a maintenance foreman of its own, reporting to the respective section head in-charge of production. Day-to-day planning of the maintenance for the section is done independently by the section head, in addition to his production responsibilities. In this organization the Chief of Maintenance is generally responsible only for the central planning and facilities like water, steam supply, plant services, machine shop for major repairs, captive equipment etc. He helps the material management for standardization, inventory control, spare part management etc. This concept of decentralization has gained so much ground in the past two decades that it is now popularized as 'Total Productive Maintenance' which is discussed more elaborately in chapter 17.

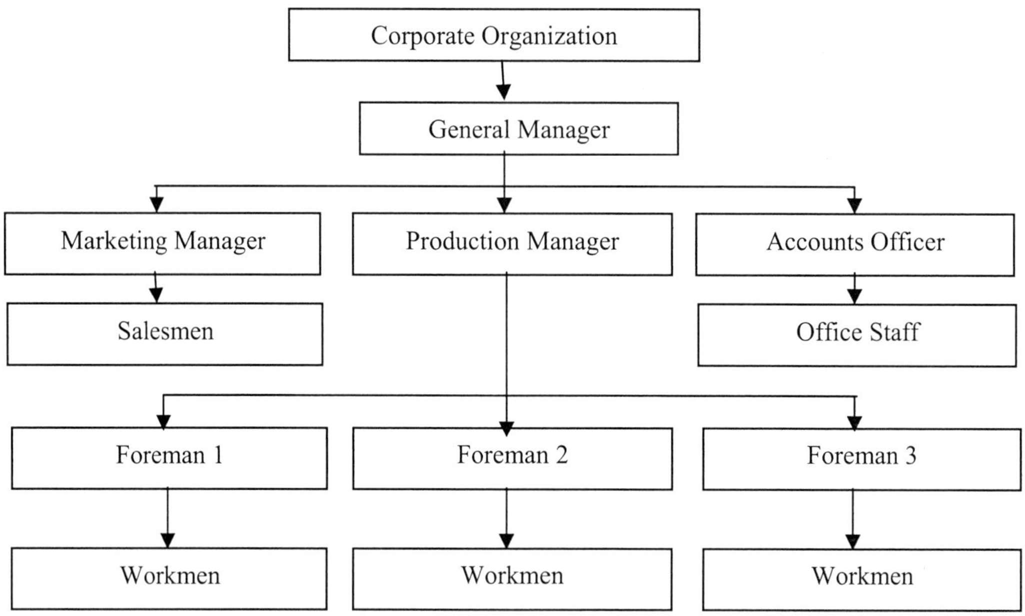


**Figure 19.5** Decentralized organization.

*Advantages of decentralized organization*

1. Maintenance is more aimed at smooth production.
2. More similarity of the operators with the behavior and maintenance requirements of their respective equipment.
3. Better rapport with the operating personnel and more chances of the production personnel helping the maintenance gang for physical assistance in case of need.
4. Better chances of opportunity maintenance, as and when the equipment is idle for other reasons.
5. Quick repair of equipment.

*Disadvantages of decentralized organization:*

1. Maintenance technology becomes stagnant.
2. Non availability of expert knowledge for the organizing and carrying out of the specialized maintenance jobs.
3. Less efficient use of the production engineer, since he has less time to devote on the production problems.
4. Risk of long term of suspension of preventive maintenance activities.

## 19.8 Matrix Type of Organization

This is partly centralized and partly decentralized (Fig. 19.5). The specialist maintenance functions like planning, preventive maintenance scheduling, overhauling etc., are centralized, while routine maintenance jobs needing semi skilled or unskilled attention like machine oiling, cleaning, electric fuse replacement etc., are decentralized. The overall responsibilities for this function still are vested centrally.

Here people with similar skills are pooled for work assignments, resulting in more than one manager. This is mostly applicable for a large organization having several major products and/or several geographical locations. Each product or location may have its own head and his own production staff, but does not justify to have a separate maintenance team under each product or location. All the maintenance team would be under a central maintenance manager, but whenever needed the team goes to the particular location or the product and for that job they fully report to the location manager. As another example, all engineers may be in one engineering department and report to an engineering manager, but these same engineers may be assigned to different projects and report to a different engineering manager or a project manager while working on that project. Therefore, each engineer may have to work under several managers to get his job done.

Rallis India group of engineering industries provides a good illustration for the matrix organization. During the sixties, it basically had 3 units, Ralliwolf making portable drilling machines etc., Rallifan making table and pedestal fans, both located in Bombay and Ralli Machines located in Surat making 3 distinct products, the ceiling fans under Rallis brand, the table top machinery under the brand of Rockwell machines and 50 cc Petrol engines under the brand of JLO. While each of the three units have independently centralized control, corporate management activities like audit have the matrix function. In Ralli machines unit of Surat, while each of the three product units have separate production managers, other functions like maintenance, industrial engineering, and purchasing are common staff functions and form matrix type of control.

Kevan Hall in his book *Making the Matrix Work* identifies 4 C factors for the success of matrix type of organization.:

- Context: Matrix managers need to make sure that people understand the reasoning behind matrix working and change their behaviors accordingly
- Cooperation: A matrix is intended to improve cooperation across the groups, but shall not lead to more meetings and slower decisions where too many people are involved.
- Control: In a matrix, managers are often dependent on strangers where they don't have direct control. Matrix managers need to directly build trust in distributed and diverse teams and to empower people, even though they may rarely get face-to-face.
- Community: The formal structure becomes less important to getting things done in a matrix so managers need to focus on the "soft structure" of networks, communities, teams and groups that need to be set up and maintained to get things done.

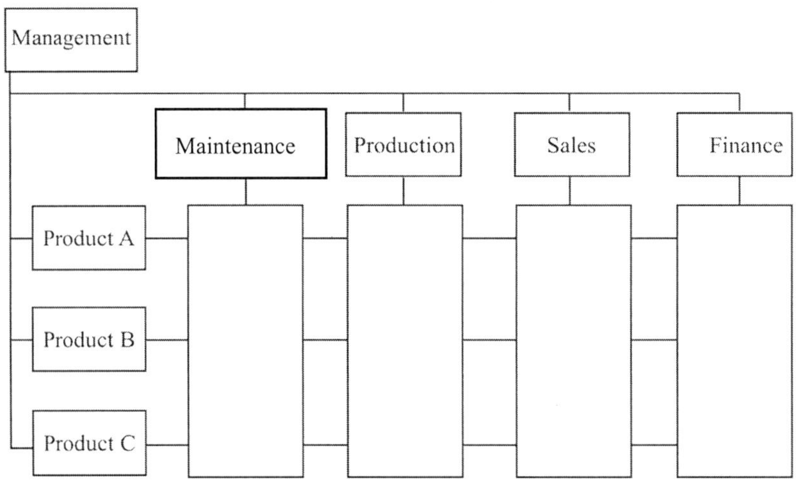


**Figure 19.6** A typical matrix of organization.
(Source: Wikipedia)

Wikipedia reports that visual representation of Matrix charts have been a challenge ever since it was invented. Most organizations use dotted lines to represent secondary relationships between people, and charting software such as Visio and OrgPlus supports this approach. Until recently, Enterprise resource planning (ERP) and Human resource management systems (HRMS) software did not support matrix reporting, but recently SAP software was developed to support matrix reporting, and Oracle eBusiness Suite can also be customized to store matrix information.

## 19.9 External Maintenance Services

A majority of small and medium plants find it uneconomical to have a full fledged maintenance department. Instead they go in for external maintenance services. Now-a-days this is becoming more popular even with medium and some large industries also for the simple reason of overall economy and less labor related issues. This can be reflected by the fact that 20 years back every public sector undertaking in Bangalore used to operate a large fleet of buses of their own to such an extent that those of 4 to 5 undertakings alone outnumbered the city transport buses. But today, roads of Bangalore present a different picture. Most of these undertakings have now reduced their fleet sizes, and are engaging contract services. So is the trend in hiring external maintenance services.

*Types of services offered by the external agencies:*

1. Janitorial services.
2. Annual contract for cleaning and periodic maintenance of equipment including supply of spare parts. This is called annual maintenance contract or AMC.
3. Annual contracts for maintaining strategic areas like scrap yard, tool room and electroplating shop and other areas normally considered as difficult areas.
4. Individual job contracts during major repairs and shutdowns to supplement the plant's maintenance force.
5. Transportation of equipment or other maintenance materials.
6. Building, owning, operating and supplying of specialized services within the plant's premises like oxygen, supply, boiler plant and captive power plant
7. Generally all manufacturers of any equipment purchased by the plant provide after sales service with a guarantee period of 1 to 5 years depending upon the type of equipment during which they provide free maintenance. This free maintenance can be extended beyond guarantee period under AMC terms.

*Advantages of engaging external services:*

While these have all the advantages of the centralized systems like skill specialization, etc, they have the additional advantage that line competition among the several private service agencies leads to better service and upkeep, they have less labor liabilities. In short

1. It is very economical, especially for small scale industries
2. Availability of experts technically qualified and experienced in specific equipment
3. Healthy competition from other agencies lead to better services.

## 19.10 Factors shall be considered for Deciding the Manpower Requirement

- Complexity of the job
- Periodicity of the activities
- Time required for each of these activities
- Quantum and type of activity
- Layout and location of the activities
- Allowances for leave, holidays, shifts etc.
- The company policy

## 19.11 Size and Type of Maintenance Organization

Manpower is the main resource for any organization and careful planning is needed to determine the size of the department. The actual type of maintenance organization depends on the following factors:

- Type of industry whether process or continuous or batch.
- Size of the industry and volume of its operations.
- Complexity and nature of the business or jobs in that type of industry.
- Age and condition of the equipment.
- Physical location of the plant's facilities and departments, whether they are spread out or compactly located within a close geographical area.
- Availability of the skilled labor and maintenance personnel employed by the company.
- Availability of external agencies for the required skills and experience needed in the field, of operation and the specialized equipment available in the plant.
- Extent of automation and built-in testing and monitoring equipment facilities provided.
- Local labor laws, industrial, culture and practices.
- Policy of the company.

## 19.12 Manpower Planning for Maintenance Activities

Manpower planning involves the following step by step procedure:

- Define overall company objectives and then convert them into manpower objectives

- Design the manpower information systems and undertake a manpower inventory
- Analyze manpower requirement
- Analyze manpower availability and improve manpower utilization
- Identify training needs and arrange for training
- Control labor costs in maintenance.

Job description for all the jobs are prepared considering the jobs to be performed, their interrelationships etc. The total crew strength requirement is calculated considering the total work content in planned maintenance as well as breakdown maintenance can be statistically estimated; the work content can be determined by any of the measurement techniques. Total crew size can be calculated based on the total work content so obtained and the total available hours per year per person. Provision also is to be made for shift relievers, leave reserves etc.

## 19.13 Conclusion

For the effective performance of the maintenance function, a well-conceived and clear definition of the relationship between the various subtasks or subsystems not only for the maintenance department but holistically among all the departments, as well as other stake holders like suppliers, shareholders etc., is essential as illustrated and discussed in this chapter.

# Chapter 20

# Equipment Record

*Chapter synopsis:* *Recording data about equipment and property including all details of the maintenance activities like specifications, purchase date, expected lifetime, warranty information, service contracts, service history, spare parts and every related detail would be of immense help to management or maintenance workers. This chapter discusses such equipment records including Asset register and how they should be maintained.*

*Keywords:* *Equipment record, Fixed Asset Register, Asset Management, Entertprise Asset Management System, Asset Knowledge Science (AKS), Asset Condition Assessment, Asset Identification Label, Asset Maintenance Management System (AMMS), Asset Management Framework, Asset Management Improvement Plan, Enterprise Asset Management (EAM) Systems, Machinery Information Management Open Systems Alliance (MIMOSA), Return on Assets (ROA), Mainpac EA, Mainpac Asset Intelligence, Unit Record card, drawing record.*

## 20.1 What is Equipment Record?

An effective functioning of PM system is a pre-requisite for the ability to plan and control the maintenance activities and will result in a higher degree of planning of maintenance jobs. This effect will often be very marked, permitting a greater degree of control of maintenance operations. Otherwise there will often be too few jobs that can be planned in advance, and too many urgent jobs. This calls for an effective recording system for all the equipment and facilities of the company that includes all essential and optional data elements to provide efficient and effective management of property.

Even though the assets like machinery or plant facilities belong to the manufacturing or other departments, since the maintenance department ensures their proper functioning, it is consequently their duty to maintain the equipment record.

## 20.2 Fixed Asset Register

*Asset Register* usually refers to a formal, structured listing of the land, machines, office equipment, buildings, patents, trademarks, copyrights, etc. at a given location, while a *Fixed Asset Register* ***(FAR)*** records all the fixed assets like land, machines, office equipment, buildings etc, (other than patents, trademarks, copyrights, etc.) held for the purpose of production of goods or rendering of services and are not held for the purpose of sale.

### 20.2.1 Objectives in Maintaining a Fixed Asset Register (FAR)

1. To plan preventive maintenance schedules for all the company assets.
2. To record the history of the maintenance activities done for each equipment for future reference as and when needed.
3. To assess the cost of maintaining certain old equipment.
4. To categorize the equipment based on their complexity of maintenance.
5. To comply with legislation governing corporations, companies, etc.
6. To allow the company to keep track of details of each fixed asset, ensuring control and preventing misappropriation of assets.
7. To keep track of the correct value of assets to enable computation of depreciation and for tax and insurance purposes.
8. To allow the company to keep track of fixed assets those are not under simple, direct control of the company. These include owned and leased assets, assets under construction, and imported assets.
9. To aid in capital budgeting.

### 20.2.2 Details to be provided in a FAR

The format and the details to be provided in a FAR generally depends upon the following

(a) Nature of assets.

  (i) If moveable assets constitute a significant portion of total fixed assets, details will be necessary on their movement from one department or cost center to another.

  (ii) *Cost of assets*: Greater control and security is required for costly equipment.

(b) Customized reports on fixed assets required by management.

(c) Disclosure norms/regulatory compliances as per statutory laws applicable to the entity.

(d) Extent of owned, and assets taken on lease/hire purchase.

(e) Requirements of insurance company.

(f) *Location of fixed assets*: If fixed assets are located at numerous locations, greater details will have to be given.

(g) *Maintenance costs*: Some fixed assets require regular servicing to keep them running in an efficient and satisfactory manner. It would be necessary to keep a tab on the maintenance costs, dates of servicing etc., during a stated period.

## 20.3 Advantages of Equipment Record

When planning a maintenance job, many benefits will be obtained from an Equipment Record.

1. Use of this practice will result in more effective management of equipment including its monitoring, performance of physical inventories, self-assessments, disposition, etc.
2. It will cut down repair times and the administration necessary for planning a repair job, since it is easier to find particulars of drawings, instructions, special tools, safety regulations, components etc.
3. The operation is less dependent on individuals and it is easier to familiarize new employee with the job.
4. It is often easier to keep records of fixed assets, insurance policy values etc., up to date.
5. Standardization is more simplified if based on the records in the technical data record.
6. When designing new equipment it is easier to benefit from experience of present equipment.
7. An equipment record can be divided into a number of levels.
   - Unit data
   - Technical data
   - Component data
   - Documentation and drawings records.

## 20.4 Enterprise Asset Management Systems

SKF in its website www.skf.com/binary/Asset-Management-Services, details the following explanations for the documentation processes.

### 20.4.1 Asset Management

Asset Management is the process of organizing an asset's life cycle to achieve the lowest life cycle cost with the maximum availability, performance efficiency, overall equipment efficiency and highest quality. In other words, asset management is the systematic process of planning and control of a physical asset throughout its life. This may include the specification, design, and construction of the asset, its operation, maintenance and modification, and its disposal. Asset management provides a strong focus on technical aspects of a facts-based, proactive management attitude with a mission, vision, and objectives derived from that of the company.

### 20.4.2 Asset Knowledge Science (AKS)

It is a process of documenting asset knowledge relevant to monitoring and diagnosing asset anomalies. The process encompasses generic literature and unique details of the asset and makes use of software called SKF aptitude Decision Support System. The goal is to provide a structure to justify which measurements are needed to detect and diagnose failures in an early stage.

### 20.4.3 Asset Condition Assessment

An Asset Condition Assessment is the process of continuous or periodic inspection, assessment, measurement and interpretation of the resultant data to indicate the condition of a specific asset so as to determine the need for some preventative or remedial action. It is a crucial part of asset management to determine remaining useful life and an asset's capability to meet performance requirements. This is in line with the condition monitoring described in chapter 6, more specifically adapted in respect to the land and buildings.

### 20.4.4 Asset Identification Label

An Asset Identification Label is an adhesive plastic tag that can be affixed to an asset to help identify it. The tags may or may not be bar coded.

### 20.4.5 Asset Maintenance Management System

An Asset Maintenance Management System (AMMS) is a system designed to program & record an organisation's maintenance activities.

### 20.4.6 Asset Management Framework

An asset management framework is a set of documents, systems and processes that addresses an organisation's asset management responsibilities

### 20.4.7 Asset Management Improvement Plan

An Asset Management Improvement Plan is a strategic plan that provides for monitoring and control of the Asset Management Improvement Activities. It provides the link between the Asset Management Strategy and the annual operations, plans and budgets. This plan will ensure that acceptable progress is made on improving asset management processes and procedures and that progress can be verified and quantified.

### 20.4.8 Enterprise Asset Management (EAM) Systems

This is a software application to support enterprise wide asset management philosophy. A computerized maintenance management system with extended functionality which typically includes timekeeping, project management, human resources issues, payroll, core financial data, etc.

### 20.4.9 Machinery Information Management Open Systems Alliance (MIMOSA)

Machinery Information Management Open Systems Alliance (MIMOSA) advocates open exchange of equipment condition related information between condition assessment, process control and maintenance information systems through published, consensus, conventions. This is to gain greatest value by combining vital condition information from multiple sources for collective evaluation, reaching accurate determinations of current condition and projected lifetime and communicating results in a useful, understandable form. MIMOSA is committed to preserving the advantages, effectiveness and rich detail contained in specialized applications such as vibration, temperature, lubricating oil and electric motor monitoring and analysis systems within an integrated enterprise information structure (See:http://www.mimosa.org).

## 20.5 Some of the other Asset Management Software

Return on Assets (ROA), Mainpac EA, Mainpac Asset Intelligence are some of the other software available on asset management.

## 20.6 Example of a Unit Record

For each machine unit or machine part a unit record is to be filled in by the operation departments, as per Fig. 20.1 shown. The equipment classification is based on the existing PM system, with numbers designating main unit and component. Equipment classes are combined or subdivided if this is justified for convenience of spare parts. An illustration of the unit record proposed by the author to a Tanzanian industry during the seventies, is given in Fig. 20.1. The subsequent paragraphs indicate the data to be filled in each column according to the number as explained therein.

**SPARE PARTS IN STOCK**

| Work Area (1) | Name of Equipment (2) | Log No. (3) |
|---|---|---|

Priority (5) ____________

(4)

| Drg. No. (6) | Spec. No. (7) | | Supplier Cat. (8) |
|---|---|---|---|
| Mfg. Drg. No. (9) | Year of mfr. (10) | | year (11) |
| Supplier (12) | Del. Dt. (13) | Job/Ref. No. (14) | Order No. (15) |
| Manufacturer (16) | Type (17) | | Sl. No. (18) |

**MODIFICATIONS EFFECTING SPARE PART STOCKING**

| Date (19) | Foreman (20) | Job Description (21) | Drg. No. (22) | Record updated (23) |
|---|---|---|---|---|
| | | | | |
| | | | | |
| | | | | |
| | | | | |
| | | | | |
| | | | | |
| | | | | |
| | | | | |
| | | | | |

**Figure 20.1(a)** Front side of the unit record card.

| Spare parts in stock | Int. No. | Qty. in stock | Qty. on m/c | Supplier | Designation/suppl.No. | Storage location |
|---|---|---|---|---|---|---|
| (24) | (25) | (26) | (27) | (28) | (29) | (30) |
| | | | | | | |
| | | | | | | |
| | | | | | | |
| | | | | | | |
| | | | | | | |
| | | | | | | |
| | | | | | | |
| | | | | | | |

SPARE PARTS TO BUY WHEN NEEDED

| Supplier spare parts | Qty. | Lead time | | | |
|---|---|---|---|---|---|
| (31) | (32) | (33) | | | |
| | | | | | |
| | | | | | |
| | | | | | |

SPARE PARTS TO BE REPAIRED AS NEEDED

| Repaired | Qty. | Supplier | Designation/suppl.No | Drg.No. |
|---|---|---|---|---|
| (34) | (35) | | | (36) |
| | | | | |
| | | | | |
| | | | | |
| | | | | |
| | | | | |
| | | | | |
| | | | | |

**Figure 20.1(b)** Reverse side of the unit record card.

The details of the data to be filled are explained below.

1. Work area, division of plant (e.g. factory, saw mill) to facilitate file division, Eg.:
   - 01 Pulp
   - 02 Lines I + II
   - 03 Line III
2. Equipment name in PM system
3. Equipment number in PM system, for example,
   - 40-026 Timber sorter, bin 1
   - 40-02601 Timber sorter, bin 1, driving motor
   - 40-02602 Timber sorter, bin 1, worm gear
4. Technical data of interest for repair and spare parts purposes.
5. *Priority classification* This is a ranking of equipment based on how fast it is required for production. If it is not permissible for a repair to be delayed due to the waiting time for aspired part the stock level will be determined by time for the part and by the time required for dismantling ahead of the repair in question. A variety of solutions are possible for different spare parts used in the same equipment item of a given priority class. The priority classes are determined by permissible waiting times as follows:
   - *Priority class* 1: 0 - 2 hr
   - *Priority class* 2: 2 – 8 hr
   - *Priority class* 3 8 - 24 hr
   - *Priority class* 4: 1 - 3 days
   - *Priority class* 5: 3 - 6 days
   - *Priority class* 6: over 6 days
6. Internal drawing number assigned by Drafting Office
7. Specification Number, referring to
   - Supplier's specification if any, of equipment design function, major components etc.
   - Internal specification, if any, of similar parts used in more than one location.
8. Supplier's catalogue containing technical specifications, if any
9. Drawing number provided by supplier/manufacturer.

10. Year and month of manufacture.
11. Historical purchase, if updated indicate year.
12. Supplier (not necessarily manufacturer).
13. Year and month of delivery.
14. Job number or order reference (to simplify location of order)
15. Order number.
16. Manufacturer (principle manufacturer or technical representative)
17. Manufacturer's type number.
18. Serial number.

To help keep record up-to-date, points 19 to 23 are completed whenever the modification or redesign of machine affects its spare parts stocking requirements (capacity, dimensions materials, urgency class etc.)

19. Date of alteration.
20. Responsible foreman.
21. Description of modification.
22. Drawing number if any, for execution job (available with drafting office or maintenance department)
23. Signature when unit record has been corrected following machine alteration.

At the backside of the Unit Record, the following are recorded

24. Spare parts to be held in stock
25. Internal Part Number not to be used, since internal numbering parts have not been implemented at present.
26. The quantity of the part to be kept in stock ( taking into account all points of use in the whole enterprise)
27. Quantity of parts used in this machine.
28. Name of the supplier of the part.
29. Name of part, supplier's order number if required.
30. Stores location.
31. Spare parts not to be kept in stock, availability being assured by stockless buying agreement with outside supplier.
32. Quantity of part that can be supplied at need.
33. Lead time under stand-by agreement.
34. For parts/components of equipment that would be repaired immediately a failure is noticed.

Repair may be done by internal or outside shop. The spare column may be used if desired for estimated repair time.

35. The quantity of the identical spare parts required for the machine.
36. The drawing number of working drawing needed for repair work.

If there is no space enough on the back of the Unit Record, addendum sheets with the same layout are used.

## 20.7 Documentation and Drawing Record

This is a special record for seeking drawing and other documentation. The Documentation and Drawing Record greatly simplifies communication between maintenance, the drafting office and the supplier.

Records of drawing numbers etc., may also be included in the technical data record and component record, but it is often considered preferable to process this record separately.

With advancement of technology most of these records are now maintained and communicated digitally.

## 20.8 Conclusion

The various aspects of the record keeping of the maintenance activity as discussed in this chapter would go a long way in computerizing the departmental activities and provide an updated reference to the auditing and other departments whenever needed.

# Chapter 21

# Work Order Systems

*Chapter Synopsis:* *To ensure that every person down the line understands the procedure for executing a maintenance job, a clear cut work order specifying the job to be undertaken is essential. This chapter explains the principles and procedures of maintenance work orders, illustrating with a flow chart incorporated in a cable manufacturing company.*

*Keywords:* *Work order, job ticket, manufacturing environment, maintenance environment, non-facilities, work order priorities, Emergency Priority, Rush Priority, Code Compliance Priority, Routine Priority, work order flow chart.*

## 21.1 What is a Work Order?

A work order or job order or job ticket or work ticket is an order created internally within the organization, received by another department or to the work force of the same department. In a manufacturing environment, the production or the concerned department raises the maintenance work order for repair or service of any equipment which has developed a defect. In some cases it is the maintenance supervisor that fills in the job order based on the written or verbal complaint from the concerned department.

In a manufacturing environment, a work order is converted from a sales order to indicate to the manufacturing supervisor the quantity, time and other details of the product or operations to be made as requested by the customer. In a maintenance environment, a work order or a service order indicates the location, date and time the service is to be carried out and the nature of work that is to be done.

More specifically the work order would state:

1. The quantity of the product to be manufactured, built or fabricated,
2. The amount of raw material to be used, its price and amount,
3. The types and number of labour required, as well as the tools needed,
4. The machine utilisation for each machine during the routing process, its rate and amount.

## 21.2 Categories of Work Orders

There can be three categories of work orders:

- The general maintenance work order for repairs or service work orders written daily or as needed by the initiator.
- Maintenance Planning and scheduling of work orders is the hub of an efficient maintenance organization. For effective maintenance, planning and scheduling work, many other systems need to work well. Most importantly equipment inspections through preventive maintenance, technical database such as bill of materials, work order history, and standard job plans. Preventive maintenance work orders are generated weekly or monthly based on preventive maintenance schedules rather than by request. The supervisor generates the work orders and sends to the concerned staff. They are returned to the originator after the completion of the job on the scheduled date. If not, any problem would be notified to him and appropriate action taken after recording the same in the work order, which is continuously tracked till the completion of the job.
- The third category is the capital oriented project request, for any modification and additions to the existing facilities. For this procedure the facilities are defined as:
    - ♦ Buildings, structures,
    - ♦ All installed equipment and machinery. These include boilers, furnaces, installed pumps and compressors, all piping work, electrical fittings, installed motors and control systems, firefighting equipment, refrigeration and A/c units etc.
    - ♦ Non-facilities are items not installed in a building or not associated with the normal functions like recreational equipment, transport logistics, staff canteen etc.

## 21.3 Objectives of Work Orders

The overall objectives of the management reporting system relate to the comprehensive maintenance system.

1. The operating units are to be served by keeping productive equipment operable, efficient and safe.
2. Materials, manpower and equipment shall be coordinated for timely availability of resources at the site.

3. An optimum level of maintenance will be established and provided in the most practical and optimally low cost manner in accordance to the management policy on economics, appearance, safety and environmental standards.
4. As a guidance for achieving improvements, management will be provided with information on status, effectiveness and cost of maintenance activity and program.

## 21.4 Work Order Priorities

www.uco.edu/administration/facilities has suggested the following priorities in issuing work orders to facilitate planning and scheduling of work to be accomplished.

### 21.4.1 Emergency Priority-1

Fire, health and safety items that demand immediate response to protect and save property and lives.

- Fire alarms
- Electrical sparks, smoke
- Downed electrical lines
- Broken water or gas lines
- Toilets overflowing
- Person trapped in elevator.

### 21.4.2 Rush Priority-2

Items that require response as soon as reasonably possible (few hours) to restore systems to normal operations to protect property or to prevent disruption to activities.

- Broken glass
- Power failure
- HVAC temperature adjustments
- Toilets/urinals defective
- Elevator repairs
- Safety auditing compliance issues, e.g. inoperative door opener.

### 21.4.3 Code Compliance Priority-3

Items that need to be corrected as soon as possible to remedy fire regulations based on inspection discrepancies, code inspector or safety findings, etc. These take priority over routine work orders.

### 21.4.4 Routine Priority-4

Other work that can be handled on a routine planned and scheduled basis.

- Painting
- Furniture/office/surplus property moves
- Set up for special events
- Cracked glass replacement.

## 21.5 General Procedure for Work Order System

The order system cited below is used in two large industries, one a process industry (cement) and the other a cable factory. In both cases the order system is designed as an integrated part of the management control system in the company

Of course, for some very large jobs, the work order form shave to be completed along with other forms. In many cases PERT and CPM is used for the planning, and the forms from the computer then act as complements to these forms as explained in paragraph 21.6.

The maintenance management control system will be described separately. All information to this system is collected via the forms shown.

In both cases the office work is performed by trained clerks, most of them former craftsmen with several years of experience from the shop floor. They act as order receivers, preparatory and planners, time estimators and schedulers. The number of men in the office is not great compared to the flow of information. The actual practical results of the system show a reduction of maintenance work time corresponding to an efficiency increase of around 20%.

The priority rules in both companies are similar. Requesting person must always give the priority to a job according to these rules. For some of them, the higher priorities, no special authorization of the job request is necessary. The lower grades must be authorized by a production engineer or a manager, depending on the nature of the requested job. If a request on a job, needing authorization, is given directly to the maintenance department, the receiving man returns it immediately or sends it to an authorized person before starting the preparation work.

The request, which has been sent to the maintenance office, is returned to the requesting person with a note about delivery time etc., from the maintenance office, and this is done as soon as the preparation of the work order has been done. In many cases also the total cost of the requested work is given on the form returned.

While the function of the order system is described in paragraph 21.1, the description of what is happening in the different steps is detailed in paragraph 21.6. The work order form/time card and the material request/report are also shown in full scale.

In many cases the number of copies of the work order form must be rather great. Each worker should have one copy, each supervisor one own copy and one control copy, the material store should have one copy and another copy is sent to the Accounts department to "open the account" and one copy is used for the report that the job is finished. Sometimes extra copies are needed. The system is, however, based on copying an original on a copying machine, which is placed in the maintenance office. Hence, any number of copies may be made and rewriting the order form is not necessary. Today except the master control copy all other copies are digitally maintained and referred to whenever required.

## 21.6 An Illustration of Work Order Flow Chart

Shown below is a case study of development of a flow chart by this author for the movement of work orders in a cable manufacturing company in Tanzania. The 5 industrial engineering symbols of operation, movement, inspection, delay and storage are indicated and the step number of each is indicated inside the symbol. The explanatory notes given later detail each and every step in the flow chart.

Infact the system of preparing multiple copies was prevalent during the 1960'-1970's especially in Bombay textile mills. Today, except the control copy all other are digitally transmitted to the concerned personnel or department.

1. The requester issues a written request on a special form (not shown) stating what is requested and the priority of the job. Original and one copy.
2. The request copy is left in the block for own notes.
3. The original is sent to an authorized person, if authorization of the request is necessary. It is always sent to the Maintenance Office, which is Order Center or Dispatch Office.
4. The request is checked and approved by the foreman in the Maintenance Office. If the requester has asked for price and/or time of delivery this information is written on the original.

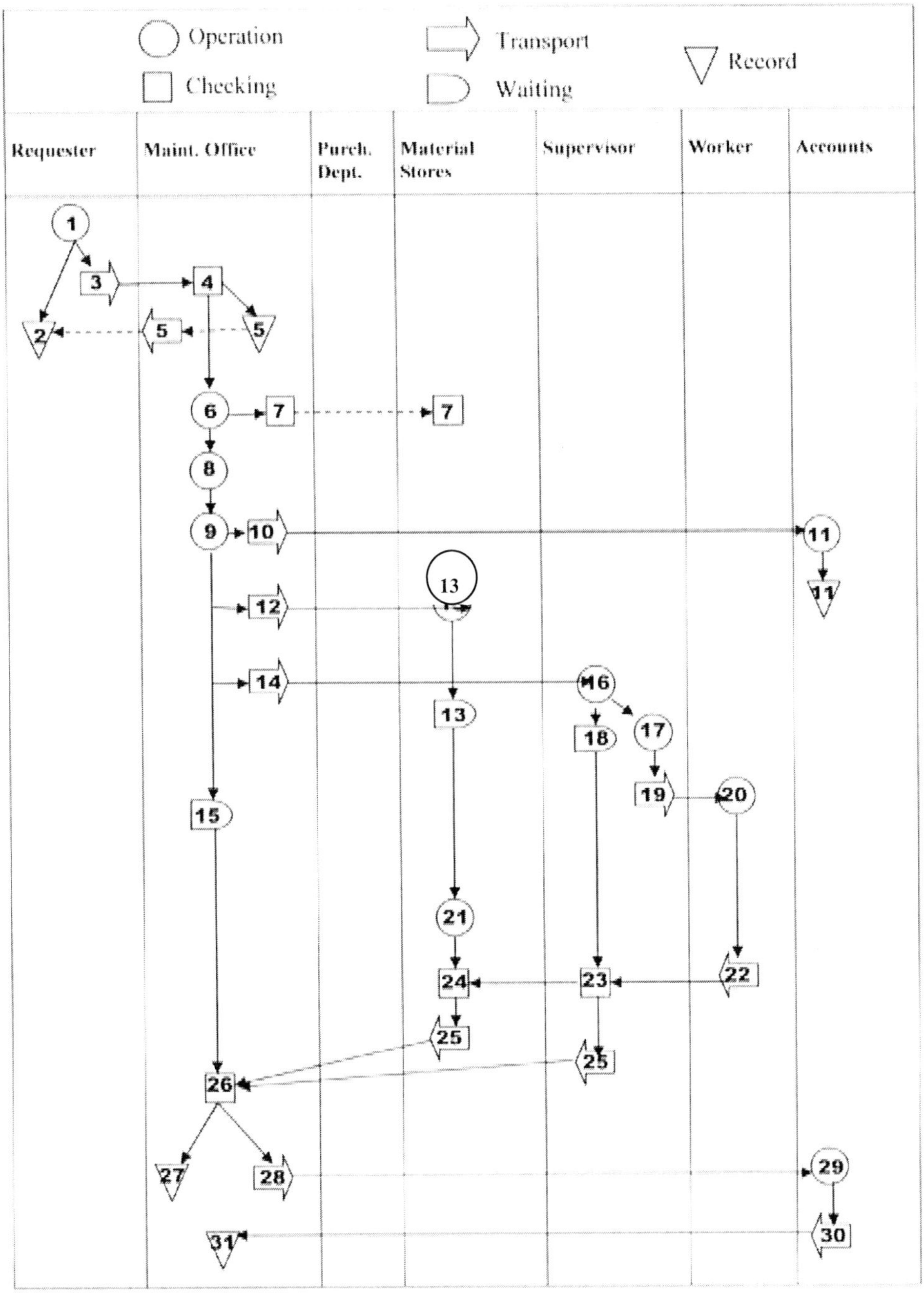


**Figure 21.1** Flow chart on work order distribution.

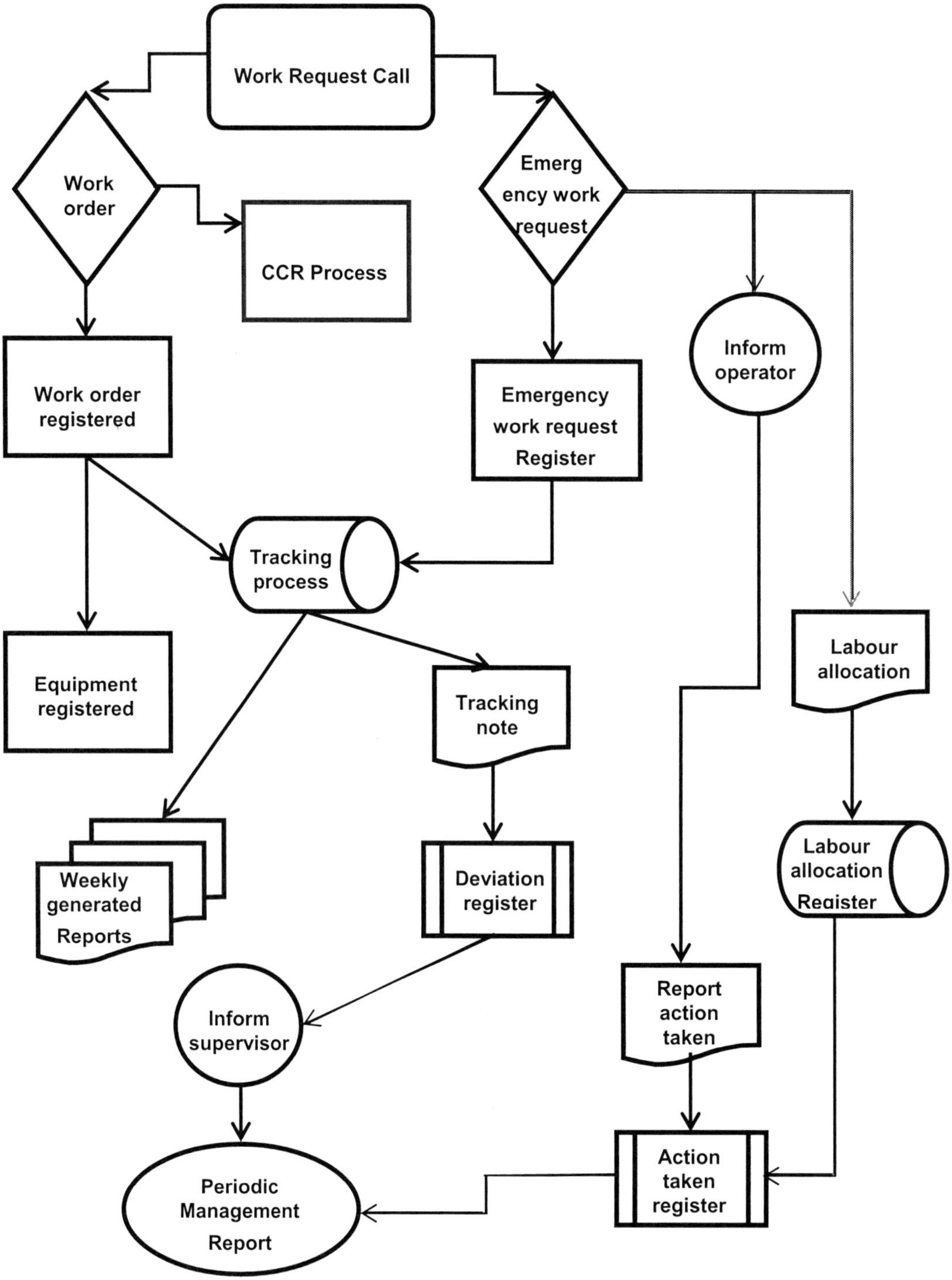


**Figure 21.2** Another illustration of a work flow chart.

5. The original is recorded in the Maintenance Office. Or sent back to the requester with information about price and/or time of delivery.

   (Note: In case a telephone request is used the original is left in the block and is never sent to the Maintenance Office).

6. The Work Order original is filled out.
   - First the identification of the object, No.1 and 2 on the form,
   - If a precalculation is wanted in advance cross in space 3 left, or after finishing the job, cross in space 3 right.
   - The type of work is entered in space 4 and the Work Order No in space 5, date or request No. 6, time of delivery No. 7, date the work can start on the machine No. 8, the requester's name, telephone No, 9, 10 and 11, the job priority No. 12.
   - Then the job preparation starts. This may be performed by the receiver or any other man in the maintenance office.
   - The operation No. and the description of each operation is entered in the space 14 and 15, tools, drawings.

7. At the same time a Material Request is made on another form. The planner asks the Material Store over the telephone if material and spare parts are available.

8. The Work Order is sent to the time estimator. Again it might be the same man.
   - Time for each operation is entered in the space 17, the number of the performing crafts group in space 18.
   - The preparation proceeds with the survey planning. For each operation the necessary time is established and in space 19 the WEEK for the performance is entered. Finally, the total load on each one of the performing groups or departments is calculated and the figures entered in space 20. Next, the cost is calculated for outside labour (one or more operations can be sent out to other workshops outside the company) in space 21, the total value of material for this order from outside suppliers is entered in space 22 and the total estimated value of material from own stores is entered in space 23. These are used for a precalculation.
   - The total number of labour hours estimated to be used by own department is entered in space 24 and the total cost is entered in space 25. These two are used for the precalculation and are not recorded in the computer. Total calculated cost added in square 32.

9. The work order original is now copied on a machine. The number of copies taken depends on the size of the job and how many departments/workers are involved. The general rules are:
   - Each supervisor should have two copies
   - Each worker should have one copy
   - Each group of workers should have one copy
   - One copy is used for "opening of the account" in the computer record
   - One copy is used for the planning control

   If the work is so large that the work description or the number of workers on the first form are not sufficient a second form can be used. This form is marked only with the work order number in space 5.
10. One copy is sent to the computer and "opens the account" for the work order number. If any information is fed to the computer before this "opening of the account" has been done, the computer refuses to record.
11. This indicates the "opening of the account".
12. The material card is sent to the store office.
13. Material is cut and collected and place in a box or a platform, marked with the work order number in space 5. The collected material then waits until a worker or a supervisor collects it for starting the work.
14. The copies to be used by the workers and the supervisors are distributed to the supervisors. The supervisor's group is filled in space 18 of the order form.
15. The original and the copy for planning control are placed - the first in a box in work order, number order, the second in a box, divided in days and weeks. It is first placed in the space, corresponding to the planned starting day or week. See point 16.
16. The supervisor now starts the planning and scheduling, finds out the first day and reports that to the survey planner, who enters the day in space 26 of both the original and the control copy.
17. The supervisor then states the name of the worker and enters his number, name, group and department number, week number, estimated hours for his part of the job, his wage type code and the time estimators code in a separate job card. One copy is used for each worker on the job.
18. The supervisor's own copy is stored in work order, number order in a box in the supervisor office.
19. One work order copy is handed out to the worker.
20. The worker starts by entering the hour in his job card.

21. The worker collects the material ordered for the job.
22. The job is finished and the worker stamps the time on the card, and the card given to the supervisor. The finishing date is entered on space 33 in the form head.
23. The supervisor checks the time card. If the job is finished he informs the material store over the intercom or telephone. He also enters the failure cause and failing component (only if the job is classified as repair) in spaces 27 and 28.
24. The store finds the material card and, calculates the cost.
25. The forms are all sent to the planning office.
26. The forms are checked against the planning control copy and all copies belonging to one job order number are collected. Calculation of time can then be made and entered in spaces 29, 30 and 31 (manual system).
27. The original and the planning copy are placed in two separate records, one on a job type record, and other on the machine number.
28. The time cards, the material card and one copy with all information sent to the computer, if Automated Data Processing (ADP) is used.
29. The wages and other costs are calculated and recorded in the maintenance control system.
30. The complete copy with the information is returned to the maintenance office, together with the material card.
31. Where both cards are recorded on the machine number. The control copy is destroyed.

## 21.7 Conclusion

While the work orders raised in the production departments for the production if components are temporary in nature and generally not retained after an year. especially because of the huge number of orders. However, in case of the work orders of maintenance jobs it is essential to keep them as a permanent records, since similar jobs might reoccur after a long period, and such record would certainly help in the decision making.

# Chapter 22

# Computerised Maintenance Management System

*Synopsis:* *A maintenance programme requires a lot of clerical work in compiling and analyzing the complex data on equipment and jobs for planning the schedules. As in any other field, use of computers in maintenance function greatly simplifies the data processing work so that the engineer can effectively do the planning and execution of the job. Computer software specific to the maintenance jobs have been developed, which are discussed in this chapter.*

*Keywords:* *Manual computation, CMMS, Tracking Work-Orders, Asset History, Specific software, MPulse, COGZ, Coherent, PMXpert, SCADA, mobile applications, web-based CMMS, Integrated Maintenance Management System (IMMS), expert system, Maintenance Management System Improvement.*

## 22.1 Introduction

Preventive maintenance programme requires analysis of several complex data of all the equipment for planning the schedules. This needs lots of clerical work in compiling and analyzing the data. In any normal company, the following are the difficulties in manually generating analysis.

- Huge volume of required data in view of too many equipment and infinite data needed to be collected for each and every equipment.
- Possibilities of human errors due to the monotonous nature in data collection, data entry and computation.
- Interpretation of data leading to improper communication to the person in-charge or
- Manual computation and analysis of simple statistical data takes a lot of time of the clerks/managers, who can otherwise concentrate on taking vital decisions.

Such a situation is greatly simplified by implementing computerized systems. Initially this computer application was utilized only for the analysis of data for making decisions. In the recent times, several computer softwares are developed in helping the maintenance staff in effectively making decisions on the procedures and schedules based on the outputs of these software. Most of the efforts of the engineers, managers and clerks in identifying the equipment and components can now be diverted to other execution works.

## 22.2 CMMS

Computerized Maintenance Management System (CMMS) can be defined as computer software that helps maintenance teams keep a record of all assets they are responsible for, to schedule and track maintenance tasks and to keep a historical record of work they performed.

The computers not only do the data processing but also do data analysis to list out the effective schedules by development of suitable software. The computerized systems plan both the current and future maintenance schedules and print the estimated job time for each element of the maintenance operations. The software formulates simulations and balances the maintenance operations in cases of production variations. Since the software provides time and cost factors for any job to a reasonable accuracy, service contracts are being undertaken as consultancy services.

In short the basic functions of CMMS can be cited as

- *Tracking Work-Orders*: Since the managers can select equipment with a problem, describe the problem and assign a specific technician to do the work.
- *Scheduling Tasks*: By maintaining a good work calendar for scheduling recurring work and sending reminders to the right people.
- *Recording Asset History*: For assets that are 10, 20, even 30 years old and have a long history of repairs. When a problem comes up, it is always useful to see how this problem was solved last time.
- *External Work:* Requests from people outside the team like a request from an assembly line operator who is hearing a strange noise from a drill.
- *Managing Inventory*: In general, maintenance department stocks more varieties of parts than the production, but a minuscule of the production inventory by value resulting in overstocking as it is considered not worth mentally to keep track on so many items. CMMS enables effective stock control even for spare items stocked in less quantity.

- *Audit and Certification*: Since a full record of assets and performance at lower energy usage at optimal cost, enables you to get favorable report from the auditors and certifiers.
- *Evidence in Legal Issues*: For example, if a company is involved in liability cases, the CMMS data base can serve as evidence that proper safety maintenance has been performed.

## 22.3 Benefits of CMMS

While each of the above point can be cited as the benefit of CMMS, in short, the benefits of Computerized Maintenance Management systems (CMMS) can be listed as:

1. A data base can be developed by the computers. This data base can also be extended to provide a complete history of the past performance of each equipment to be available when needed.
2. This information helps the management in making decision whether to repair or replace.
3. A well understood and interpreted standardization in systems and procedures can be established.
4. Time standards for repetitive maintenance operations can be developed to estimate the time factors with reasonable accuracy.
5. This will enable to prepare cost budgets for the maintenance departments, which earlier were not possible or difficult.
6. Detailed instructions, checklists and guidelines can be printed for the use of the staff.
7. The engineers and managers can concentrate more on the effective execution of the operational outputs of the software and work for maximum effectiveness.
8. This simplifies record keeping too since the input data can automatically be edited and tabulated in standardized record formats.
9. Less work outages, with less unexpected breakdowns
10. Better accountability by checking if a technician did their work on time and per scheduled process, and get alerted for any deviations.
11. A full record of assets and performance at lower energy usage at optimal cost, enables you to get favorable report from the auditors and certifiers.
12. Savings on purchases by a rational inventory planning well in advance to buy parts at spare-time pricing, rather than at a higher rate by panic buying.
13. The repetitive malfunctions can be analyzed and an early action can be taken to prevent them.

14. The frequency of the periodic reporting can be increased by computerization. The reports which were possible only once in a quarter can now be done every week.

## 22.4 Specific Software Available for Maintenance Function

CMMS can either be developed in house or by engaging outside consultants or purchasing readymade software. Some computer manufacturers offer computerized maintenance systems free of cost or at nominal costs. Besides, there are several software companies that develop software suitable for the maintenance functions. These can either be industry specific or for general application, suitable for a majority of industries. Some of these software are illustrated below with their trade names. Further details of these can be got from their websites by typing their names.

1. **MPulse** is a Work Order Management system that schedules preventive maintenance, manages requests for repairs, creates work orders, tracks work history with inventory and maintenance costs, and provides detailed management reports.
2. **COGZ,** a CMMS maintenance software integrates preventive maintenance work orders and scheduled work orders, with inventory control, and purchase order management.
3. **Coherent™** is an integrated suite of tools that comprises of a flexible and customizable facilities and maintenance management software solutions.
4. **PMXpert** is a user-friendly Preventive Maintenance "CMMS" Software program usable in all types of facilities, like machine shops, medical centers, hotels, shops, and fleet companies.
5. **SCADA** (**supervisory control and data acquisition**) is a type of computer based industrial control system (ICS) to monitor and control industrial processes, specifically for large scale processes that include multiple sites, and large distances.
6. **Maintenance Management System Improvement (MMSI)** is an expert system where all data on the maintenance history of each and every asset is made available in the form of computerized reports. Its benefits would be:
    1. To provide the Division of Maintenance with access to real-time data,
    2. To allow more effective monitoring and control of materials by improving the accuracy of data that is entered and stored,
    3. To make it easier to generate reports, as Division of Maintenance staff are able to electronically share data,

4. To eliminate duplicate data entry. End-users do not need to enter the same information on multiple systems,
5. To reduce the amount of time end-users spend on administrative tasks, such as recreating information that has already been collected,
6. To make it easier for users to effectively coordinate large projects involving multiple crews,
7. To allow users to easily determine where equipment is located and who is using it,
8. To provide a central repository for information at all levels, better enabling Headquarters to justify the need for more resources.

## 22.5 Future of CMMS

**Mobile applications:** With the advancement of technology it has now become very common for maintenance personnel to perform a majority of their maintenance activities using iphones. Since maintenance workers spend most of their time outside the office, CMMS is being available right on their mobile phone with great benefits. The technician can record what they are doing as they are doing it, recording pictures of the work and requesting help onsite. MEX-App and MC Express are some of the software specially developed for mobile applications. In recent times WhatsApp has become very common for instant communication between two individuals or within a group. The field staff can take screenshots of the actual situation and send message to the boss sitting in the office and get instructions within seconds.

**Easy to Use CMMS:** While more advanced scheduling software can be used in the office, Easy to Use CMMS just for data recording while on the job can be developed to simplify its use without carrying a thick manual.

**Web-based CMMS Systems:** While older CMMS systems are generally downloadable and need to run on a server at the client company, the new CMMS companies mostly focus on providing a private CMMS system for their clients which runs online. Many companies prefer these since the IT, security and backup needs are taken care by the CMMS-maker. Many organizations are already using Web-based computerized maintenance management system (CMMS) software on their phones or tablets.

## 22.6 System Integration

CMMS packages often link them to the manufacturing and process control systems for a better coordination and control, like SCADA, and the Scheduled maintenance based on number of units of output or condition of machines. Other aspects where the integration can take place are

- Scheduling
- Asset management
- Fixed assets with details the machines being maintained and their associated tooling
- Parts inventory/Parts locations/purchasing
- Costing
- OEE Systems – to reduce maintenance response time to failure (RFT)

## 22.7 Conclusion

The computerization of maintenance jobs and the large number of software developed purely for the maintenance jobs as discussed in this chapter, establish the benefits of such programs unlike the concept we held five decades back.

## 22.8 Websites on CMMS

Some of the websites that can be referred for CMMS are:

1. http://en.wikipedia.org/wiki/Computerized_maintenance_management_system
2. www.mapcon.com/us-en/cmms-software-products -
3. www.free-cmms.sourceforge.net
4. www.wbdg.org/om/cmms.php
5. www.softwareadvice.com/cmms
6. www.faciliworks.com
7. www.mpulsesoftware.com
8. www.cogz.com
9. www.commsouth.in/coherent-facilities-management-software.php
10. www.pmxpert.com
11. http://www.maintenanceconnection.com
12. www.ferret.com.au/mex-cmms-app-for-ipads-and-iphones
13. www.reliableplant.com
14. www.mapcon.com/us-en/mobile-cmms-software

# Chapter 23

# Replacement of Machinery and Equipment

***Chapter synopsis:*** *Replacement Policy is a set of decision rules for the replacement of facilities that wear out, deteriorate, become obsolete, or fail over a period of time. It is essential to formulate the optimal policy and guidelines in procuring new equipment in replacement of the existing, instead of spur-of-the-moment decisions necessitated by the sudden failures of the machinery that become irrepairable after usage for extended periods. This chapter presents the methods and procedures used in studying the economics of replacing equipment and machinery, illustrated by examples.*

***Keywords:*** *Replacement policy, Beyond Economic Repair (BER), Level of Repair Analysis (LORA), make or buy decision, deterioration, obsolescence, inadequacy, traditional evaluation methods, payback period, total life average, average rate of returns, discounted cash flow method, net present worth (NPW), profitability index, internal rate of returns, MAPI method, defender, challenger, adverse minimum, time adjusted average, operating inferiority, annual inferiority gradient, MAPI urgency rating, Markov Analysis, replacement study, required return, minimum attractive rate of return, cutoff rate of return, depreciation, sinking fund depreciation, straight-line depreciation, future worth, salvage value, Estimated Replacement Value (ERV).*

## 23.1 Why Replacement?

Components of a system function for a time and then fail. When the part is said to be "beyond economic repair" (BER), the central problem is to decide whether a strategy of scheduled replacement is preferable or the running of all components until failure. We had earlier seen that all machines and equipments deteriorate with use and exposure, which reflects in

- Inability to take specified load,

- Inability to produce to the same degree of quality,
- Reduction in speed mostly due to increased friction,
- Increase in hazards to safety of the employees,
- Reduction in the useful life of various components,
- Increased vibration and noise etc., in the machine due to worn out components like bearings.

Analysis of economic repairability can be formally evaluated using Level of Repair Analysis (LORA).

## 23.2 Basic Strategies of Repair/Replacement

When a machine stops due to a failure of a certain part there can be three strategies of repair:

- Remove only the defective part, repair it and refit to the machine.
- Remove the part and replace it by a new part.
- Remove a group of similar parts related to the defective part and replace all of them presuming the time and cost of the disassembly is more than the cost of the parts, and also if by condition monitoring, you expect that these parts to fail in a short time, like the bolts of an tool post or the bulbs in a control panel.

Which of these strategies to be adopted depends basically on overall cost consideration, in addition to the difficulty of the disassembly, the cost of the spares etc. Sometimes the convenience of the operative too becomes a factor. For example, if a street bulb gets fused out, the EB lineman replaces all the bulbs of a certain group of lights, if they are known to have served for a fairly long period.

## 23.3 When to Recondition instead of Replacement

(a) For imported machinery, reconditioning is generally more preferred atleast for the time being.

(b) If the equipment and/or the spares are patented.

(c) Even for local equipment the major spares are so expensive that it would be cheaper to recondition them atleast to increase its life time.

(d) If the spares available otherwise cause compatibility problems with the equipment, then it would be wiser to recondition the existing spare or modify the newly purchased spare to suit the equipment.

(e) Critical components like shafts, gears are generally made to high precision to last longer than other parts of the machine. Minor adjustment or re-machining might render them re-usable as good.

(f) The budget limitations to the department may make it difficult to replace each and every time a part or equipment fails, especially in case of unplanned or emergency breakdowns.

(g) During most of the preventive maintenance schedule for the disassembly, the additional labor for reconditioning would be much less than the cost of disassembly.

(h) If the existing machine is custom built to suit the process, it would be highly expensive to get a new replacement machine of the same specifications.

(i) After the completion of the expected trouble free life time the reconditioning/ replacement of critical components can be planned, as a routine.

(j) When outside expert subcontractors who do a perfect reconditioning job are available, it is better to opt for reconditioning.

(k) In any case a meticulous cost-benefit analysis has to be made for *'make or buy decision'* viz repair or replace decision.

## 23.4 When to Replace Machinery and Equipment

Replacement models generally are concerned with comparing the increasing operating costs (and possibly decreasing revenues) associated with aging equipment against the net proceeds from alternative equipment. In certain cases when the machine itself gets replaced due to factors cited below.

1. *Deterioration of the equipment* due wear and tear, misalignment of the shafts, increase in the running friction etc. These lead to:
   (a) Decline in the performance and productivity compared to a new machine of similar specifications.
   (b) Increases in accident proneness,
   (c) Increase in maintenance costs,
   (d) Reduction in product quality,
   (e) Increased operation cycle times,
   (f) Decreased operational efficiency,
   (g) Increase in labor costs and overall operation costs.
2. Obsolescence in technology
   (a) Machines with advanced technology and higher productivity at the same or marginal increase in the cost being available in the market.
   (b) Procurement of machines with higher technology would have a positive impact on the reputation of the company among the competitors.

(c) Increase the profitability of the operation.

(d) The obsolete machine can be disposed off at its depreciated value.

(e) If the original equipment manufacturer (OEM) has discontinued the equipment making, the availability of spares impossible or difficult.

3. Inadequacy

(a) The need for increasing the production capacity necessitates new machinery with higher production capacity to replace the existing machines, even though they are not obsolete,

(b) In ever diversifying environment, industries always need to introduce new products and new processes. The existing machines may be inadequate necessitating replacement by more versatile machines.

4. Working conditions

(a) If the existing machines are accident prone, they may have to be replaced immediately to increase the employee motivation.

(b) Availability of new machines that are less noisy or smoky makes it preferable for replacement

## 23.5 Factors for Investment Decision on Facilities

1. Technical factors
2. Plant Efficiency
3. Machine capacity
4. Process capability Index (cpk) of the machine
5. Maintenance requirements
6. Operative skills needed
7. Availability of spares
8. Safety for operatives.

## 23.6 Phases of Replacement Analysis

The four phases of Replacement analysis are:

1. *Creative phase* involving searching for data on several alternatives from available data or research and tabulating or listing the data in these alternatives in logical and easily comprehendible manner.

2. *Definitive phase* involving quantification of these alternatives into definite data for future evaluation. Some alternatives can be straight away eliminated if they are beyond the scope of the present project or against the company policy or any such reason that should be clearly defined and specified in the remarks.
3. *Analysis phase* involving conversion of the data on these alternatives into comparable common denominators like present cash outlay, capital recovery period, finances, effect on the employees etc. Items not definable in quantitative terms must be described qualitatively for comparison and evaluation.
4. *Decision phase* when these alternatives thus described and tabulated must be reviewed and compared with quantitative overall cost effectiveness and the qualitative benefits to the organization and the employees. Judgement is added where facts are missing.

## 23.7 Methods of Evaluation

The methods for equipment replacement analysis can be grouped into traditional methods and discounted cash flow methods as under.

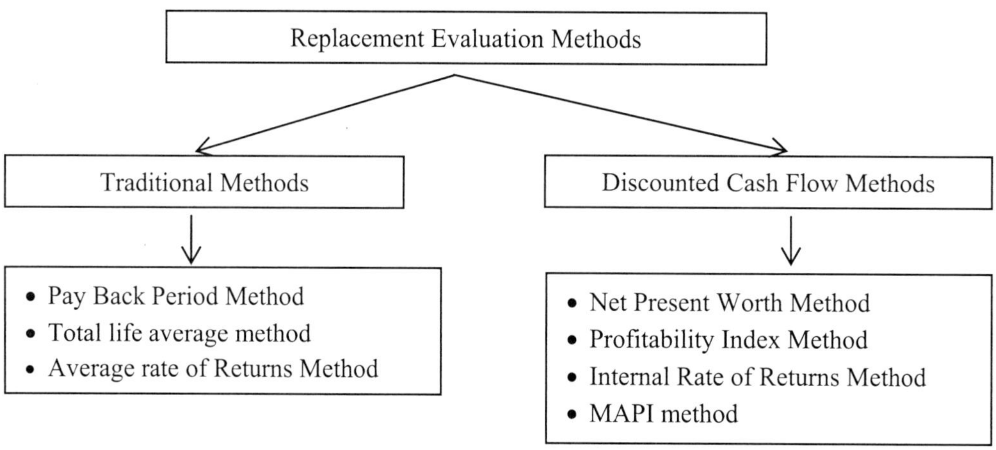


## 23.8 Traditional Methods

### 23.8.1 Payback Period Method

Payback period is defined as the number of years required to recover the original cash investment. In other words it is the period of time at which a machine, facility, or other investment has produced sufficient net revenue to recover its investment costs.

If P is the payback period, $C_i$ and $C_s$ are the initial cash investment and the final scrap value at the end of the period, and $C_a$ is the average annual cash flow, then

$$P = \frac{(C_i - C_s)}{C_a}$$

If the payback period calculated as above is less than the minimum acceptable then accept to procure the new equipment.

### 23.8.2 Total Life Average Method

Here the total cost involved in purchasing, operating and maintaining equipment are added and the total is divided by the number of its life years, to get the total cost per year; this figure is compared with that of all the alternatives to select the best. In principle this is the converse of the Payback period method.

If $C_a$, $C_e$, $C_o$, and $C_m$ are the average cost, equipment cost, operating cost and maintaining cost and L is the expected life of the equipment, then

$$C_a = \frac{(C_e + C_o + C_m)}{L}$$

### 23.8.3 Average Rate of Returns Method

Here the investment proposal judged based on their relative profitability. If I, P and $C_a$ are the total investment, the total number of years the project is expected to cover and the average cash flow per year (which is equal to the total net profit after tax for all the years divided by the number of years) respectively, the Average Rate of Returns is given by

$$ARR = \frac{(C_a \times P)}{I}$$

*Disadvantages of the traditional methods:*

- They fail to consider the inflation and the resultant reduction of the value of the money over time
- They ignore the length of the project
- They donot consider the concept of cash flow, but follow the accounting of the income.

## 23.9 Discounted Cash Flow Methods

### 23.9.1 Net Present Worth (NPW) or Net Present Value (NPV)

NPV compares the value of a rupee today to the value of that same rupee in the future, taking inflation into account. For example, if the inflation rates 10%, Rs. 100

today will be worth only Rs 90 after one year. Its value would further fall down to Rs. 81, 72 and 64 after 2, 3 and 4 years respectively.

The net present value NPV is given by:

$$NPV(i,N) = \sum_{t=0}^{N} \frac{R_t}{(1+i)^t}$$

where

t is the time of the cash flow

i is the inflation rate and

$R_t$ is the net cash flow

Conversely the future value of today's equipment is given by the formula

$$F = P(1+i)^n$$

where,

F is the future worth of an equipment

P is its present value

i is the average rate of inflation or the interest rate

n is the number of years.

For example buying an equipment for Rs 10,000 today, it is as good as buying it for $10000(1+0.10)^{10}$ = Rs 25,940 after 10 years at an inflation rate of 10%.

In case of comparison of alternatives, accept the acceptance whose net present value of the annual cash flows are higher.

### 23.9.2 Profitability Index Method

Profitability Index is the ratio of the present value of each cash inflow to the present value of each cash outflow.

$$P_i = \frac{PV_i}{PV_o} \times 100$$

where, $P_i$ is the profitability index while $PV_i$ and $PV_o$ are the present values of the total cash inflows and outflows. The equipment that has higher profitability index is accepted, all things being equal.

### 23.9.3 Internal Rate of Returns Method

The internal rate of return (IRR) or economic rate of return (ERR) is a rate of return used in capital budgeting to measure and compare the profitability of investments. It is that rate at which the sum of the discounted cash inflows is equal to the sum of the discounted cash outflows. It equates the present value of the cash inflows to the present value of the cash outflows.

If the IRR is greater than the cost of capital involved in procuring the equipment, accept the proposal, otherwise reject it.

### 23.9.4 Differences between Net Present Worth and Internal Rate of Returns Methods

1. The NPW takes the interest rates as a known, predetermined value while IRR takes an unknown value.
2. NPW seeks to determine the amount that can be invested in a given project so that its anticipated earnings will be enough to repay this amount with interest at the market rate. IRR seeks to find the maximum rate of interest at which the funds invested in the project could be repaid with the cash inflows arising out of the project.
3. In short, NPW refers to reinvestment of the inflows at discounted rate, while IRR refers to the reinvestment of the cash flows at maximum interest rate. Hence NPW is more reliable than IRR.

## 23.10 MAPI Method

The Machinery and Allied Products Institute (MAPI) has suggested a procedure for equipment replacement analysis and capital investment analysis, called MAPI method. It calculates the relative profitability of the reinvestment in relation to the cash release, revenues and costs of next year. In MAPI method, the existing equipment is called the *defender* while the proposed equipment is called the *challenger*. The other terminologies used are:

*Capital cost (P)*, which is the average annual cost of capital recovery with interest.

*Time adjusted average* is the uniform annual amount of either the capital cost or the operational inferiority based on the selected life and interest rate.

*Operating inferiority* is the amount by which a machine is operationally inferior to the best alternative available. It accumulates in a machine over a period of time due to wear and tear and hence cannot produce as much product quality as it could when it was new.

*Adverse minimum*, DAM for defender and CAM for challenger, is the lowest sum of the time adjusted average of capital cost and operating inferiority obtainable from a machine.

*Annual inferiority gradient* (g) is the annual amount of inferiority that the challenger (the new machine) accumulated during its life period.

*MAPI urgency rating (D)* is the percentage of annual rate of return expressed in terms of net savings available from the machine divided by the net investment needed. This is given by the formula

$$\text{MAPI urgent rating D} = \frac{BG + VK - ES - EK}{IK}$$

where (based on the next year),

BG is the change in operating profit, i.e., the changes in the inputs and outputs through the replacement investment.

VK is the avoided capital consumption (depreciation and maintenance).

ES is the resulting tax burden, for example, 30% of (BG + VK) at 30% tax

IK is the investment cost, i.e. the avoided cost less repairs in all remaining years of use of the old system and their residual value.

EK is the emergent capital consumption

D is the urgency scale higher than the discount rates, the replacement investment is beneficial to the MAPI method.

Among several possible investments the one with the highest urgency scale is to be selected.

## 23.11 Markov Analysis

Markov Analysis, named after Russian mathematician Andrei Andreyevich Markov, is a method forecasting the value of a variable whose future value is independent of its past history. In finance, it introduces a method for forecasting random variables. It is helpful in estimating the proportion of a company's accounts receivables that will become bad debts and forecasting future brand loyalty of current customers. In maintenance it is helpful in evaluating alternative machinery and in replacement analysis. This is discussed more in paragraph 9.14 also.

## 23.12 Conclusion

The principle of discounted cash flow suggested by Machinery and Allied Products Institute is essentially a cost analysis for the repair or replacement of an equipment. The checklists given in the appendix would help the maintenance engineer to take decisions in adverse conditions.

# Appendix I
## Checklist for Equipment Management
### (Source: Wikipedia)

1. **Policy and objectives**

   1.1 How is equipment management integrated into company policy?

   1.2 Are equipment management policy and objectives set and prioritized correctly?

   1.3 Are the management guidelines and evaluation criteria good?

   1.4 Are long-term and annual plans integrated?

   1.5 Are policy and objectives thoroughly understood and implemented?

   1.6 Are accurate checks done?

   1.7 Are the year's results considered in formulating goals, objectives, and plans for the next year?

2. **Organization and operation**

   2.1 Are the organization and personnel assignments right for managing the equipment?

   2.2 Is a good organization in place for promoting TPM?

   2.3 Is the TPM promotion organization in close contact with production lines?

   2.4 Are all the necessary departments participating fully in TPM?

   2.5 Is effective co-ordination being done between the head office and the factories?

   2.6 Are there any impediments to the exchange and use of information?

   2.7 Are relations good with any subcontractors responsible for equipment, dies, tools, and maintenance work?

3. **Small-group activities and autonomous maintenance**

   3.1 Is the organization of the small group right?

   3.2 How are small-group activity objectives set?

   3.3 Do the groups meet frequently and are their meetings lively?

   3.4 Are there lots of good suggestions, and are they handled properly?

   3.5 How is goal attainment ascertained?

   3.6 Do operators take the initiative in maintaining their equipment?

4. **Training**

   4.1 Do the different departments understand TPM?

   4.2 Are the training programs broad enough and their curricula right?

   4.3 Do the training programs follow the curricula?

   4.4 Do people take part in outside training programs?

   4.5 How many people have technical or expert certifications?

   4.6 How knowledgeable and skilled are people in maintenance techniques?

   4.7 How is skill assessment done?

   4.8 Is there any way to make sure the training is having an effect?

5. **Equipment maintenance**

   5.1 Is 5-S being implemented?

   - Is the machinery clean and free of dust, filings, oil, and other waste material?
   - Have policies been instituted to deal with dirt, places that cause foiling, and places that are difficult to clean, inspect, and lubricate?
   - Are lubrication labels, gauge limitations, bolt tighten-to-her marks, and other visible indicators used?
   - Are all of the tools, materials, gauges, and other things stored neatly and kept clean?

   5.2 Are equipment diagnosis techniques used in these cases?

   - Cracking, corrosion, and loosening.
   - Abnormal vibration, heating, and noise.
   - Water, oil, gas, and air leakage.

   5.3 Are power lines, water lines, hydraulic lines, and other lines neatly and properly handled?

   5.4 Are oils properly selected and properly replaced or filtered at the appropriate intervals?

6. **Planning and management**

   6.1 Are the appropriate efforts being made to improve maintenance techniques and efficiency?

   6.2 Are equipment standards properly set and enforced in a planned manner?

6.3 Are monthly and annual maintenance plans drawn up and implemented properly?

6.4 Are the purchasing plans for spare parts and other maintenance equipment properly drawn up (e.g., how much of what to buy from where) and are such things cared for?

6.5 Are equipment blueprints well cared for?

6.6 Are the dies, tools, and gauges properly cared for?

6.7 Are goods records being kept on equipment wear and equipment failures that mandate stoppages or other maintenance efforts?

6.8 Are maintenance records used to improve processes?

6.9 Are the right maintenance techniques properly applied?

**7. Equipment investment plans and maintenance planning**

7.1 Is equipment investment coordinated with the development of new products and new processes?

7.2 Is equipment investment cost-effective?

7.3 How is the plant investment budget drawn up and controlled?

7.4 Are maintenance planning suggestions duly reflected in equipment investment standards?

7.5 Are reliability and maintenance duly considered in selecting, designing, and placing equipment?

7.6 Is equipment closely monitored in the start-up stages?

7.7 Is the company good at developing its own dies, tools, and equipment?

7.8 Are policies promptly instituted to keep major problems from recurring?

7.9 Are plant assets properly managed?

**8. Production volume, scheduling, quality, and cost**

8.1 Is equipment control closely integrated with production volume and scheduling?

8.2 Is equipment control closely integrated with quality control?

8.3 Are maintenance budgets drawn up and managed properly?

8.4 Is energy and other resource conservation practiced?

**9. Safety, sanitation, and environmental conservation**

9.1 Are sound policies in place for safety, sanitation, and environmental conservation?

9.2 Are the right organizations in place for safety, sanitation, and environmental conservation?

9.3 Are safety, sanitation, and environmental conservation methods known and practiced?

9.4 Is equipment investment integrated with safety, sanitation, and environmental conservation considerations?

9.5 Are safety and sanitation polices paying off?

9.6 Do environmental policies meet all of the legal requirements?

**10. Results and Assessments**

10.1 Are result properly measured?

10.2 Are policies being implemented and objectives met?

10.3 Is maintenance paying off in terms of enhanced productivity and other management aims?

10.4 Does the company make an effort to publicize its activities and its successes?

10.5 Does the company know where the problems are?

10.6 Have plans been drawn up for future progress?

## Appendix II

## Some Definitions from the website of American Inst. of Industrial Engineers

*Replacement Study:* An economic analysis involving the comparison of an existing facility and one or more facilities with equal or improved characteristics proposed to supplement or displace the existing facility.

*Rate of Return (Internal Rate of Return):*

1. The interest rate earned by an investment.
2. The interest rate at which the present worth equation (or the equivalent annual worth or future worth equations) for the cash flows of a project or project increment equals zero.
3. As used in accounting, often it is the ratio of annual profit, or average annual profit, to the initial investment or the average book value
4. As per 'Your Dictoionery.com" - The discount rate at which the present value of future cash flows equals the current market price of the investment or opportunity. Also called dollar-weighted rate of return.

*Required Return:* The minimum return or profit necessary to justify an investment. Often it is termed interest, expected return or profit, or charge for the use of capital. It is the minimum acceptable percentage, no more and no less.

*MAPI method*: A procedure for equipment replacement analysis developed by George Terborgh for the Machinery and Allied Products Institute. It uses a fixed format and provides charts and graphs to facilitate calculations. A prominent feature of this method is that it includes explicitly an allowance for obsolescence.

*Minimum Attractive Rate of Return:* The effective annual rate of return on investment, either before or after taxes, which just meets the investor's threshold of acceptability. It takes into account the availability and demand for funds as well as the cost of capital. Sometimes termed the minimum acceptable return.

*Economic Life:* The period of time, extending from the date of installation to the date of retirement from the intended service, over which a prudent owner expects to retain an equipment or property so as to minimize cost or maximize net return.

*Depreciation:*

1. Decline in value of a capitalized asset.
2. A loss of value due to physical or economic reasons.

3. A form of capital recovery, usually without interest, applicable to property with two or more years' life span in which an appropriate portion of the asset's value periodically is charged to current operations.
4. In accounting, depreciation is the allocation of the book value of this loss to current operations according to some systematic plan. Depending on the existing income tax laws, the amount and timing of the charge to current operations for tax purposes may differ from that used to report annual profit and loss.

*Sinking Fund Depreciation:*

1. A method of computing depreciation in which the periodic charge is assumed to be deposited in a sinking fund that earns interest at a specified rate. The sinking fund may be real but usually is hypothetical.
2. A method of depreciation where a fixed sum of money regularly is deposited at compound interest in a real or hypothetical fund in order to accumulate an amount equal to the total depreciation of an asset at the end of the asset's estimated life. The depreciation charge to operations for each period equals the sinking fund deposit amount plus interest on the beginning of period sinking fund balance.

*Straight-Line Depreciation*: A method of computing depreciation wherein the amount charged to current operations is spread uniformly over the estimated life of an asset. The allocation may be performed on a unit of time basis or a unit of production basis or some combination of the two.

*Future Worth:*

1. The equivalent value at a designated future date based on the time value of money.
2. The monetary sum, at a given future time, which is equivalent to one or more sums at given earlier times when interest is compounded at a given rate.

*Economic Life*: That period of time after which a machine or facility should be retired from primary service and/or replaced as determined by an engineering economy study. The economic impairment may be absolute or relative.

*Physical Life:* That period of time after which a machine or facility can no longer be repaired or refurbished to a level such that it can perform a useful function.

*Service Life:* That period of time after which a machine or facility cannot perform satisfactorily its intended function without major overhaul.

*Obsolescence*:

1. The condition of being out-of-date. A loss of value occasioned by new developments which place the older property at a competitive disadvantage. A factor in depreciation.
2. A decrease in the value of an asset brought about by the development of new and more economical methods, processes, and/or machinery.
3. The loss of usefulness or worth of a product or facility as the result of the appearance of better and/or more economical products, methods, or facilities.

*Present Worth (Present Value):*

1. The monetary sum which is equivalent to a future sum or sums when interest is compounded at a given rate.
2. The discounted value of future sums.

*Present Worth Factor:*

1. Mathematical formulae involving compound interest used to calculate present worth of various cash flow streams. In table form, these formulae may include factors to calculate the present worth of a single payment, of a uniform annual series, of an arithmetic gradient, and of a geometric gradient.
2. A mathematical expression also known as the present value of an annuity of one (The present worth factor, uniform series, also is known as the annuity fund factor).

*Salvage Value*:

1. The cost recovered or which could be recovered from a used property when removed from service, sold, or scrapped. A factor in appraisal of property value and in computing depreciation.
2. Normally, an estimate of an asset's net market value at the end of its estimated life. In some cases, the cost of removal may exceed any sale or scrap value; thus net salvage value is negative.
3. The market value of a machine or facility at any point in time.

*Estimated Replacement Value (ERV):* The cost of capital works required to replace specific existing assets, including necessary facilities (such as buildings, energy, and other supply systems) with new assets capable of producing the same quantity and quality of output. This is a key value often used in benchmarking activities at the highest level. It is not the book value, nor the current cost accounting value, nor costs to build a state of the art replacement. ERV is an estimate of the *current* replacement cost (normally the insurance value).

To calculate the ERV, determine if the original equipment investment figures reasonably agree with equipment actually in use. Then identify clusters of equipment by the year in which they were acquired. Consider each cluster of investment and escalate it to a current value using a selected inflation index (e.g., the Bureau of Labor Statistics Construction Cost Index (BLS CCI)). Then the indexed clusters of investment are totalled to get the current value of plant and/or equipment. The indexed value of the plant could be compared with other plants recently built, adjusting for size and available insurance values. Even when a company is self-insured, there is normally an established "insurance value" to help define the financial exposure of the company risks. An insurance underwriter typically prepares these values, even if the plant is self-insured. Underwriters follow a procedure very similar to the one described.

ERV is often used to normalize maintenance cost as a convenient basis for comparing plants of a similar type that vary in size. The rationale for using the ERV rather than the original cost of the plant is to account for construction cost escalation over time (inflation). Two relatively new plants built 10 years apart could have original costs that vary by 50 to 100 percent. Maintenance cost/ERV can be used to set long-term goals, and targets for plant reliability. World-class plants tend to fall in the range of 1 to 2.5 percent.

# Chapter 24

# Environmental Issues

***Chapter synopsis:*** *The definitions of the basic environmental issues, sustainable development and the international initiatives taken in this direction are discussed in this chapter. The role of the maintenance engineer in sustaining the environment and the principles of green design are also highlighted.*

***Keywords:*** *Environmental management system, environmental audit, sustainable development, global environmental issues, Magna Carta of environment, World Environment Day, United Nations Conference on Environment Development, Rio Declaration, Kyoto Protocol, green house gases, ozone depletion potential, National Productivity Council, Oil spills, inflatable boom, environmental impact assessment, green design, environmentally conscious manufacture, environmental ethics, cost oblivious approach and cost benefit approach.*

## 24.1 Introduction

Industrial Revolution and the consequent development of new chemical industries and other manufacturing processes involving chemicals have caused substantial pollution of the environment. This pollution was considered as a necessary evil of the economic development of a country and was not taken seriously till the middle of 20$^{th}$ century.

Although nature was the main focus of the nineteenth and twentieth century philosophies, contemporary environmental ethics emerged as an academic discipline only in the 1970's. The questioning and rethinking of the relationship of human beings with the natural environment reflected an already widespread perception in the last thirty years.

Multiplication of this problem resulting in issues like depletion of ozone layer has created a high level awareness as well as a movement among all concerned. This movement sought to highlight the role of the engineer as the key figure in the development of processes that result in pollution, and that he is also responsible to develop counter processes to control pollution at the same time, achieving the desirable production targets. This created an awareness of what the engineer must do and commit himself in controlling the degradation of environment. Among all fields of engineering it is the maintenance field that can play a larger role in sustaining environment and hence this chapter highlights his role after explaining the basic environmental issues and concepts.

## 24.2 Basic Environmental Concepts

Before explaining the ethical aspects of environment, it is better to have the basic concepts of the environmental issues as given by the following definitions.

### Definitions

*Environment* is the surroundings in which an organisation operates, including the air, water, land, natural resources, flora, fauna, humans and their interrelationship. Surroundings in this context extend from within the organisation to the global systems.

*Environmental Management system* is the overall management system that includes organisational structure, planning activities, responsibilities, practices, procedures, processes and resources for developing, implementing, achieving, reviewing and maintaining the environmental policy.

*Environmental management system audit* is a systematic and documented verification process of objectively obtaining and evaluating evidence to determine whether an organisation's EMS conforms to the EMS audit criteria set by the organisation or to the international standards and for communicating the result of this process to the management.

*Environmental policy* is a statement from the organisation of its intentions and principles in relation to its overall environmental performance that provides framework for action and for the setting of its environmental objectives and targets.

## 24.3 Sustainable Development

- The medieval concept of industrial development was that if development has to take place a little of environment has to be sacrificed. If use of chemicals is necessary for a better process, discharging the effluents of these chemicals into open is inevitable.
- During the mid-twentieth century global consciousness on the need to sustain the environment despite the development has been created.

- In simple terms, ***sustainable development*** is the development process without any destructive activity, thus integrating development with environment.
- It is the process of meeting the basic needs of the present generation without effecting the life and scope of the future generations.

## 24.4 Global Environmental Issues

- Global warming (or greenhouse effect)
- Acid rain
- Ozone layer depletion
- Transboundary movement of hazardous wastes
- Ocean contamination
- Decrease in diversity of wild life
- Deforestation
- Desertification
- Pollution of rivers, lakes and ponds
- Water related diseases among humans as well as animals.
- Increasing respiratory illnesses
- Unmanageable solid and hazardous waste generation.

Apart from the above, some of the recent environmental issues faced in Indian context are:

- Artificial ripening of mangoes by harmful chemicals as unearthed in May 2005.
- Bio-medical disposal, as highlighted in the press during August 2005.
- Toxic wastes from the bleaching and dying units in South India, that became a major issue in 2004-2005, etc.

## 24.5 International Initiatives on Environmental Issues

At the United Nations Conference on Human Environment held at Stockholm on 5$^{th}$ June 1972, the famous declaration called the *Magna Carta on Environment*, was made and the following resolutions were made.

- Man has the fundamental right to freedom, equality and adequate conditions of life in an environment of quality that permits a life of dignity and well being.
- Man bears a solemn responsibility to protect and improve the environment for the present and future generations.

The U.N. General Assembly adapted the above resolution in December 1972 and declared 5th June as *World Environment Day*.

The other global initiatives are:

- ✓ Charter on Economic Rights and Duties of States in 1974.
- ✓ United Nations Habitat Conference on Human Settlement held at Vancouver in 1976.
- ✓ United Nations Desertification Conference at Nairobi in 1977.
- ✓ Inter-Governmental Environmental Education Conference at Georgia in 1977.
- ✓ United Nations World Water Conference for 'Clean drinking water and sanitation for all' in November 1980.
- ✓ World Conference on Environment Development held in 1984 where the expression of *'Sustainable Development of Environment'* was conceived. An Earth Summit for the 1992 in Rio de Janeiro was planned.
- ✓ Montreal Protocol in 1987.
- ✓ Clean Air Act amendments of USA in 1990.
- ✓ Copenhagen Amendments in June 1992.
- ✓ United Nations Conference on Environment Development (UNCED) Earth Summit on Environment was held at Rio De Janeiro in 1992, the most strategic international effort for the preservation of environment, where the Rio Declaration on Environment and Development was signed by all participating countries. The details of Rio Declaration is given in Annexure I.
- ✓ Kyoto Protocol in 1997, which is one of the latest and most significant document, required the industrialised nations to reduce emission of carbon dioxide and other green house gases (GHGs) by an average of 5.3% below the 1990 level by 2012. There was a consensus of the scientific community, including that in US, that the climatic change is a man-made phenomenon and there is an urgent need to cut down the carbon emissions. This had come into force as late as 20th February 2005 after all the countries have ratified it except USA, which was blamed for its big brotherly attitude by all other nations including European Union. Even during the latest G-8 conference conducted in Gleneagles in Scotland during May-June 2005, the US administration initially refused to accept the above consensus but in deference to the pressure from the European countries, conceded to the above consensus. But till day there has not been any positive step taken to achieve Kyoto protocol.

Nevertheless, in general, most of these initiatives as cited above have emphasised and succeeded to some extent in the following:

- ✓ Categorising the ozone depleting substances like CFC's and halogens according to their ODP (Ozone Depletion Potential) and gradual reduction of the production of these items.
- ✓ Reduction in the use of high ODP materials in automobile and refrigeration industries.
- ✓ Removal of air conditioners, foam insulation materials and CFC's from cars before crushing them.
- ✓ Use of warning labels on the products containing High ODP materials.
- ✓ Banning of non-essential products using ozone depleting substances.

## 24.6 Environmental Scenario of India

The National Productivity Council, in one of its seminars on Environmental Management Systems, has indicated the following scenario in India with respect to environment pollution.

- India is one of the top twelve among impure water countries.
- 75% of our rivers, lakes and pond waters polluted.
- 75 million working days per year are lost due to water related diseases.
- India is World's third largest generator of organic water pollutants, discharge load being 17,60,000 kg/day, which is 8% of the global discharge.
- Phasing out of ozone depleting substances like CFCs, is not effective so far.
- Overall environmental compliance by industries is only 45%.
- Cleaner production technology and techniques are not adapted in industry.
- Reactive approach of industry for pollution and its related problems is not defined.
- Lack of awareness of green consumerism and green markets.
- Poor awareness among public on environmental issues.

## 24.7 Oil Spills

Not with standing the above scenario, a recent news item reported that Chennai Port has adapted an inflatable oil boom, that would contain the oil that has spilled into the waters and prevent from further spreading, before it could be skimmed off, or otherwise removed. As a matter of fact such a boom was witnessed by the author at San Diego Port as illustrated in the photograph below.

**Figure 24.1** The inflatable boom operating from a tugboat at San Diego port. (Photograph taken by the author).

## 24.8 Environmental Impact Assessment

As a result of the Bhopal Gas Tragedy, Indian environmentalists have given serious thought to the process of Environment Impact Assessment (EIA) to fulfil the following objectives.

- To identify adverse environmental problems that may be expected to occur,
- To incorporate into the development of action oriented measures,
- To identify the environmental benefits and drawbacks of the project, as well as its economic and environmental acceptability to the community,
- To identify the critical environmental problems that require further study and monitoring,
- To examine and select the optimal alternative from the various relevant options available,
- To invite the public in the decision-making process related to the environment,
- To assist all stakeholders and those involved in the development and environmental matters to understand their roles, responsibilities and overall relationships with one another.

## 24.9 Engineer's Role in Environment Protection

1. As an experimenter involving environmental issues, the engineer must be aware of his role in Environmental protection. He should be human-centric and eco-centric.
2. He should have full knowledge and confidence in his projects and should be meticulously careful to foresee the environmental effects of the project activities.
3. He should have sincere concern about the environment during the project planning and execution stage and ask the following questions for himself:
    - How does the industry effect the environment?
    - How far such ill effects, if any, can be controlled?
    - Is political or physical regularisation needed?
    - Whether reasonable protective measures are available for immediate implementation?
    - Whether the engineer as an individual can ensure safe and clean environment?
4. He should preplan all the activities and processes and the control systems without frequent replanning or redesigns.
5. He should plan for safe exits.
6. He must budget the funds required for these control systems including the safe exits etc.

## 24.10 Environmental Engineering

Environmental Engineering is the application of engineering principles during all stages of setting up a manufacturing unit with a view of protecting and enhancing the quality of environments and enhancing the public health and welfare. These include green design, green manufacture and development of environment friendly systems for disposal of solid, liquid and gaseous wastes.

## 24.11 Green Design

Green Design involves systematic consideration of environmental objectives and factors during all stages of developing products, services or production processes. This also includes the planning for recyclability, biodegradability and other eco-friendly disposals of the products after their life cycle. Other names for Green Design are Design for Environment (DOE), Environment Engineering etc.

## 24.12 Principles of Green Design

- Consider the physical and chemical structure of the material before selection
- Avoid toxic materials
- Evaluate manufacturing processes
- Design for longer life
- Design for ease of assembly and disassembly
- Incorporate source reduction
- Design for interchangeability
- Design for recycling
- Review packaging.

Avoid design and use of throw-away-after-use materials like the ultra-cheap carry bags made of plastic.

## 24.13 Environmentally Conscious Manufacture

Environmentally conscious manufacture (ECM), also called *Green Manufacture,* is to reduce the harmful effects of manufacturing and by developing processes and methods of manufacture suitably to reduce the generation of harmful wastes, emissions and other hazardous effects of and to reduce the energy consumption. The principal issue is reduction or elimination of occupational hazards. One striking example is the elimination of the use of lead in soldering or as an additive in petrol or paints in view of its toxicity. Another example is the mandatory reduction in the manufacture of asbestos sheets.

## 24.14 Basic Approaches for Resolving Environmental Problems

(a) ***Cost oblivious approach***: All efforts are to be made to make the environment as clean as possible, whatever, may be the cost to do so.

  - No level of environmental degradation is accepted.
  - This approach is somewhat similar to that of *rights and duty ethics.*
  - Though ideal, this approach has two obvious problems.
    - It is difficult to define exactly what is *as clean as possible* and
    - In the highly competitive world of Indian industry, where every rupee counts, industries try to use the above vagueness as a loop hole and only try to do minimal expenditure to provide short run measures enough to create an impression among the public that they are protecting the environment which may not be true in the long run.

(b) *Cost Benefit Approach*: The problems are analysed in terms of the benefits derived by reducing the pollution problems. The costs and the benefits are weighed to determine the optimum combination.

- Here the target is not to achieve a completely clean environment but an economically viable environment protection.
- This can also be compared to *Utilitarian theory.*

However, this approach too suffers from 4 major difficulties.

- It is difficult to assess the true cost of human life or loss a species or environmental protection.
- It is difficult to assess accurately the costs and benefits, and much guess work or factor of safety has to go into the calculations.
- This approach does not necessarily specify who should bear the cost and who should get the benefit.
- The cost benefit analysis does not take morality and ethics into account. The decision is simply based on mathematical simulations and calculations and there is no room for a discussion whether what is done is right or wrong

(c) *A combination of professional and personal ethics*: Unlike professional decisions like bridge designs, projects involving environment effects the engineer is involved personally even as a member of the public. Hence the need to apply his decision making with reference to his personal ethics.

Whatever may be the approach it is essential that the engineer applies both his professional and personal ethics at the same time meticulously following laws and regulations of the state.

From the perspective of human health, the engineer's responsibility to protect is clear, which must be balanced between the consideration of the well being of his employer, the public, the community.

Reference may be made by the maintenance engineer to the books of Professional Ethics and Human Values, which specify the risk benefit analysis an engineer should perform before undertaking a project to control environmental pollution caused by his company. This analysis requires a comprehensive estimation and evaluation of the risks taken and the benefits that accrue by these risks. It entails careful quantification of the costs associated with a particular program for reducing or avoiding the risks. This is something like the cost-benefit analysis. Both deal with the resultant costs incurred by the company in the form of losses / damages or the costs incurred in reducing the defects or improving the operating methods are estimated.

In risk-benefit analysis, the risks due to the hazardous elements of the design or the operation are assigned a rupee or a dollar value and compare both the rupee or the dollar value of the benefits accrued by taking the risk. Then the most beneficial proportion of the risk is chosen.

## 24.15 Conclusion

Having understood the scenario of the environmental issues and the evolution of the environment management systems, it is hoped that the engineer perceives any process or the industrial project from the environment perspective and applies his mind in achieving his objective of the project with minimal environmental degradation.

# Chapter 25

# Industrial Safety

*Chapter Synopsis: The primary goal of spreading awareness in Industrial safety is to manage risk, eliminating or reducing it to acceptable levels. This is significant in cases of press shops where the maximum number of accidents involving loss of hands and toes are known to have occurred in industry. This chapter lists some of the factors that should be kept in mind in designing safety features in an industry.*

***Keywords:*** *Machine guard, non-incombustible guards, interlocking mechanism, gears, friction drives, sprockets, automatic feeding, railed foot walk, personal protective equipment , eye protection, welding and grinding goggles, National Safety Council, safety engineer, Central Safety Committee, Workmen Safety Committees, safety inspection, plant safety tours.*

## 25.1 Introduction

In conformation of certain university syllabi that includes Industrial safety as a topic of Maintenance Engineering and Management, this chapter indicates some aspects of safety factors that should be followed to ensure safe operating environment in an engineering factory especially a press shop. Nevertheless the subject of Industrial safety itself is very vast with several separate books, which may be referred for further information.

The ultimate objective of an organization depends significantly on how safe are the working conditions provided in the plant. Safety is not only compatible with all objectives of an organization but is also a positive aid. The successful operation of safety measures depend on good planning, design, installation, processing and operation of safety schemes. A safe factory would be far more productive than an unsafe factory. There is significant interplay between the production departments and safety problems in one production department would have a chain reaction in other departments, quite often resulting in production delays. Fig. 25.1 illustrates safe working habits in sustaining accident-free environment.

## 25.2 Essentials of a Machine Guard

1. The guard shall be so designed and constructed to prevent accidents both to the operator at his regular work or to others, who may fall or slip or carelessly touch the machine.
2. Should not interfere with production (if so, it is liable to be bypassed deliberately).
3. In general, should be attached to the machine. If attached to floor, use suitable connections that would interfere least.
4. The guarded part must be easily accessible for oiling, inspection and repair besides normal working by a door or removable section, hinged or fastened to the machine.
5. The guard should not interfere with the sweeping and cleaning of the neighbouring floor. Usually at 6 feet above the ground.
6. Should be strong enough to resist injury and keep its shape.
7. Non-combustible guards like CI sheet metal wire mesh etc. are preferred.
8. When subjected to acid fumes, wooden guards are necessary.
9. Should have interlocking mechanism where possible, to prevent operation without guards.
10. The guard shall be designed to prevent wear on the parts guarded, e.g. a solid gear enclosure.

## 25.3 Mechanical Power Transmission

A. *Gears, Friction Drives, Sprockets etc*:

1. All gears and friction drives shall be completely enclosed or atleast have a continuous band guard, provided with side flanges, extending inwards beyond the roots of the teeth.
2. All sprockets must be completely encased.
3. Chains shall be guarded as prescribed for belts.
4. Where there is spoke hazard, gears and sprockets shall be encased on the exposed sides.
5. CI guards are preferable, as they fit smugly, present a good appearance and protect from dust. Sheet metal makes the guards cheaper and can be fabricated in a variety of sizes and shapes.
6. In large guards, angle iron is used to make the joint between the flat side and curved part of the guard. For small guards, these joints may be done by cutting projections, bent, spot welded or riveted.

B. *Friction Clutches*: Any revolving part of the clutch having dangerous projections within 7 inches of the floor shall be completely and substantially encased. This also includes any transmission shafting or any start-stop devices that revolves even when the machine is stopped.

C. *Belts, Chains and Ropes*:

1. If vertical or inclined, complete guarding upto 7 feet from the floor.
2. If both tracks are horizontal, and within 6.5 feet, the guard shall be at bottom and till 15 inches above the belt, the sides shall be guarded.
3. If both tracks are horizontal, and above 7 feet, the guard shall be at bottom and till 6 inches above the belt, the sides shall be guarded.
4. For pulleys below 6.5 feet, complete casing (or railing for large diameter) shall be provided.

D. *Shafting*:

1. For vertical or inclined, complete casing and provision of troughs below.
2. Horizontal below 7 feet, complete casing and provision of troughs below.
3. Shaft ends or hubs of diameter greater than 2 feet shall be cordoned off.

E. *Projections of Revolving Parts*: Revolving parts like set screws, keys, oil caps located over the rim of the pulley rim, couplings with protruding bolts and nuts, broken pulleys or gears should be carefully designed and encased.

F. *Belt Shifters:* Each set of tight and loose pulleys should have belt shifter levers to shift the belts from one pulley to another, without being touched by hand.

G. *Individual and Group Drives:* Group shafting is adapted where there are compact groups of machines which are to be run continuously or simultaneously. They should be arranged to prevent any interface between the pulleys or the belts or the gears.

H. *Wood Working Machines*

1. Automatic feeding is safer than hand feeding.
2. *Hoods* should be provided on wood working machines to remove the saw dust, chips or shavings. Otherwise they
   - Clog to the teeth and reduce efficiency.
   - Obstruct work.
   - Dangerous to operator.
   - Make the floor slippery and increase fire risks.
3. Floors should be kept neat

4. Saw guards for circular saws, rip saws, swing saws, band saws, travelling cut off saws.
5. Jointers, planers, shapers, matchers, molders, stickers etc.

**I.** ***Metal Working Machines like Lathes, Presses, Drills, Planers and Grinders***

**J.** ***25.3 and Elevators: EOT Cranes should have***

1. Railed foot walk with a height more than 18 inches shall be provided along the bridge
2. Fixed ladder between cab and foot walk
3. Fixed ladder from floor to cab
4. Bumpers at each end of rails
5. Warning gong or other signals
6. Railing or other enclosures on all sides of cab
7. Fenders on bridge wheels
8. Hoist limit switches with dynamic braking.

**K.** ***Engine Flywheels and Pits:*** Where high speed flywheels are used, they should properly guarded and accidental access during motion prevent.

**L.** ***Boilers:*** All boilers and pressure vessels should be operated and maintained Per OSHA regulations.

**M.** ***Portable Ladders***: All portable ladders must always be removed from the place of use and safely kept aside.

## 25.4 Characteristics of Safe Industrial Buildings

A. Clearance between the building and rail track shall be minimum 12 inches
   - Between wall and ends of crane bridge
   - Between EOT crane and the existing or future OH pipes, shafting etc.
   - Between stairs and the floors above and below.

B. Accessibility of valves, switches, oil cups, lamps etc. should either by location or by providing runways, stairs etc.

C. Careful planning for the crossings of railroads, roads and pedestrian crossings.

D. Proper lighting is necessary for safety, health and efficiency.

E. Atleast two safe exit points are to be provided. Stair cases enclosed in fire resisting walls are preferred. Horizontal exits run through fire resisting walls or bridges or balconies to other rooms or sections.

F. All stairs must be provided with railings.

G. Floors shall be of such materials that will not become slippery by usage or become uneven. Ceramic stairs or mosaic stairs are dangerous.

H. Floor opening shall be guarded with railing.

## 25.5 Eye Protection

A. As per US bureau of standards, common eye hazards are

1. Large flying objects like chips
2. Dust and small flying particles like in grinding
3. Dust and wind like automobile smoke
4. Splashing metal like in babbiting
5. Gases, fumes and liquids like in acid handling
6. Excessive dust sand flying particles like in sand blasting
7. Reflected light or glare like in snow covered grounds
8. Injurious radiant energy of moderate intensity like oxy-Acetylene welding
9. Injurious radiant energy of high intensity like electric welding or furnace proximity

B. The lenses of chipping goggles must withstand 10 blows of 16 grams of steel, falling 54 inches from above or blown 36 inches from below.

C. Goggles used for grinding may be lighter.

D. Antirust and antacid head covers must be corrosion resistant and must fit the face properly.

E. Goggles for welding must not retransmit visible and invisible rays as per norms.

F. Other protections

1. Leggings - Steel capped toes
2. Protective cloths
3. Gloves
4. Apron
5. Full clothing.

## 25.6 Supervision and Training

A. *50% to 90% of reduction in accidents is possible by:*

1. Mechanical guarding
2. Redesign of machinery and review of operating methods
3. Creation of awareness

4. Strict supervision and enforcement
5. Training of workmen both in general work habits of carefulness and safe methods in specific jobs.

B. *National Safety Council has recommended the following 10 steps:*

1. Secure cooperation of manager
2. Secure cooperation of the supervisor
3. Appoint a safety engineer
4. Appoint a central safety committee
5. Secure cooperation of foremen and sub-foreman
6. Organize foreman's safety meetings
7. Appoint foreman's inspection committee
8. Hold safety meetings for all employees
9. Appoint a workman's safety committee
10. Start use of safety bulletin boards.

C. *Role of Workers:*

1. Ensure that he works as per the acceptable safety practices.
2. Realize that others are concerned about his safety as much as and as sincerely as himself.
3. Realize that the extra time spent on safe practices is worth the trouble.
4. Serve on the safety committees and give safety suggestions.
5. Undertake a job only after fully understanding it.
6. Cooperate with the concerned personnel fully in the investigation of accidents and implementation of safe practices.

D. *Role of Supervisor in Industrial Safety:*

1. Accept safety as much part of the job as the production, quality etc., functions.
2. Recognize the relationship between safety and good house keeping.
3. Give adequate safety instructions to new employees as well as to old employees on new jobs.
4. Take corrective action when safety rules are violated and make subordinates take safety rules seriously as another company rules.
5. Cooperate with the safety engineer and safety committees.

E. *Role of Safety Engineer:*

1. Make regular visits and discover unsafe practices, unsanitary conditions and non-use of safety guards.

2. Keep in touch with the factory manager and secure their cooperation.
3. Recommend methods safe guarding and follow up methods.
4. Develop and conduct safety training programs.
5. Safety Job Hazard Analysis for each job and exhibit them in bulletin boards.
6. Supervise the bulletin board services and safety educational activities.
7. Investigate into all serious accidents and near accidents.
8. Keep records and prepare reports on accident experience.

F. *Role of Workmen Safety Committees:*

1. Make regular visits and discover unsafe practices, unsanitary conditions and non-use of safety guards.
2. Instruct and warn workmen against unsafe acts.
3. Investigate into all serious and near accidents and submit reports.

G. *Role of Central Safety Committee:*

1. Discuss on controversial issues and formulate safety methods.
2. Review reports of engineers, foremen and workers.

H. *Role of Top Management in Safety:*

1. Assume full responsibility and accountability for the safety in the company.
2. Identify executives responsible for safety.
3. Authorize necessary expenditure for safety programs.
4. Pass on safety policies as formulated by the safety department or the safety committees. Reference may be made to the policies of other organizations if necessary and if possible.
5. Plan and organize regular safety inspections.
6. Participate in the safety programs where necessary.

## 25.7 Safety Inspection

### 25.7.1 Principal Function

The principal function of the plant safety engineer or any other officer responsible for safety is to pinpoint the unsafe conditions and ensure their rectification before any eventuality occurs. A planed inspection of all equipment and facilities with respect to safety is essential to any safety program.

### 25.7.2 Objectives

1. To locate faults and rectify them before an accident occurs.
2. To ensure that all the equipment and facilities are constantly under safe conditions and they do not lead to serious damages to property and life.
3. To keep the working condition in the factory under acceptable standards.
4. To enthuse a sense of confidence in the employees that the management is serious about the safety factor. The inspection should not only regular and as planned, but should be participated by the Chief Executive, worker representatives etc.
5. To create a sense of participation among the employees and to ensure their cooperation in the promotion of safety.
6. To ensure that all jobs are performed as per the established safe practices and that unsafe work habits are controlled.
7. To ensure that the safe work methods and working conditions are promptly reported.
8. To pave the way for effective enforcement of safety rules.

### 25.7.3 Purpose of Safety Inspections

1. To see that equipment, machine tools, and parts are in good condition and not be worn or damaged to cause accidents.
2. To ensure that all the above are properly safe guarded.
3. To ensure material used in the manufacturing process do not create an uncontrollable health or explosion hazard.
4. To ensure that all safe exits are properly maintained and cleared of any obstructions.
5. To ensure that the personal protective devices, fire equipment, first aid equipment, machine guards are used properly.
6. To ensure that the aisles, floors, stairs, elevators, ramps etc. are being maintained in safe condition.
7. To check illumination and noise levels.
8. To check all work practices to ensure that they confirm to safety standards.

### 25.7.4 What to Inspect

A: *Work Area:*

1. House keeping
2. Job layout

3. Aisles and exits
4. Stairs, ramps and ladders
5. Floors and other working levels.

B: *Material Handling*:

1. Piling and storage
2. Loading and unloading
3. Racks, platforms and bins
4. Hand signals
5. Cranes, chains, hoists, slings etc.

C: *Hand Tools:*

1. Condition of hand tools
2. Improper use of hand tools
3. Difficulties if any, in securing hand tools
4. Housekeeping as related to hand tools.

D: *Machines used:*

1. Lack of physical safeguards
2. Physical condition of safeguards
3. Actual protection offered by the safeguards
4. Location of controls
5. Mechanical condition of the machinery and equipment with respect to safety.

E: *Personal protective equipment*:

1. Plant wide safety tours conducted by the management, safety personnel and others.
2. Safety inspections of sections of the plants by the teams of line supervisors and safety personnel.
3. Regularly scheduled inspection of the specific equipments and machines by trained personnel.
4. Special safety appraisal procedures like those for new or relocated equipment.
5. Safety tours conducted by employee union.
6. Inspection conducted by external organizations like insurance companies, industrial hygienists, fire prevention specialists.
7. Statutory inspection conducted by government through factory inspectors etc.

8. Constant safety vigilance by the supervisor.
9. Inspection of some equipment like tool boxes by the contractor himself.
10. Surprise safety checks and inspection tours by the chief executive, employee union etc.

## 25.8 Plant Safety Tours

Emphasis should be placed in periodic safety tours around the factory premises joined by the chief executive to create enthusiasm and confidence in the workers and secure their whole hearted cooperation. It is also essential to publicize the corrective action taken as a result of these tours. This is similar to Gemba Walk in Japanese terminology, which is described as under.

> *Gemba walk is the practice of senior managers to tour the several places of the factory along with the concerned operatives with the basic purpose of identifying the areas of improvements. Each and every Muda (a Japanese word referring to 'waste') noticed or suggestions offered by anyone for improvements would be recorded and analyzed and posted on the notice board to motivate the operatives. This is generally done during the afternoon hours or Sundays to create a friendly and holiday atmosphere which may be followed by snacks or lunch. Preferably the Chief Executive Officer accompanies them to instill interest among the workers.*

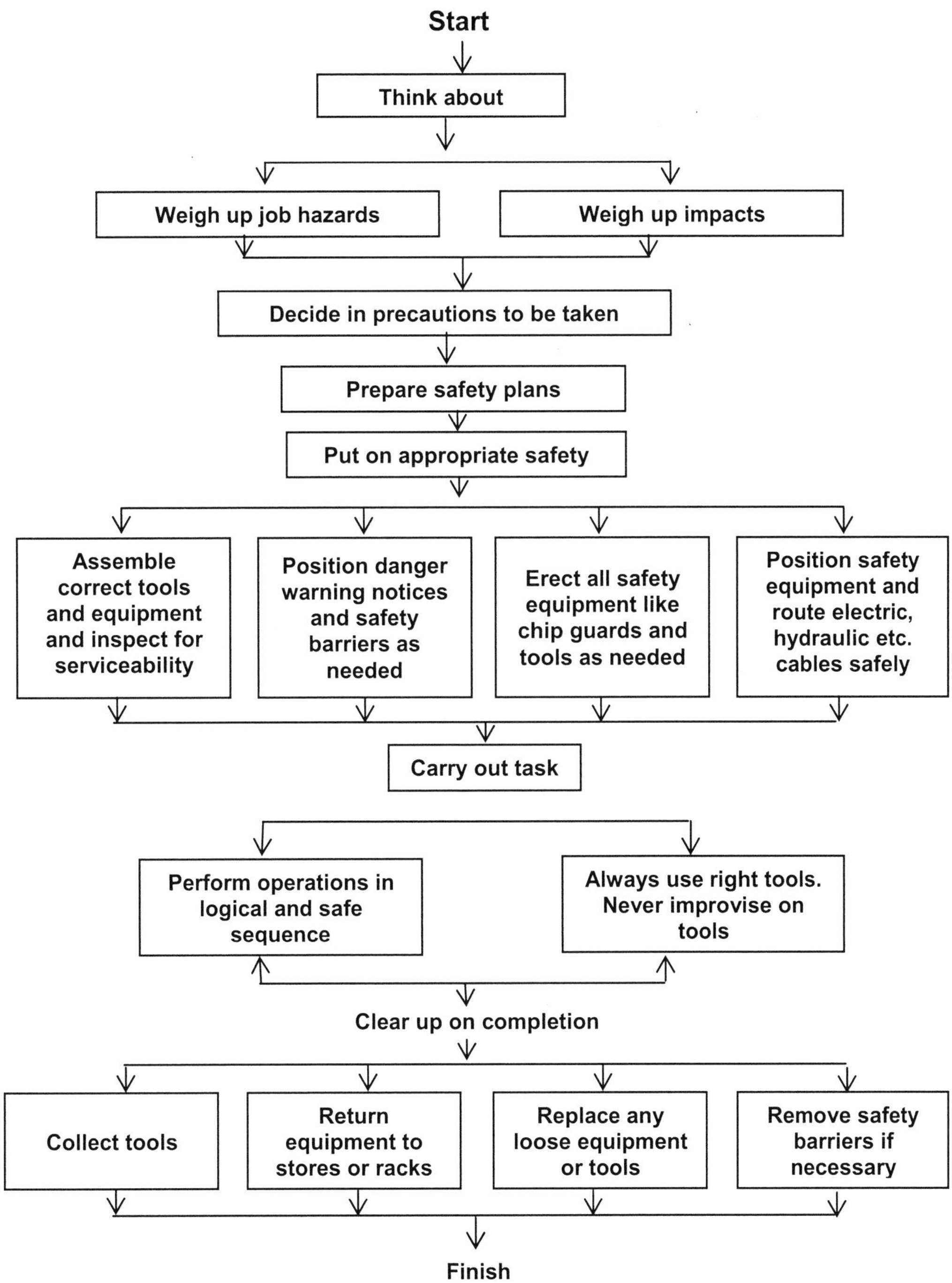


**Figure 25.1** Illustration of safe working habits.

## 25.9 Conclusion

Contrary to the traditional belief that workers are accident prone and accident occur due to human errors, ILO has emphasized that in most cases the accidents are caused due to equipment faults or conditions beyond the control of the operative. Hence a clear cut understanding of the anatomy of accidents and how they can be prevented is essential. This chapter hence cites some of the safety standards for industrial equipment. The next chapter aptly describes systematic method of analyzing how the accidents are likely to happen and take preventive measures.

## Appendix

Illustration of a typical authoritative memorandum from the Factory Inspectorate to a small industry failing in industrial safety.

### MINISTRY OF LABOR - FACTORY INSPECTORATE

To

ABC Automotive Parts Company

Your reference:

Our reference:

Dear Sirs,

At a recent visit of the Inspector of Factories to your premises, it was observed that the matters below required your attention. I am seriously concerned by the poor standard of working conditions and safety in your factory. All of these matters have previously been brought to your notice in my earlier letter and I must therefore advise you that the continued failure to comply with the law will result legal proceedings being taken that may result in withdrawal of the license issued to you.

## 1.0 HEALTH

### 1.1 Cleanliness

Workrooms and canteen are not kept in a clean state. Whitewashing of walls and ceilings of workrooms was long overdue. Also the floors of the canteen and heat treatment departments had not apparently been cleaned at least once a week.

### 1.2 Temperature

Effective provision was not made for securing and maintaining a reasonable temperature in the Thread Rolling, Cold Rolling and Capstan Departments.

### 1.3 Sanitary Conditions

Sanitary conveniences were not kept clean and effective provision is not made for lighting them.

## 2.0 SAFETY

The following were not securely fenced.

1. Flat belt and pulley drives of lathes no. 6 and 7 in the capstan department.
2. Stock bars projecting around the headstocks of lathes nos. 3, 8 and 9 in the same department.

3. Vee belt and pulley drives of two hacksaws and the compressor guards should be extended to provide total enclosure.
4. Vee belt, pulley drives, low shafting and bevel gears of the rumbling barrels are to be fully enclosed.

## 2.1 Floors

1. The floor of the Capstan Department was not kept clean from oil and swarf and this likely to cause persons to slip.
2. Gangways in the General Stores and Dispatch Department were not kept clear of obstructions such as stillage and containers.

## 2.2 Electricity Regulations

Regulation I - There was much temporary wiring on the premises. You are reminded that this should comply fully with these regulations and in particular with Regulation No. 1, which requires that all apparatus and conductors shall be so constructed, installed, protected, worked and maintained so as to prevent dangers as far as it is reasonably practicable.

Regulation 2- The covers of several fuse boxes were not securely closed.

## 2.3 Horizontal Milling Machine Regulations

Regulation 3- The cutter of the Cincinnati Milling Machine was not fenced by a strong guard enclosing the cutting surface except such part as is necessarily exposed for milling operation.

## 2.4 Woodworking Machinery Regulations

Regulation 10- The raving knife at the circular saw was not capable of the required adjustment (i.e. to within not more than 0.5 inches from the teeth of the saw). There was no adjustable top guard for the saw.

## 3.0 AIR RECEIVERS

The Air receiver had not been thoroughly cleaned and thereafter thoroughly examined by a competent person within the previous 26 months and a report in Form 59, containing the prescribed particulars obtained. The cleaning and examination should be carried out forthwith and the required report obtained. Please forward a copy of the report which will be returned to you.

Yours faithfully,

District inspector of Factories.

# Chapter 26

# Job Hazard Analysis

*Chapter Synopsis:* *Any accident, however minor it may be, causes not only production hold ups but results in injuries to the worker and hospitalization expenditure to the Company. This chapter advocates the principle of 'prevention is better than cure' whereby every job is analyzed for the safety aspects and a well-planned and executed Job Hazard analysis is developed during the design stage prior to the execution of a process.*

*Keywords:* *Accident free work environment, hospitalization, production holdups, job hazard, OSHA regulations, definitions for JHA, Risk assessment, ALARP, work place safety, method study, SREDDIM, selection of a job for JHA, checklist, work habits, creative examination, protection equipment health hazards, JSA software, ESS portal.*

## 26.1 Why Accident Free Work Environment?

One of the major functions of the process engineer is to ensure that each operation is performed in the equipment without any mishap or accident. It must be noted that any accident, however minor it may be, causes not only production hold ups but in injuries and hospitalization expenditure to the worker, thereby annulling whatever savings effected by the improvements in productivity achieved by the method improvement study undertaken. A preplanned and well executed job hazard analysis during the design stage prior to the execution of a process would go a long way in harnessing productive hours which otherwise would have been lost due to accidents. It is the author's experience that in many medium and small scale industries he has visited, it is only after the occurrence of a mishap that the cause is investigated and job modified to avoid reoccurrence of the accident. The principle of 'prevention is better than cure' is advocate by analyzing the job for the safety aspects at the design stage itself, giving due consideration for the possible hazard content from the view point of machine features, handling methods needed to be adapted etc. By this production losses due to accidents can be reduced if not eliminated.

Though JHA is discussed in general, the principles and procedures are applicable for all situations in industry.

## 26.2 What is a Hazard

A hazard is the potential for an accident. In practical terms, a hazard often is associated with a condition or activity that, if left uncontrolled, can result in an injury or illness. It is for this reason that every job that has been under operation or in the process of being designed has to be analyzed meticulously with regard to the safety aspects, so as to eliminate or reduce the hazards associated with the job.

## 26.3 OSHA Definition on JHA

A job hazard analysis (JHA) is a procedure which helps to integrate accepted safety and health principles and practices into a particular task or job operation. In a JHA, each basic step of the job is to identify potential hazards and to recommend the safest way to do the job. Other terms used to describe this procedure are job safety analysis (JSA) and job hazard breakdown.

## 26.4 OSHA Regulations

- The Occupational Safety and Health Act (OSHA) of 1970 provides for the standards formulated by the Dept. of Labor for lifting the health and safety to a higher plateau throughout American Industry.
- Where the specific standards are not available, the employer is obliged to develop these standards relating to the conditions, practices, means, methods, operations and processes used as the General Duties of the employer to safeguard the employees. For tasks that are complex, unusual, difficult, require the interaction of many people or systems or involve new tools or methods, a JSA should be performed.

The Job Hazard Analysis forms a part of these General Duties.

## 26.5 How Accidents are Caused?

The pre-1920 concept of accidents is:

- Industrial injuries result only from industrial accidents.
- Accidents are invariably caused by unsafe acts of persons or by exposure to unsafe mechanical conditions.
- Unsafe actions and conditions are caused by fault of persons, and
- Faults of persons are created by environment or acquired by inheritance.

However later thinking gives a holistic approach than blaming the operator squarely. ILO elaborates this view as:

- Man is not a machine. His performance is not fully predictable and he sometimes makes mistakes. A mistake may be made by:
  - The architect who designed the factory,
  - The contractor who built it,
  - The machine or the process designer, the manager, the engineer, the chemist, the electrician, the foreman, the maintenance man, the operator - in fact everyone who has anything to do with the design, construction, installation, management, supervision and use of the factory or anything in it.

## 26.6 Definition and Purpose

Job Hazard Analysis is an organized manner in which each job is analyzed critically, element by element either at the process design stage or at the operation stage, to achieve the following objectives

- To identify the potential hazard in each element of the job
- To develop solutions to eliminate or control the hazards
- To specify safe practices, safe equipment and the dos and don'ts for each stage of the operation.

OSHA and Wikipedia define JHA as follows:

A Job Hazard Analysis is a technique that focuses on job tasks as a way to identify hazards before they occur and the relationship between the worker, the task, the tools, and the work environment. Ideally, after you identify uncontrolled hazards, you will have to take steps to eliminate or reduce them to an acceptable risk level

*- Occupational Safety and Hazards Administration(OSHA)*

Job Safety Analysis is also called Job Hazard Analysis, one of the safety management tools that can be used to define and control the hazards associated with a certain process, job or procedure. Job Safety Analysis is a term used interchangeably with Job Hazard Analysis and Risk Assessment. The purpose of a JSA is to ensure that the risk of each step of a task is reduced to ALARP (As Low as Reasonably Practicable).

*- Wikipedia*

## 26.7 Methodology

While a majority of books or papers treat Job Hazard Analysis adapting several methodologies, this book prefers to adapt a methodology similar to *Method Study*, which the author had successfully adapted in his several projects. The relationship between method study and JHA can be better understood by the equation.

**Method Study : Critical Examination Technique:**
**Work Place Safety : Job Hazard Analysis**

This methodology involves the steps **SREDDIM,** meaning,

- Select,
- Record,
- Examine,
- Develop,
- Define,
- Install and
- Maintain.

Now let us see how the established method study procedure can be applied for successful development of JHA.

### 26.7.1 Select

Though as a principle, all the 100% of the jobs should be analyzed, the factors contributing for the priority selection are given below to which priorities may be assigned.

- Major accidents
- Frequent accidents at a single work station
- Heavy loss of production hours due to injuries
- Complaints from the supervisors or workers about excessive hazards in specific jobs
- Introduction of new processes or operations.

Wikipedia gives detailed explanation to the above five as:

- *Accident frequency and severity*: Jobs where accidents occur frequently or where they occur infrequently but result in disabling injuries.
- *Potential for severe injuries or illnesses*: The consequences of an accident, hazardous condition, or exposure to harmful substance are potentially severe.
- *Newly established jobs*: Due to lack of experience in these jobs, hazards may not be evident or anticipated.
- *Modified jobs*: New hazards may be associated with changes in job procedures.

- *Infrequently performed jobs*: Workers may be at greater risk when undertaking non-routine jobs, and a JSA provides a means of reviewing hazards.

*Jobs appropriate for job hazard analysis can be summarized as follows*

- Jobs with the highest injury or illness rates;
- Jobs with the potential to cause severe or disabling injuries or illness, even if there is no history of previous accidents;
- Jobs in which one simple human error could lead to a severe accident or injury;
- Jobs that are new to your operation or have undergone changes in processes and procedures; and
- Jobs complex enough to require written instructions.

It may be mentioned here that while the first four factors are generally realized and adapted in today's industry, the last one, viz performing the job hazard analysis during the design stage is rarely done. This chapter hence emphasizes this requirement. Though it discusses in general, the principle and practices are applicable for all the factors.

Questions to be asked to identify potential hazards

- Can any part of the body get caught in or between objects?
- Do tools, machines, or equipment present any hazards?
- Can the worker make harmful contact with moving objects?
- Can the worker slip, trip, or fall?
- Can the worker suffer strain from lifting, pushing, or pulling?
- Is the worker exposed to extreme heat or cold?
- Is excessive noise or vibration a problem?
- Is there a danger from falling objects?
- Is lighting a problem?
- Can weather conditions affect safety?
- Is harmful radiation a possibility?
- Can contact be made with hot, toxic, or caustic substances?
- Are there dusts, fumes, mists, or vapours in the air?

### 26.7.2 Record

Break up each and every job into elements and record them in relevant charts and diagrams as listed below:

Some of the useful charts are:

- Outline process charts
- Flow process charts
  - Man type
  - Material type
  - Equipment type
- Two handed process charts
- Flow charts
- Process charts
- Multiple Activity Charts
- Chronocyclographs come handy when analyzing fast movements which you feel have unsafe elements.

*Remember :*

1. Involve the direct employees, since they have a unique knowledge of the job and the potential hazards. This will also help in minimizing oversights and motivate the workers to accept the solutions. Remember it is the Japanese attitude of treating the operator as the central figure, that helped Japanese industry to succeed.
2. Brainstorm with them for ideas to eliminate or control those hazards.
3. Review the accident history with the employees and supervisors.
4. Conduct a preliminary Job Review.
5. If any hazards exist that pose an immediate danger to an employee's life or health, take immediate action to protect the worker. This will demonstrate your commitment for safety.
6. List, rank, and set priorities for hazardous jobs.
7. Outline the steps or tasks.

### 26.7.3 Examine and Evaluate

Critically analyze each and every element with respect to the potential hazards and list them in charts. The analysis can be done

- By machine or equipment type
- By job, assembly or product flow
- By profession category like plumbing, painting etc.
- By the extent of movement involved like machine operating or tote box carrying.

*Principles of Creative Examination*

1. Divide and conquer
2. Set quotas and deadlines for yourself
3. Let loose your mind
4. Two heads are better than one
5. Question each and every detail

The last point is the cream of creativity. Rudyard Kipling said in his famous poem;

"I had six stalwart serving men,
They taught me all I know,
Their names were What and Where and When
and Why and How and Who "

Again as per our Hindu philosophy,

*Esa sarveshu bhuteshu gudho atma na prakasate,*
*Drsyate tvagryaya buddhya sukshmathi suksmam darshibi'*

Atma or the truth is in all beings, but hidden and therefore is not manifest. It can be realized however, by the concentrated reasoning of those who have trained themselves in perceiving subtle and more subtle truths.

This is the basis for all research analyses and the basic principle behind all the engineering techniques, whether method study or reengineering or any analytical or research process. We call it ***creativity***.

*How to Identify Workplace Hazards?*

Your goal is to discover the following, and document the answers to these questions in a consistent manner

- What can go wrong?
- What are the consequences?
- How could it arise?
- What are other contributing factors?
- How likely is it that the hazard will occur?

Also identify

- Where it is happening (environment),
- Who or what it is happening to (exposure),
- What precipitates the hazard (trigger),

- The outcome that would occur should it happen, and
- Any other contributing factors.

### 26.7.4 Develop

Through the critical examination techniques as illustrated above, develop ideal alternatives not only to specify safe practices, but also to enable the operative and everyone associated with the job, to get the awareness of the potential hazards and how to adapt the safe practices.

The following are the points to be borne in mind.

- ➢ Develop an entirely different way of doing the same job
- ➢ Revise the job procedure
- ➢ Correct the defective equipment like the guards
- ➢ Change the physical conditions like posture and position of controls
- ➢ Train the workers and the supervisor.

*Safety Factors in Developing*

1. Keep materials handling to a minimum.
2. Provide safe working surfaces on floors, stairs, platforms, gangways etc.
3. Provide adequate space for machinery and equipment.
4. Provide safe access to every place to which the workers have to go.
5. Provide for adequate safe exits in case of fire etc.
6. Provide for safety of maintenance and repair personnel.
7. Provide for safe transport facilities.
8. Isolate dangerous processes like spray painting or processes with high fire risk.
9. Always buy machinery with built in safety devices.

### 26.7.5 Install and Maintain

Like any method improvement study, it is not enough in developing or implementing the proposals. Ensure the following:

- ➢ Publish the analysis charts on the notice board or in the form of posters in simple and specific words.
- ➢ Train each operator in the specified procedure.
- ➢ Ensure strict adherence to these safety instructions.
- ➢ In case of any queries from the workers, clarify them clearly and fully.
- ➢ If any accident still happens, start investigations by checking if any of these safety instructions are violated or overlooked.

Wikipedia suggests that after the JSA worksheet is completed, the work group that is about to perform the task should have an on-the-job discussion on the hazards and controls, delegate responsibilities, ensure that all equipment and PPE described in the JSA are available and that contingencies such as firefighting are understood, communication channels and hand signals are agreed etc. Then, if everybody in the work group feels that it is safe to proceed with task, work should commence.

## 26.8 Some Suggestions for making a Job Hazard Analysis

1. Make a list of all the jobs to be analyzed. Assign priorities and set target dates of completion.
2. Analyze the job while it is being performed.
3. List all the tools used in each step.
4. Make sure that the potential hazards are described in a way as to indicate the step that should be taken by the workers to avoid them.
5. Give full specifications of the safety equipment when indicated or suggested.
6. Formulate and use a checklist to ensure that all the instructions are followed.

## 26.9 Use and Follow up of Job Hazard Analysis

1. Check the analysis with the supervisor, the worker and other concerned personnel and encourage them to make additions and modifications to our suggestions.
2. Always explain the 'why' of everything to the workers.
3. Make sure that a concerned person understand the chart and realize the potential hazards.
4. Discuss freely with the workers regarding the unsafe work habits. Point out with humorous undertone, but never criticize the habit, not even sarcastically. Let him feel the importance of correcting the habits.
5. Make sure that the analysis is available for ready reference. Never consider such documents as confidential and inaccessible affair. Keep the salient features exhibited on or near the machine.

## 26.10 Advantages of Job Hazard Analysis

1. Acts as a reference not only for safety aspects but as future method improvement studies.
2. Makes the supervisors to realize the seriousness of unsafe practices that are otherwise treated as minor, and also helps them to monitor the worker's safe practices.

3. Provides a steady comparison between the existing and the proposed systems so that
   (a) The management can be concerned about the need for the new system.
   (b) The related engineers, supervisors and workers can discuss the issue together and arrive at further improvements.
   (c) The benefits of safe practices can be presented effectively.
4. It facilitates the analysis of the hazard potential of a job before the accident occurs.
5. Ensures to secure good cooperation from the labor unions.
6. Aids enforcement of stricter safety rules.
7. Helps in planning safety education and training.
8. Increases the morale of the workers.
9. Reduces the loss of production hours and increases overall productivity.
10. Reduces the hospitalization expenditure for the management.

## 26.11 Checklist for Job Hazard Analysis

*Work Habits:*

- Lifting and handling
- Body posture
- House keeping
- Horse play
- Care of tools

*Availability of protecting equipment:*

- Gloves
- Respirators
- Eye protection
- Safety shoes

*Health hazards:*

- Solvents
- Dust, fumes and gases
- Chemicals
- Hot atmosphere

- Glare
- Ambient noise

*Adherence to work rules:*

- Red and yellow tags
- Non-smoking areas
- Treatment of injuries

*Clothing:*

- Loose hair and/or head covering
- Ties and long sleeves
- Rings, bracelets etc.
- Slippers

*Equipment features:*

- Unprotected moving parts
- Sharp corners and edges
- Projections hampering normal movement
- Emergency stop switches beyond normal work area
- Insufficient space or need for inconvenient posture while handling or operating
- Absence of yellow demarcating lines

## 26.12 Software for JHA

The Employee Self Service portal www.ess-home.com/products suggest a software for JHA, which is detailed below. It

- Identifies hazards and required control measures for each task step.
- Associates personal protective equipment (PPE) to each task step to help prepare written PPE programs, and pinpoint root causes associated with PPE.
- Defines standard phrases for describing hazards and control measures and duplicate JHAs for similar jobs.
- Creates a list of required PPE for any employee, based on job title, location and department.
- Attaches Windows supported media files, images, and documents to JHAs, including observational videos, facility diagrams or photographs showing potential hazards.

- Documents environmental or ergonomic conditions that may be present during a job or activity.

The benefits & features of the software can be summarized as follows

| Benefits | Features |
|---|---|
| **Recognizes and defines job hazards** | Identifies hazards and required control measures for each task step. Document environmental or ergonomic conditions |
| **Institutes safety processes for hazard reduction** | Associates personal protective equipment (PPE) to each task step and pinpoint root causes associated with PPE. |
| **Develops consistent and thorough JHA auditing processes** | Attaches Windows supported media files, images, and documents to JHAs, including observational videos, facility diagrams or photographs showing potential hazards. Records all parties involved in JHAs, including those interviewed, observed, or otherwise concerned. |
| **Documents non-regulatory factors** to identify high risk functions or areas | Associates JHAs with job titles, departments, and locations to track potential areas or job roles where job hazards may be more likely to occur. |
| **Defines proactive measures** | Associates training requirements to the JHA and those will appear as required training in the affected employee's record in the Training Management Software |
| **Saves time** by eliminating duplicated effort | Defines standard phrases for describing hazards and control measures and reapply JHAs for similar jobs. Uses duplicate function to copy over full JHA content for similar tasks or for jobs at another facility |

## 26.13 Conclusion

While this chapter explains the hazard analysis with respect to the existing jobs, it also emphasizes the need for such analysis at the time of process planning stage itself, with due consideration of the possible hazards. Since a majority of today's text books do not cover this aspect, it is hoped that this chapter would trigger of a series of experiences and case studies from practicing engineers.

## 26.14 References

1. Modern Safety Practices by De Reamer
2. Materials Handling by Immer, McGraw Hill Publ.
3. Industrial safety by Paramanayagam, Publ. by Central Labor Inst., India.
4. Industrial Engineering Handbook ed. by H.B. Maynard, 2nd Edition,

5. Production Handbook ed. By Carson, Publ. by Arnold Press
6. Management Handbook for Plant Engineers, ed. By Bernard Lewis
   (a) Chapter 1 - Plant Protection and Safety by Richard J. Healy,
   (b) Chapter 2 - Safety by William Miller.
7. Job Hazard Analysis by James Roughton, and Nathan Crutchfield
8. Encyclopedia of Occupational Health and Safety, ILOPubln, 1972
9. Accident Prevention - A Workers' Educational Manual, ILOPubln, 1972.
10. Kinetic Methods of Manual Handling in Industry ILOPubln, 1967
11. *Job hazard Analysis*by Prof. D. R. Kiran– Manufacturing Technology & Management, Sept 2000, p 27 to 29
12. www.setonresourcecenter.com/safety
13. OSHA Answers: Job Safety Analysis www.ccohs.ca/oshanswers/hsprograms
14. Job Hazard Analysis Software: www.ess-home.com/products/...suite/job-hazard-analysis/
15. Osha Job Hazard Analysis:www.docstoc.com/docs/.../OSHA-JOB-HAZARD-ANALYSIS
16. Safety by Williams Heinrich. Section 10, Chapter 10 -
17. Industrial Safety Handbook ed. by Handley

## Appendix

An illustration of the format for Job Hazard Analysis for a drilling operation

| Company's name | | Date | Page ... of ... |
|---|---|---|---|
| Component | Operation | | Machine and location |
| **KEY ELEMENT** | **MACHINE OR EQUIPMENT** | **POTENTIAL HAZARD** | **SAFE PRACTICE** |
| 1. Collect material for operation | Tote Box | (a) Drop box on foot<br>(b) Back strain<br>(c) Overloaded boxes<br>(d) Slipping enroute<br>(e) Bump against some object enroute | (a) Have firm grip on box<br>(b) Use hard toe shoes<br>(c) Inspect handle of box before lifting<br>(d) Load 25 pieces only into tote box.<br>(e) Ensure the path is fee of oil, swarf, obstructions etc. |
| 2. Clamp one piece in the jig and setup drill | Bench Drill | (a) Machine defective<br>(b) Loose clamping<br>(c) Chuck wrench not removed<br>(d) Adjusting while machine is running<br>(e) Crushed finger while clamping | (a) Report defective jigs and fixtures and discard if necessary<br>(b) attach DO NOT USE tag<br>(c) Ensure proper clamping of the work<br>(d) Ensure proper fitting of drill bit collect<br>(e) Ensure presence of no loose objects<br>(f) Stop machine for all adjustments<br>(g) Use proper blocks or clamps for fixtures. |
| 3. Drill hole | Bench Drill | (a) Loose clothing, hair etc. getting caught in the machine<br>(b) Flying tool<br>(c) Spinning work piece<br>(d) Chips injuring eyes and other parts of the body<br>(e) Broken drill bit<br>(f) Pinch points at belt | (a) Avoid loose clothing, loose hair, rings etc.<br>(b) Be attentive<br>(c) Follow instructions<br>(d) Stop machine and report if work is spinning even if properly clamped<br>(e) Use chip breaker tools<br>(f) Use eye goggles<br>(g) Do not attempt to apply more force than needed for drilling<br>(h) Get worn out belts replaced<br>(i) If machine is defective attach DO NOT USE tag. |
| 4. Unclamp job and carry to inspection table | Tote box | Same as in 1 | Same as in 1 |

# Annexure - I

# University Syllabi Referred to in this Book

1. **ANNA UNIVERSITY-**
   **MBA P03 - ADVANCED MAINTENANCE MANAGEMENT**

**Unit 1 - Maintenance Concept**
Objectives and functions of Maintenance, Types, Maintenance Strategies. Organization for maintenance. Five Zero Concept

**Unit 2 - Failure Data Analysis**
MTBF, MTTF, Useful Life – Survival Curves – Repair Time Distribution Breakdown time distributions, Poisson, Exponential a: nd Normal distribution. Availability of repairable Systems – MaintainabilityPrediction – Design for Maintainability

**Unit 3 - Maintenance Planning**
Overhaul and Repair: Meaning and Difference, optimal overhaul/Repair / Replace maintenance policy for equipment subject to breakdown. Replacement Decisions: Optimal interval between preventive replacement of equipment subject to breakdown, group replacement

**Unit 4 - Maintenance Systems**
Fixed Time Maintenance, Condition based Maintenance, Operate to Failure, opportunity maintenance, Design out maintenance, total productive maintenance.

**Unit 5 - Advanced Techniques**
Reliability Centered Maintenance (RCM) – Total Productive Maintenance (TPM) - Philosophy andImplementation. Signature Analysis – MMIS – Expert Systems –

Concept of Tero technology. Reengineering maintenance process. Total productive Maintenance, Concept of reliability, reliability improvement, Concept of maintainability and Maintainability improvement.

## 2. ANNA UNIVERSITY - PE 803 / ME 2037 - MAINTENANCE ENGINEERING

**Unit - 1 - Principle and Practices of Maintenance Planning**

Basic Principles of maintenance planning – objectives and principles of planned maintenance activity – reliability & machine availability – MTBF, MTTR and MWT – factors of availability – maintenance organisation – maintenance economics.

**Unit - 2 - Maintenance Policies – Preventive Maintenance**

Important and benefits of sound maintenance systems – maintenance categories – comparative merits of each category-preventive maintenance, maintenance schedules, repair cycle – principles and methods of lubrication – TPM

**Unit - 3 - Condition Monitoring**

Condition monitoring-cost comparison with and without CM – on-load testing and off-load testing – methods and instruments for CM – temperature sensitive tapes – pistol thermometers – visual inspection aids – borescopes, endoscopes, fibrescopes, stroboscopes- corrosion coupons-wear debris analysis-

**Unit - 4 - Repair Methods for Basic Machine Elements**

Repair methods for beds, slideways, spindles, gears, leadscrews, and bearings-failure analysis-failure and their developments-logical fault location methods – sequential fault location – midway check method

**Unit - 5 - Machine Erection and commissioning (for PE 803, PROD. ENGG. STREAM))**

Equipment needed for erection, erection procedure, commissioning – check list – safety – equipment records job order systems – use of computers in maintenance

**Unit - 6 - Repair Methods for Basic Machine Elements -2 (for ME 2037, MECH. ENGG. STREAM)**

Repair methods for Material handling equipment - Equipment records –Job order systems -Use of computers in maintenance.

## 3. JAWAHARLAL NEHRU TECHNICAL UNIVERSITY –HYDERABAD 58018: MAINTENANCE AND SAFETY ENGINEERING

**Unit - 1 - Introduction**

Need for Maintenance, Facts and Figures, Modern Maintenance, Problems and Maintenance Strategy for the 21st Century, Engineering Maintenance Objectives, Maintenance Equipment, Life Cycle, Terms and Definitions.

**Unit - 2 - Maintenance management and Control**

Maintenance Manual, Maintenance, Facility Evaluation, Functions of Effective Maintenance Management, MaintenanceProject Control Methods, Maintenance Management Control Indices.

**Unit - 3 - Types of Maintenance**

Preventive Maintenance, Elements of Preventive Maintenance Program, Establishing Preventive Maintenance Program, PM Program Evaluation and Improvement, PM Measures, PM Models, Corrective Maintenance Types, Corrective Maintenance Steps and Downtime Components, Corrective Maintenance Measures, Corrective Maintenance Models.

**Unit - 4 - Inventory Control in Maintenance**

Inventory Control Objectives and BasicInventory Decisions, ABC Inventory Control Method, Inventory Control Models, Two-Bin Inventory Control, and Safety Stock, Spares Determination Factors, Spares Calculation Methods

**Unit - 5 - Quality and Safety in Maintenance**

Needs for Quality Maintenance Processes, Maintenance Work Quality, Use of Quality Control Charts in Maintenance, Work Sampling, Post Maintenance Testing, Reasons for safety Problems in Maintenance, Guidelines to Improve Safety in Maintenance Work, Safety Officers' Role in Maintenance Work, Protection of Maintenance Workers.

**Unit - 6 - Budget**

Preparation methods and Steps, Maintenance Labor Cost Estimation, Material Cost Estimation, Equipment Life Cycle, Maintenance Cost Estimation, Maintenance Cost Estimation Models.

**UNIT VII: RCM**

Reliability, Reliability Centered Maintenance (RCM), Goals and Principles. RCM Process and Associated Questions, RCM Program Components, Effectiveness Measure Indicators, RCM Benefits and Reasons for its Failures, Reliability Versus Maintenance

Reliability in Support Phase, Bathtub Hazard Rate Concept, Reliability Measures and Formulae, Reliability Networks, Reliability Analysis Techniques.

**UNIT VIII: Maintainability**

Maintainability Importance and Objectives, Maintainability in Systems Life Cycle, Maintainability Design Characteristics, Maintainability Functions and Measures, Common Maintainability Design Errors.

## 4. VISWESWARAYYA TECHNICAL UNIVERSITY -MBA SYLLABUS 08MBAPM457 IA: MAINTENANCE MANAGEMENT

**MODULE 1**

Importance of maintenance – objectives of maintenance – structure of the plant – reasons for and nature of maintenance – production – maintenance systems

**MODULE 2**

Types of maintenance – maintenance systems – planned and unplanned maintenance – breakdown maintenance – corrective maintenance – opportunistic maintenance – routine maintenance – preventive maintenance – predictive maintenance – condition based maintenance systems – design-outmaintenance – selection of maintenance systems.

**MODULE 3**

Maintenance planning and scheduling – establishing a maintenance plan – items to be maintained

**MODULE 4**

Safety precautions – characteristics of items to be maintained – classification of items – maintenance procedure – guidelines for matching procedures to items – universal maintenance procedure – establishing a new maintenance schedule

**MODULE 5**

Maintenance organization – resource characteristics – resources structure – maintenance control – administrative structure – training of maintenance personnel

**MODULE 6**

System operations and documentation – documenting maintenance operations – record keeping – data collection and analysis – failure statistics – planning and scheduling plant shutdowns

**MODULE 7**

Depreciation & Machine Life, Replacement policies, spares and types of spares, spares planning, maintenance control, network techniques in maintenance activities, evaluation of maintenance performance.

**MODULE 8**

Total productive maintenance – development and scope – basic systems of TPM – procedures and steps – productivity circles – TPM as a part of TQM – benefits of TPM.

## 5. BHOPAL UNIVERSITY: BE(MECH) - M 451 - MAINTENANCE ENGINEERING & MANAGEMENT

**UNIT - I**

Introduction - Fundamentals of Maintenance Engineering. Maintenance Engineering its importance in material & energy conservation, inventor y control, productivity, safety, pollution control etc. Safety Regulations, pollution problems, human reliability, total quality management (TQM), total productivity maintenance (TPM), environmental issues in maintenance, ISO 9000.

**UNIT - II**

Maintenance Management - types of maintenance strategies, Planned and unplanned maintenance, breakdown, preventive &predictive maintenance -their comparison, advantages & disadvantages. Limitations, computer aided maintenance, maintenance scheduling, spare part management, inventory control, and organisation of maintenance department.

**UNIT - III**

Tribology in Maintenance, friction wear and lubrication, friction & wear mechanisms, prevention of wear, types of lubrication mechanisms, lubrication processes. Lubricants - types, general and special purpose, additives, testing of lubricants, degradation of lubricants, seal & packings.

**UNIT- IV**

Machine health monitoring - Condition based maintenance, signature analysis, oil analysis, vibration, noise and thermal signatures, on line & off line techniques, Instrumentation & equipment used in machine health monitoring. Instrumentation in maintenance, signal processing, data acquisition and analysis, application of intelligent systems, data base design.

**UNIT - V**

Reliability, availability & maintainability (RAM) Analysis - Introduction to RAM failure mechanism, failure data analysis, failure distribution, reliability of repairable and non-repairable systems, Improvement in reliability, reliability testing, reliability prediction, utilization factor, system reliability by Monte Carlo Simulation Technique.

## 6. CHHATISGARGH SWAMI VIVEKANANADA TECHNICAL UNIVERSITY - MAINTENANCE & RELIABILITY ENGG

**UNIT - I**

Maintenance Engineering Objective and functions, organization and administration, economics and maintenance policies. Types of maintenance systems-planned, unplanned, preventive, predictive, conditional monitoring, total predictive maintenance.

**UNIT - II**

Failure Analysis Analysis of source identification, classification and selectivity of failures, catastrophic, wear out and cumulative failures, failure rate Mortality distribution, statistical and reliability concept of failure analysis, equipment replacement policy.

**UNIT - III**

Reliability Engineering Concept, bath tub curve, elements, Hazard Models- constant, linearly increasing, weibull. System Reliability - Series configuration, parallel configuration, mixed configuration, reliability improvement – Improvement of components, Redundancy – element, unit, standby, repairable and non repairable systems, reliability, availability, maintainability, MTBF, MTTR, reliability allocation for simple series system.

**UNIT - IV**

Maintenance Management Maintenance planning, maintenance scheduling, work orders, work measurement, maintenance cost budgeting, store and spare control, maintenance planning and control techniques, Incentives for maintenance work.

**UNIT - V**

Maintenance of Mechanical System Introduction, Bearings, Friction Clutches, Couplings, Fastening Devices, Chains, Gear Drives, Support Equipment, Cooling Towers.

## 7. GUJARAT TECHNICAL UNIVERSITY - MAINTENANCE AND SAFETY ENGINEERING

**1. Introduction:**

Importance of maintenance, functions of maintenance, type of maintenance, including total productive maintenance and its implementation, organization of maintenance.

**2. Wear and service life of equipment:**

(i) Methods of assembly and fitting - assembly of keyed joints, splined joints, fixed joints, assembly of ball and roller bearings, repairs and assembly of gears.

(ii) Wear of machines - types and reasons of wear, defects due to wear of equipment, corrosion and its prevention.

(iii) Recovery and strengthening of machine elements various methods of recovery and and increasing service life.

**3. Maintenance of Production Equipment:**

Maintenance and repair of shafts, bearings, spindles, couplings and clutches, gears, bed services and link mechanisms.

**4. Restoring The Guide Ways Of Machine Tools:**

Test of repaired equipment, fault-tracing sequence in fault tracing, drawing decision tree.

**5. Planning and Scheduling Maintenance Work:**

Factors involved in effective planning of maintenance work, Various methods of scheduling work, Categorization of plant/equipment for the purpose of priorities, VAIN analysis.

**6. Preventive Maintenance:**

Philosophy of PM, methods & schedules. Maintenance cost & replacement economics, Types of cost, Maintenance cost, Methods of cost comparisons, Factors in equipment Replacement, MAPI methods, Economics, Concept of maintainability.

**7. Safety Engg.:**

Background of Industrial safety, Accident Causation, Industrial hazards, Accident investigation, prevention, Safety education, Safety consideration in design of equipment, Legal aspects of Ind.. Safety

## 8. RAJASTHAN TECHNICAL UNIVERSITY - 8ME4.1: RELIABILITY AND MAINTENANCE ENGINEERING

**Unit 1 - Introduction**

Maintenance Objectives and Functions; Maintenance Organisation and Administration of Maintenance Systems. Need of planned maintenance. Maintenance policies; Breakdown, time based maintenance: Block replacement, age replacement and periodic replacement policy. Corrective and preventive maintenance. Maintenance planning, Scheduled maintenance. Cost of maintenance versus Cost of equipment and production delays. Inspection: Inspection intervals, Inspection reports, card history system.

**Unit 2 - Predictive maintenance**

Equipment wear records, standards. Equipment used in predictive maintenance. Computerized maintenance, Total Productive Maintenance. Methods of condition monitoring, Non-destructive testing, Liquid Penetrate, Magnetic particles, Ultrasonic testing, and Vibration analysis. Oil analysis, Radiographic testing.

**Unit 3 - Reliability**

Definition, failure data analysis, Mean failure rate, mean time to failure (MTTF), mean time between failures (MTBF), hazard rate, Bathtub curve. Use of Weibull probability chart foe assessing characteristics life, guarantee period etc.

**Unit 4 - System reliability**

Series, parallel and mixed configuration; Simple problems. Reliability improvement: Techniques, use of Pareto analysis-Design for reliability, redundancy unit and stand by redundancy, Optimization of reliability.

**Unit 5 - Spare Parts Management**

Spare parts, features and categorization of spares, cost considerations, Techniques of cost reduction; Selective controls used in spare parts control; ABC analysis, FSN, XYZ, VED and other approaches. Inventory control of spares.

## 9. RAJIV GANDHI TECHNICAL UNIVERSITY - ME-801 (C) RELIABILITY AND MAINTENANCE

**Unit - 1 Basic Concepts of Reliability**

Probability distributions used in maintenance engineering- Binomial, Poisson, Exponential, Normal, Log-normal, Gamma and Weibull distribution; failure rate, hazard rate, failure modes, MTTR, MTBF, MTTF

**Unit - 2 System Reliability Models**
System reliability-n-component series systems, m-component, parallel systems and combined system; standby systems; K-out-of-m systems; redundancy techniques in system design; event space, decomposition (Key Stone), cut and tie sets, Markov analysis, reliability and quality, unreliability, maintainability, availability

**Unit - 3 Maintenance Concepts and Strategies**
Introduction, maintenance functions and objectives, maintenance planning and scheduling, maintenance organization. General Introduction to Maintenance Types: Breakdown, emergency, corrective, predictive, and preventive; maintenance prevention; design-out maintenance, productive maintenance, shutdown maintenance and scheduled maintenance.

**Unit - 4 Condition Based Maintenance**
Principles of CBM, pillars of condition monitoring, CBM implementation and benefits; condition monitoring techniques- visual monitoring, vibration monitoring, wear debris monitoring, corrosion monitoring, performance monitoring

**Unit - 5 Reliability Centered Maintenance** (RCM):
Concept, methodology, benefits;

**Total Productive Maintenance:** Evolution of TPM, TPM objectives, concept, pillars of TPM.

**Failure Modes and Effects Analysis** (FMEA)/ Failure Modes, Effects and Criticality Analysis (FMECA): Overview, elements of FMECA, applications and benefits, risk evaluation, risk priority numbers, criticality analysis, process FMEA, qualitative and quantitative approach to FMECA; design FMEA and steps for carrying out design FMEA

## 10. UNIVERSITY OF PUNE - RELIABILITY ENGINEERING

1. **Fundamental concepts:**
   Reliability definitions, failure, Failure density, Failure Rate, Hazard Rate, Mean Time To Failure, MTBF, maintainability, availability , pdf, cdf, safety and reliability, Quality, cost and system effectiveness, Life characteristic phases, modes of failure, Areas of reliability, Quality and reliability assurance rules, product liability, Importance of Reliability,

2. **Probability theory:**
Set theory, laws of probability, total probability theorem, probability distributions-binomial, normal, poisson, lognormal, weibull, exponential, standard deviation, variance, skewness coefficient, chebyshev inequality, central limit theorem.

3. **System reliability and modeling:**
Series, parallel, mixed configuration, k- out of n structure, complex systems-enumeration method, conditional probability method, cut set and tie set method, Redundancy, element redundancy, unit redundancy, standby redundancy- types of stand by redundancy, parallel components single redundancy, multiple redundancy. Markov analysis.

4. **Maintainability and Availability:**
Objectives of maintenance, types of maintenance, Maintainability, factors affecting maintainability, system down time, Availability - Inherent, Achieved and Operational availability, reliability and maintainability trade-off.

5. **System reliability Analysis:**
Reliability allocation or apportionment, Reliability apportionment techniques – equal apportionment, AGREE, ARINC, feasibility of objectives apportionment, dynamic programming apportionment, Reliability block diagrams and models, Reliability predictions from predicted unreliability, minimum effort method.

6. **Strength based reliability:**
Safety factor, safety margin, Stress strength interaction,

7. **Failure Mode, Effects and Criticality Analysis:**
Failure mode effects analysis, severity/criticality analysis , FMECA examples, RPN, Ishikawa diagram for failure representation , fault tree construction, basic symbols development of functional reliability block diagram, Fau1t tree analysis, fault tree evaluation techniques, minimal cut set method, Delphi methods, Monte carlo evaluation.

8. **Design of Mechanical components and systems:**
Material strengths and loads, Reliability testing and reliability growth testing.

## 11. UTTTAR PRADESH TECHNICAL UNIVERSITY - EME-052 : MAINTENANCE ENGINEERING & MANAGEMENT

### Unit - I

Introduction, operating life cycle, reliability, Failure data analysis, failure rate curve, hazard models, elements in series, parallel, mix, logic diagrams, improving reliability, redundancy-element, unit, standby, maintainability, availability, reliability and maintainability trade off.

**Unit - II**

Maintenance Strategies: Break down maintenance, planned maintenance, strategies, preventive maintenance, design out maintenance, planned lubrication, total productive maintenance, zero break down, preventive inspection of equipment used in emergency.

**Unit - III**

Replacement planning maintain or replace decision, replacement of items that deteriorate identical equipment, replacement of items that fail without deterioration individual, group replacement, replacement in anticipation of failure.

**Unit - IV**

Break down maintenance planning, assignment model, waiting time models expected waiting time, minimum cost service rate, PERT.

**Unit - V**

Maintenance Management, production maintenance system, objectives and functions, forms, policy, planning, organization, economics of maintenance, manpower planning, materials planning, spare parts planning and control, evaluation of maintenance management.

## 12. UTTTAR PRADESH TECHNICAL UNIVERSITY (Different Course)

**UNIT I**

Introduction to maintenance: Need for maintenance. Types of maintenance, breakdown, corrective and preventive maintenance. Maintenance planning, Scheduled maintenance. Cost of maintenance versus Cost of equipment and production delays. Inspection: Inspection intervals, Inspection reports, card history system.

**UNIT II**

Predictive maintenance. Equipment wear records, standards. Equipment used in predictive maintenance. Computerized maintenance. The role of computers in a maintenance programme. Types if lubrication system and selection of lubricants. Non-destructive testing: Liquid Penetrate, Magnetic particles, Ultrasonic testing, and Vibration analysis. Oil analysis Radiographic testing.

**UNIT III**

Maintenance of mechanical drives: Bearings: Overheating, noise, vibration. Chain drives: Normal wear in chain drives. Tension in chain drives, Sprockets: Lubrication,

Belt drives: Tension in belts, slip &creep. Gears: Normal wear in gears, Lubrication & alignment problems. Shock, overloading. Couplings: Rigid flexible couplings.

**UNIT IV**

Maintenance of fluid power systems: Pumps: Noise & heat, Compressors: Heating, noise, vibration problems. Maintenance of Control valves.

**UNIT V**

**Life Testing-Reliability:** Life testing-Objectives-failure data analysis, Mean failure rate, mean time to failure, mean time between failures, hazard rate, system reliability: series & parallel and mixed configuration-simple problems. Maintainability and availability. Reliability of acceptance sampling based on reliability test-O. C Curves.

**Quality and Reliability:** Reliability improvements-techniques, use of Pareto analysis-Design for reliability, redundancy unit and stand by redundancy, Optimization of reliability.

## 13. PUNJAB TECHNICAL UNIVERSITY - IE 514: MAINTENANCE & RELIABILITY ENGINEERING

1. **Maintenance in context**: maintenance and profitability, terro-technology, application of terro-technology.
2. **Principles**: the structure of plant, reason for nature of maintenance work, the production maintenance system a dynamic model.
3. **Establishing a maintenance plan**-preliminary consideration: items, classification of items, maintenance procedure, guidelines for machine procedures to items.
4. **Maintenance planning and control:** Basic requirements, Management information, labour costs, computer based Management information system, work planning and work control, basic rules for success.
5. **Introduction:** Reliability concepts and patterns of failure, reliability Management, reliability, for system effectiveness.
6. **Reliability and hazard rates:** Failure data, reliability function, failure rate and hazard rate, common distributions in failure mechanisms – experimental, Welbull, gamma, Normal, log normal, extreme value, model selection for components failure, failure analysis.
7. **Reliability prediction and analysis:** reliability prediction based on exponential distribution, system reliability analysis – block diagram method, fault tree and success tree methods, event tree method, failure model, failure mechanism.

8. **Reliability design:** Design for reliability, design process, assessment methodology, reliability allocation, reliability improvement, selection of components to improve system reliability.

## 14. NPC'S 2 YEAR PG. PROGRAMME IN PLANT ENGINEERING - Subject B 1 MAINTENANCE MANAGEMENT:

### 1A. Maintenance Systems

Systems concept in maintenance; production performance vis-à-vis capability performance; availability performance; definition of preventive maintenance, corrective maintenance; Objectives and functions of maintenance; Categorization of equipment and maintenance planning; Maintenance information system; Paperwork and procedures in maintenance; Maintenance performance evaluation and performance indices; Renewal Theory; Equipment replacement analysis; Development of maintenance personnel, Maintenance games; Maintenance system in deferent types of industries, i.e., Engineering, Process, Transport etc.; Plant visits; Total productivity ,maintenance, Reliability centered maintenance; 5-S Housekeeping practices.

### 1B. Reliability, Maintainability and Logistics Engineering

Life testing methods; Experimental design, data collection and analysis, estimation of parameters for exponential, normal, Weibull and Gamma failure distributions; Reliability – basic concepts; Reliability calculations for series, parallel and parallel series systems; Reliability calculations of complex systems with multi-mode failures; Reliability allocation, reliability prediction techniques; Concept of maintainability; Interaction between reliability, maintainability and availability; Design considerations for maintainability; Logistics engineering – concept and scope ; Six stages of logistics engineering; Life cycle cost; System effectiveness; Use of reliability, maintainability and availability; Concepts in system element selection, equipment tools spare parts and other resources.

### 1C. Spare Part Management

Definition of inventory, types of inventory, storage and issue records; Inventory costs, economic order quantity, inventory models; Maximum, minimum and safety stock; Classification and coding.

Introduction and scope of spare parts management in Indian Industries; factors effecting the spare part inventory; Classification and codification of spare parts; Inventory costs and economic order quantity; Different stocking policies for spare parts; Service level & safety stock for different categories of spare parts; capital spares and insurance parts: Systems and procedures in spare part management; Management reports for effective inventory control; Standardization, variety reduction and import substitution of spare parts; Development and scope of spare part bank.

## 15. INDIAN INSTITUTE OF PLANT ENGINEERS - DIPEM SEM-II C -- INDUSTRIAL MAINTENANCE

**UNIT - I: Introduction to theory and practice of Maintenance**

Need for proper maintenance in case of plant and machinery-Philosophy/ mechanism of breakdown-Concept of maintenance-out design and advantages; Haphazard/ breakdown Maintenance and its characteristics cost of down time. Need for maximization of equipment availability.

**UNIT - II: Types of maintenance**

Corrective maintenance, Preventive maintenance/Planned maintenance, Predictive maintenance, Zero hours maintenance, Periodic maintenance (time bound maintenance)

**UNIT - III: Total productive maintenance-concepts, advantages, steps**

Condition based maintenance, concepts, advantages, steps-Maintenance repair and operations, Computer based maintenance management systems, Estimating repair and maintenance costs-Maintenance Engineer's tool box-Work simplification in maintenance

**UNIT - IV: Maintenance of various equipment and systems**

Elevators and special lifts, Air-conditioning equipment,Ventilation and exhaust systems, Dust collecting and air cleaning equipment,Plain bearings, roller bearings, antifriction bearings-Flexible couplings, Chains and Gears for power transmissions-

Belt drives-Cranes, Winches & pulley hoists, ropes, chains, slings their maintenance and calibration-Reciprocating compressors, valves, Pumps, motors, Lubricating devices & systems

**UNIT - V: Legal maintenance**

Equipment & devices under pressure- Storage tanks of certain chemicals, Installation of high and medium voltage equipment-Cooling towers, Elevators & Certain lifts, Vehicles-Fire prevention facilities

## 16. DR. MGR UNIVERSITY, FACULTY OF MANAGEMENT MBA
## Subject no. MMG 13E85 Maintenance Management

**Unit - I Production and Distribution Process**

Process Functions - Functional Block Diagram - Reliability Engineering - Failure Histogram - Probability Density Function - Hazard rate and failure pattern - The Nature and role of maintenance - Maintenance: Its role and scope in total organization context, Centralized and decentralized maintenance organization structure. Design of maintenance organization in multi echelon repair inventory systems

**Unit - II Maintenance Systems**

Design and its selection - Break down maintenance - Planned and Unplanned maintenance - Routine maintenance - Remedial Maintenance - Predictive maintenance - Preventative maintenance - Corrective Maintenance - Total Productive Maintenance - Design maintenance - Contract Maintenance - Benchmarking maintenance

**Unit - III Pareto's Principles for repetitive breakdown analysis**

Spares Management- Planning consideration for each type of activities- Human Resource management for maintenance- Selection- Training-

**Unit - IV Maintenance Budget- Budgetary control**

Scheduling maintenance costs-control of maintenance expenditure -Maintenance effectiveness- monitoring of maintenance performance- Replacement Technique

**Unit - V Maintenance Quality improvement**

ISO 9000 and Its relation to Maintenance- Techniques for continuous improvement in maintenance -Reliability enhancement programme- FMCEA, RCM & POM

# Annexure - II

# Syllabi of International Universities

## 1. University of Wisconsin, Whitewater

*Course No.:450/650: Design, Maintenance, and Management of Recreation Facilities*

**Course content and schedule**

| | |
|---|---|
| Week 1 | Introduction to facility planning, design, and management process |
| Week 2 | Feasibility study, needs assessment, relationship to existing facilities |
| Week 3 | Architectural guidelines/requirements, federal, state, and local regulations |
| Week 4 | Demographic data and site development |
| Week 5 | Master planning, initiating design and construction, program statement development |
| Week 6 | Specification sheets, selection and working with architect, construction phase, and building committee |
| Week 7 | Midterm examination, Principles and concepts of indoor and outdoor facilities |
| Week 8 | Facility complexes, health and fitness facilities, sport facilities |
| Week 9 | Recreation centers, parks, Aquatic facilities |
| Week 10 | Technical considerations, natural vs. artificial surfaces, lighting, HVAC, pedestrian and motor traffic patterns |
| Week 11 | Spectator/non-participant amenities, ancillary spaces, participant amenities |
| Week 12 | Health and safety considerations, general risk management |
| Week 13 | ADA, Title IX, and OSHA, facility and equipment/negligence and product liability |

Week 14 Facility and personnel management

Week 15 Presentation of graduate student projects

## 2. KTH Royal Institute of Technology, Stockholm, Sweden

*Course No.: MG2300 :: Maintenance Management*

*Management and Organization:* This course module treats the contemporary managerial (organizational and economical) topics in maintenance activities. The curriculum covers also the maintenance activities in the development and procurement of new production equipment, translation of production requirements into maintenance objectives, and how to achieve the objectives with optimized resources. Also covered is how the maintenance experience can be used during the design phase, and how to define the future maintenance needs of a company. European standards within maintenance are discussed, as well as laws and regulations regarding labor, liability, warranties, environment, energy, etc.

*Reliability performance of production plants*: This course module covers knowledge about how to guide, control and develop the availability performance activities, in order to assure the performance of the production, the quality of the products, the safety regulations and the environment conditions. After studying this module, the student will have good knowledge of all the availability performance activities that shall be taken into account during the entire life cycle of the production system.

*Maintenance methods and techniques:* This part of the course covers definition and development of the maintenance work plan. The theories and methods that are used to optimize the mix between corrective maintenance, preventive maintenance, strategies, and modifications are discussed – how to choose the right methods for the best cost effectiveness.

*Maintenance information systems:* This part covers the different methods and systems used in the decision making process, how to assure that the maintenance activities are cost effective, and are supporting the company profit. Further, the means how to specify the system requirements and how to develop and use the information systems for planning, control, feedback analysis and improvements are discussed.

## 3. Southeast Missouri State University

*Course No.: FM554 :: Facilities Operation & Supervision*

***Course Content or Outline: Hrs***

A. Roles of the Facility Manager 6

B. Facility Manager Leadership / Management / Ethics 3

C. Communication 3
D. Quality Assessment & Innovation 6
E. Finance & Facilities Management 3 Midterm Exam
F. Real Estate & Facilities Management 3
G. Planning & Project Management for Facilities 3
H. Operations & Maintenance 6
I. Human & Environmental Factors 6
J. Technology Applications 6
K. Final Exam

## 4. Sacramento State College of Continuing Education

*Course name:: Sustainable Facility Management Green Business Operations Certificate Program*

a. Facility Management roles and responsibilities
b. Standards and regulations for sustainable operations
c. Sustainable site management practices
d. Sustainable design of facilities & systems
e. Commissioning and Re-commissioning
f. Operations & maintenance practices
g. Energy management
h. Waste management
i. Water management
j. Indoor air quality
k. LEED evaluation & certification

## 5. Linnaeus University, state university, Småland region of Sweden

1. The maintenance organization: its tasks, roles and management 1
2. The maintenance process and processes for maintenance support 1
3. Maintenance philosophies and policies 1
4. Methods and tools for maintenance resource allocation 1
5. Key performance indicators for maintenance performance measurement 1
6. Methods for maintenance improvement projects 1
7. Methods and tools for the management of maintenance related risks

## 6. Kansaa University

*Module 1 — Introduction to Managing Maintenance*

(This module is based on *The Handbook of Maintenance Management*, Second Edition by Joel Levitt, published by Industrial Press, NY)

Session 0 — Introductory Session

- A "get-to-know-you session," where the course format will be explained and general questions answered. It is recommended that all students be present online at this session.

Session 1 — Evaluating Your Operation

- Introducing assessment tools
- Identifying dominant patterns in departments

Session 2 — Maintenance Metrics and KPIs (Key Performance Indicators)

- Understanding formulas, measures, and techniques to evaluate effort
- Anticipating maintenance efforts

Session 3 — CMMS (Computerized Maintenance Management System)

- Reviewing the technology
- Outlining capabilities of popular systems

Session 4 — Maintenance Contracting and Outsourcing

- Building partnerships with other groups
- Utilizing readers as a resource

Session 5 — Maintenance Quality Improvement

- Reviewing the field of Quality Improvement
- Applying Deming's ideas to maintenance

*Module 2 — Planning and Scheduling Maintenance Activity*

(This material is based on *Maintenance Planning, Coordinating and Scheduling,* Second Edition by Don Nyman and Joel Levitt, published by Industrial Press, NY)

Session 1 — What is planning and why plan maintenance jobs?

- How to sell planning & scheduling to your management
- Where does the ROI (Return on Investment) come from?

- The nature of maintenance activities
- Metrics of planning and scheduling

Session 2 — Planning

- What are the steps in effective planning?
- Job safety analysis (JSA)
- Complete description of the planner's job
- The complete planning process

Section 3 — Estimating, Crew Size & Manpower Requirements

- Estimation and slotting
- What are the contents of a complete planned job package?

Section 4 — Preparatory activity to scheduling

- Calculating available hours from maintenance workers
- Coordination with operations and coordination meeting

Section 5 — Scheduling Maintenance Work

- Job loading until available hours are used up
- Create a credible schedule that takes all the constraints into account
- Job execution and feedback with job closeout and follow-up

*Module 3 — Preventive and Predictive Maintenance*

(This material is based on *Maintenance Planning, Coordinating and Scheduling,* Second Edition by Don Nyman and Joel Levitt, published by Industrial Press, NY)

Section 1 — Basic Understanding of PM (Preventive Maintenance)

- True breakdown costs
- Past sins
- What is PM?
- Understanding what PM is supposed to accomplish and the critical event
- Maintenance improvement curve
- Short repairs and high productivity
- Four types of task lists
- PM frequency and its effect on breakdown

Section 2 — TLC (Tighten, Lubricate and Clean)

Section 3 — Guidelines for involvement with predictive techniques

- P-F Curve

- Questions to ask before PdM (Predictive Maintenance)
- Oil analysis
- Vibration analysis
- Temperature measurement
- Ultrasonic inspection

Section 4 — Reliability enhancement programs

- RCM (Reliability Centered Maintenance)
- PMO
- Task list development

Section 5 — Management of PM activity

- Access to equipment
- Interruptive maintenance and non-interruptive maintenance
- Metrics
- Outsourcing PM

# Bibliography



| | Author | Title | Publisher | Year |
|---|---|---|---|---|
| 1 | Alan Davies | Handbook of Condition Monitoring | | |
| 2 | ALIRATI | Industrial Maintenance Management | SAF-ALIRATI | 1980 |
| 3 | Armstrong | Condition Monitoring | BSIRSA | 1988 |
| 4 | Birolini A | Reliability Engineering, Theory and Practice | Springer, 1999 | |
| 5 | Borris, Steven | Total Productive Maintenance | McGraw Hill, N Y | |
| 6 | BSI | BS. No. 3138:1978 | BSI | |
| 7 | Campbell, Dixon & Picknell | Strategies for Excellence in Maintenance Management | Productivity Press | |
| 8 | Clifton R.H. | Principles of Planned Maintenance | Edward Arnold Publishers | |
| 9 | Coxmoor | Wear Debris Analysis | | |
| 10 | Davies | Handbook of Condition Monitoring | Chapman & Hall | 1996 |
| 11 | Deming | Out of the Crisis | MIT press, Boston | |
| 12 | DeReamer, Russel | Modern Safety Practices | John Wiley & Sons | |
| 13 | Durgesh Chandra | Design Out Maintenance and Instrument Aids | Universal Book Corporation | |
| 14 | Ebeling, Charles E | Reliability & Maintainability Engineering | Tata McGraw Hill | 2000 |
| 15 | Garret & Silver | Production Management Analysis | Harcout Brace | |
| 16 | Gopalakrishnan & Banerjee | Maintenance and Spare Part Management | Prentice Hall India | 2002 |
| 17 | Handley, William | Industrial Safety Handbook | McGraw Hill | 1977 |
| 18 | Hartman, Edward G | Successfully Installing TPM in a Non-Japanese Plant | Productivity Press. | |
| 19 | HB Maynard | Industrial Engineering Handbook | McGraw Hill | |

| | Author | Title | Publisher | Year |
|---|---|---|---|---|
| 20 | Heintzelman, John E | The Complete Handbook of Maintenance Management | Prentice Hall | |
| 21 | Higgins, Lindlay R | Maintenance Engineering Handbook | McGraw Hill Inc | 5th ed |
| 22 | Himbury, S. | Kinetic Methods of Manual Handling in Industry | ILO | 1972 |
| 23 | Ifthikaruddin, Pandian et al | Principles of Environmental Science Engineering | Thirumla Publishers | 2004 |
| 24 | IIPE, Chennai | Advances in Plant Engineering and Management | Professional Graphics | 1996. |
| 25 | ILO | Accident Prevention | ILO | 1972 |
| 26 | ILO | Maintenance and Repair of Motor Vehicles | ILO | 1973 |
| 27 | Immer, John | Material Handling | McGraw Hill | 1953 |
| 28 | Ireson & Grant | Handbook of Industrial Engineering & Management | Prentice Hall | |
| 29 | ISNT, Chennai | Weld Inspection by NDT Techniques | Professional Graphics | 1998 |
| 30 | Japanese Institute of Plant Maintenance | TPM for Every Operator | Productivity Press. | |
| 31 | Jardine AKS | Maintenance, Replacement and Reliability | Pitman Publishing | |
| 32 | Kelly and Harris | Management of Industrial Maintenance | Newness, Butterworth, UK | |
| 33 | Kelly, Anthony | Maintenance Planning & Control | East West Publishers | 2000 |
| 34 | Khanna, O.P. | Industrial Engineering & Management | Dhanpatrai | |
| 35 | Kiran D.R. | Professional Ethics and Human Values | McGraw Hill | |
| 36 | Kiran D.R. | Tota; Quality Management | BSPublicaions | |
| 37 | Knight , Charles | Organization and Administration of Maintenance Function | McGraw Hill Inc | |
| 38 | Leflar, James A | Practical TPM | Productivity Press | |
| 39 | Lewis, Bernard T | Management Handbook for Plant Engieers | McGraw Hill | 1977 |

| | Author | Title | Publisher | Year |
|---|---|---|---|---|
| 40 | Mahajan | Industrial Engg. & Prod uction Management | Dhanpatrai Publishers | |
| 41 | Miller&Blood | Modern Maintenance Management | | |
| 42 | Mills & Stralingsford | Office Organization and Methods | Isaac Pitman | |
| 43 | Mishra, RC & K Pathak | Maintenance Engineering & Management | Prentice Hall India | 2002 |
| 44 | Mobley, Keith | Maintenance Handbook | | |
| 45 | Morrow | Maintenance Engg Hand Book | | |
| 46 | Muther, Richard | Systematic Layout Planning | Cahners Books | |
| 47 | Newall & Roger Vergin | Production & Operations Management - A System Concept | Houghton Miffin & Co. | |
| 48 | Oberman, Ya | Material Handling | Mir Publishers | 1988 |
| 49 | QCFI, Hyderabad | Total Quality Management | QCFI | |
| 50 | Raman Sivakumar | Introduction to Environmental Science & Engineering | Vijay Nicole | |
| 51 | Resheter, Ivanov & Fadeev | Reliability of Machines | Mir Publishers | |
| 52 | Roylance BJ and Hunt TM | Increasing Reliability through Oil Analysis | | |
| 53 | Seiichi, Nakajima | Introduction to TPM | Productivity Press | |
| 54 | Spreigel & Lansburgh | Industrial Management | Asia Publishers | 1966 |
| 55 | Srivatsava, S.K. | Induatriaql Maintenance Management | S Chand & Co. | 1998 |
| 56 | Tselang, Marthand | Industrial Engineering & Production Management | S Chand & Co. | |
| 57 | White EN | Maintenance Planning, Control and Documentation | Gower Press | 1979 |
| 58 | Wilde, Ray | Production & Operations Management | Hold Reinhart & Winston | |
| 59 | Wireman, Terry | Preventive Maintenance | Prentice Hall | |


Apart from above, reference can be made to web sites and more references indicated at the end of chapters like Wear Debris Analysis as well as at appropriate paragraphs.

# Index

## C



## D

## E



## F

# M

# Q



# R